T0396586

IUTAM Symposium on Dynamics Modeling and Interaction Control in Virtual and Real Environments

IUTAM BOOKSERIES

VOLUME 30

Aims and Scope of the Series

The IUTAM Bookseries publishes the proceedings of IUTAM symposia under the auspices of the IUTAM Board.

For other titles published in this series, go to
www.springer.com/series/7695

Gábor Stépán, László L. Kovács, and András Tóth
Editors

IUTAM Symposium on Dynamics Modeling and Interaction Control in Virtual and Real Environments

Proceedings of the IUTAM Symposium on Dynamics Modeling and Interaction Control in Virtual and Real Environments, Held in Budapest, Hungary, June 7–11, 2010

Editors

Gábor Stépán
Budapest University of
Technology and Economics
Department of Applied
Mechanics
Muegyetem rkp. 5
1111 Budapest
Hungary
Tel.: +36 1 463 1369
Email: stepan@mm.bme.hu

András Tóth
Budapest University of
Technology and Economics
Department of Manufacturing
Science and Engineering
Egry J. u. 1 T/46.b
1111 Budapest
Hungary
Tel.: +36 1 463 2516
Email: toth@manuf.bme.hu

László L. Kovács
Budapest University of
Technology and Economics
Department of Applied
Mechanics
Muegyetem rkp. 5
1111 Budapest
Hungary
Tel.: +36 1 463 1369
Email: kovacs@mm.bme.hu

ISSN 1875-3507 e-ISSN 1875-3493

ISBN 978-94-007-1642-1 e-ISBN 978-94-007-1643-8

DOI 10.1007/978-94-007-1643-8

Springer Dordrecht Heidelberg London New York

Library of Congress Control Number: 2011926532

Typeset & Cover Design: Scientific Publishing Services Pvt. Ltd., Chennai, India.

Printed on acid-free paper

Springer is part of Springer Science+Business Media (www.springer.com)

Preface

This volume contains the invited papers presented at the IUTAM Symposium on Multibody Dynamics and Interaction Control in Virtual and Real Environments held in Budapest, Hungary, June 7–11 2010.

The symposium aimed to bring together specialists in the fields of multibody system modeling, contact/collision mechanics and control of mechanical systems. The offered topics included modeling aspects, mechanical and mathematical models, the question of neglections and simplifications, reduction of large systems, interaction with environment like air, water and obstacles, contact of all types, control concepts, control stability and optimization.

Discussions between the experts of these fields made it possible to exchange ideas about the recent advances in multibody system modeling and interaction control, as well as about the possible future trends. The presentations of recent scientific results may facilitate the interaction between scientific areas like system/control engineering and mechanical engineering.

A Scientific Committee was appointed by the Bureau of IUTAM including the following members:

G. Stépán (Hungary, Chairman)
F.L. Chernousko (Russia, IUTAM Representative)
P. Bidaud (France)
W. Harwin (United Kingdom)
J. Kövecses (Canada)
F. Pfeiffer (Germany)
B. Siciliano (Italy)
H. Yabuno (Japan)

The committee selected the participants to be involved and the papers to be presented at the symposium. As a result of this procedure, 45 active scientific participants from 17 countries accepted the invitation and 39 papers and 4 keynote lectures were presented in 8 sessions.

Papers on dynamics modeling and interaction control were naturally selected into the main areas: mathematical modeling, dynamic analysis, friction modeling, solid and thermomechanical aspects, and applications. The presentations were almost equally divided between experimental and computational work and several ones addressed the question of modeling and simulation of complex multibody systems such as gearboxes and vehicles. There were separate sessions on theoretical aspects with a number of lecturers applying fractional order calculus in dynamical modeling of oscillatory systems and investigating chaotic solutions that may arise in case of many contact tasks. In addition, a separate session was offered to friction modeling that was also addressed by several authors during dynamics modeling of biologically inspired robots. A significant outcome of the meeting was the opening towards applications that has a key importance in the future of nonlinear dynamics.

Rather than presenting the papers in the order they were delivered at the Symposium, the contributions are grouped according to topic and application. The ordering is not unique as some papers may fit in several categories.

The scientific presentations were devoted to the following topics, and ordered in this book as follows:

- Dynamics modeling and control of robots (7 talks)
- Applications and control of bio-inspired robots (3 talks)
- Vehicle dynamics and control (3 talks)
- Mathematical modeling of oscillatory systems (7 talks)
- Biomechanics and rehabilitation (6 talks)
- Micro-electromechanical Systems (6 talks)
- Modeling dry friction (4 talks)
- Thermoelasticity aspects (3 talks)

The four keynote lectures indicate the multi-disciplinary character of the Symposium while they span the important topics of dynamics modeling and interaction control in virtual and real environments:

- Kouhei Ohnishi (Keio University, Yokohama, Japan)
  Real World Haptics and Telehaptics
- József Kövecses (McGill University, Montreal, Quebec, Canada)
  Approaches to Lagrangian Dynamics and Their Application to Interactions with Virtual Environments
- Michael Beitelschmidt (Technical University of Dresden, Germany)
  Real Time Simulation and Actuation of Shifting Forces of a Gearbox
- Philippe Bidaud (Institute of Intelligent Systems and Robotics, Paris, France)
  Stability analysis and dynamic control of multi-limb robotic systems

The editors wish to thank both the keynote lectures and participants, and the authors of the papers for their contributions to the important fields of multibody dynamics, interaction control and the closely related branch of disciplines reflected by the titles of topical sessions.

The success of the symposium would not have been possible without the work of the Local Organizing Committee. The members of that were:

Gábor Stépán (Chairman)
László L. Kovács (Secretary)
András Tóth (Project manager)

In addition thanks are due to Springer Science+Business Media for efficient cooperation and to Ms. Nathalie Jacobs for her help and encouragement to the publication of this volume.

Budapest
October 2010

Gábor Stépán

Contents

Dynamics Modeling and Control of Robots

Applications and Control of Bio-Inspired Robots

Vehicle Dynamics and Control

Mathematical Modeling of Oscillatory Systems

Biomechanics and Rehabilitation

Micro-electromechanical Systems

Modeling Dry Friction

Thermoelasticity Aspects

Addresses of Authors

(Session chairmen are identified by an asterisk)

Fabiola ANGULO
National University of Colombia
Department of Electrical and Electronics
Engineering & Computer Sciences
Campus La Nubia, Manizales, 170003,
Colombia
Tel: +5768879400 Fax: +5768879498
fangulog@unal.edu.co

Carsten BEHN
Ilmenau University of Technology
Department of Technical Mechanics
Max-Planck-Ring 12
Ilmenau, 98693, Germany
Tel: +493677691813 Fax: +493677691823
carsten.behn@tu-ilmenau.de

András BIBÓ
Budapest University of Technology and
Economics, Department of Structural
Mechanics
Műegyetem rkp. 3
Budapest, H-1111, Hungary
Tel: +3614631432
biboa@freemail.hu

István BÍRÓ
University of Szeged
Faculty of Engineering
Mars tér 7
Szeged, H-6724, Hungary
biroistvan60@gmail.com

Igor BOCK
Slovak University of Technology
Faculty of Electrical Engineering
Department of Mathematics
Ilkovicova 3
Bratislava, 81219, Slovakia
Tel: +421260291204 Fax: +421265420415
igor.bock@stuba.sk

Quentin BOMBLED
University of Mons, Department of
Theoretical Mechanics, Dynamics
and Vibrations
Boulevard Dolez 31
Mons, 7000, Belgium
Tel: +3265374215 Fax: +3265374183
quentin.bombled@umons.ac.be

Felix L. CHERNOUSKO*
Russian Academy of Sciences
Institute for Problems in Mechanics
pr. Vernadskogo 101-1
Moscow, 119526, Russia
Tel: +74959309795 Fax: +74997399531
chern@ipmnet.ru

Malgorzata CHWAL
Cracow University of Technology
Institute of Machine Design
Warszawska 24
Cracow, 31-155, Poland
Tel: +48126283386 Fax: +48126283360
mchwal@pk.edu.pl

Thomas GORIUS
University of Stuttgart
Institute of Engineering and
Computational Mechanics
Pfaffenwaldring 9
Stuttgart, 70569, Germany
gorius@itm.uni-stuttgart.de

James ING*
University of Aberdeen
School of Engineering
Fraser Noble Bld., Meston Walk
Aberdeen, AB24 3UE,
United Kingdom
j.ing@abdn.ac.uk

Alexander V. KARAPETYAN*
Lomonosov Moscow State University
Faculty of Mechanics and Mathematics
Leninskie gory 1
Moscow, 119991, Russia
Tel: +74959393681 Fax: +7495 9392090
avkarapetyan@yandex.ru

Piotr KEDZIORA
Cracow University of Technology
Departments of Mechanical Engeenering
Warszawska 24
Cracow, 31-155, Poland
Tel: +48126283409 Fax: +4812 6283360
kedziora@mech.pk.edu.pl

István KEPPLER
Szent István University, Institute of
Mechanics and Machinery, Department
of Mechanics and Engineering Design
Páter Károly utca 1.
Gödöllő, H-2103, Hungary
keppler.istvan@gek.szie.hu

Go KONO
Keio University
Department of Mechanical Engineering
3-14-1, Hiyoshi, Kanagawa
Yokohama, 223-8522, Japan
Tel: +81455661823 Fax: +81455661495
gogogoh20@hotmail.com

László L. KOVÁCS*
Budapest University of Technology and
Economics
Department of Applied Mechanics
Műegytem rkp. 5
Budapest, H-1111, Hungary
Tel: +3614633678 Fax: +3614633471
kovacs@mm.bme.hu

Masaharu KURODA
National Institute of Advanced Industrial
Science and Technology (AIST)
Hetero Convergence Research Team
1-2-1 Namiki
Tsukuba, 305-8564, Japan
Tel: +81298617147 Fax: +81298617842
m-kuroda@aist.go.jp

Walter LACARBONARA
Sapienza University of Rome
Department of Structural Engineering
via Eudossiana 18
Rome, 00184, Italy
Tel: +390644585293 Fax: +39064884852
walter.lacarbonara@uniroma1.it

Fumiya MATSUMOTO
Waseda University, Applied Mechanics
and Aerospace Engineering
3-4-1 #58-218 Okubo Shinjuku
Tokyo, 169-8555, Japan
11.08.gen@ruri.waseda.jp

Jacob P. MEIJAARD
University of Twente
CTW/WA
Drienerlolaan 5
Enschede, 7522 NB,
The Netherlands
J.P.Meijaard@utwente.nl

Ákos MIKLÓS
Budapest University of Technology and
Economics
Department of Applied Mechanics
Műegytem rkp. 5
Budapest, H-1111, Hungary
Tel: +3614633678 Fax: +3614633471
miklosa@mm.bme.hu

Lars MIKELSONS
University of Duisburg-Essen
Department of Mechatronics
Lotharstraße 1
Duisburg, 47057, Germany
mikelsons@imech.de

Aleksander MUC*
Cracow University of Technology
Institute of Machine Design
Warszawska 24
Cracow, 31-155, Poland
Tel: +48126283360
olekmuc@mech.pk.edu.pl

Pawel OLEJNIK
Technical University of Lodz,
Department of Automation and Biomechanics
1/15 Stefanowski
Lodz, 90-924, Poland
olejnikp@p.lodz.pl

Gerard OLIVAR*
National University of Colombia
Department of Electrical and Electronics Engineering and Computer Sciences
Campus La Nubia, Manizales,
170003, Colombia
golivart@unal.edu.co

Viviane PASQUI*
Pierre and Marie Curie University
Institute of Intelligent Systems and Robotics
4 Place Jussieu, CC 173
Paris, 75005, France
Tel: +33144276343 Fax: +331445145
pasqui@isir.upmc.fr

Shih PO-JEN
National University of Kaohsiung
Civil and Environmental Department
700 Kaohsiung Uni. Rd.
Kaohsiung, 811, Taiwan
Tel: +88675916592 Fax:+88675919376
pjshih@nuk.edu.tw

Vladimir PUZYREV
Donetsk National University
Universitetskaya 24
Donetsk, 83001, Ukraine
vladimir.puzyrev@gmail.com

Subramanian RAMAKRISHNAN
University of Cincinnati
School of Dynamic Systems
598 Rhodes Hall
Cincinnati, 45221, United States
Tel: +16466702185
subbu@umd.edu

Zaihua WANG*
Nanjing University of Aeronautics and Astronautics
Institute of Vibration Engineering Research
Yudao Street 29
Nanjing, 210016, China
zhwang@nuaa.edu.cn

Walter V. WEDIG*
Karlsruhe Institute of Technology
Institute of Engineering Mechanics,
Chair for Dynamics/Mechatronics
Kaiserstraße 10, Bld. 10.23
Karlsruhe, D-76131, Germany
Tel: +49 721 60842658
Fax: +49 721 60846070
wwedig@t-online.de

Hiroshi YABUNO*
Keio University
Department of Mechanical Engineering
3-14-1, Hiyoshi, Kouhoku
Yokohama, 223-8533, Japan
Tel:+81455661823 Fax:+81455661495
yabuno@mech.keio.ac.jp

Alexandra A. ZOBOVA *
Lomonosov Moscow State University
Faculty of Mechanics and Mathematics
Leninskie gory 1
Moscow, 119991, Russia
Tel: +74959393681 Fax: +74959392090
azobova@mail.ru

Ambrus ZELEI
Budapest University of Technology and Economics
Department of Applied Mechanics
Műegytem rkp. 5
Budapest, H-1111, Hungary
Tel: +3614633678 Fax: +3614633471
zelei@mm.bme.hu

Dynamics Modeling and Control of Robots

These papers explore dynamics modeling tools and control concepts proposed for some new and special robotic applications including personal, entertainment, micro-scale and space robots. Dynamics modeling aspects and control of an underactuated service robot platform developed within the frame of the ACROBOTER project (FP6-IST-6 45530) are discussed in the papers by L.L. Kovács and A. Zelei. The paper by T. Gorius presents simulation and experimental results obtained with the prototype of an interactive 3D pendulum presented at the Word Exhibition EXPO 2010 in Shanghai. The nonlinear control of a micro-cantilever probe of an atomic force microscope is investigated by H. Yabuno, while the presented work of F. Matsumoto addresses the dynamics and trajectory generation of a space robot model.

The ACROBOTER Platform – Part 1: Conceptual Design and Dynamics Modeling Aspects

László L. Kovács, Ambrus Zelei, László Bencsik, and Gábor Stépán

Abstract. This paper presents the conceptual design and the dynamics modeling aspects of a pendulum-like under–actuated service robot platform ACROBOTER. The robot is designed to operate in indoor environments and perform pick and place tasks as well as carry other service robots with lower mobility. The ACROBOTER platform extends the workspace of these robots to the whole cubic volume of the indoor environment by utilizing the ceiling for planar movements. The cable suspended platform has a complex structure the dynamics of which is difficult to be modeled by using conventional robotic approaches. Instead, in this paper natural (Cartesian) coordinates are proposed to describe the configuration of the robot which leads to a dynamical model in the form of differential algebraic equations. The evolution of the ACROBOTER concepts is described in detail with a particular attention on the under–actuation and redundancy of the system. The influence of these properties and the applied differential algebraic model on the controller design is discussed.

1 Introduction

Obstacle avoidance is an important problem in service and mobile robotics. Robots operating in indoor environments have to overcome various static obstacles on the floor, e.g., chairs, tables, doorsteps and even the edges of carpets in a room. Thus floor based domestic robots need to have strategies to detect and avoid these randomly placed objects.

A new direction in the development of indoor service robots is the use of robotic structures that can move on the walls and/or on the ceiling of a room. An advantage

László L. Kovács
HAS-BME Research Group on Dynamics of Machines and Vehicles,
H-1521 Budapest, Hungary
e-mail: kovacs@mm.bme.hu

Ambrus Zelei · László Bencsik · Gábor Stépán
Department of Applied Mechanics,
Budapest University of Technology and Economics, H-1521 Budapest, Hungary
e-mail: zelei@mm.bme.hu, l.bencsik@hotmail.com, stepan@mm.bme.hu

G. Stépán et al. (Eds.): Dynamics Modeling & Interaction Cont., IUTAM BOOK SERIES 30, pp. 3–10.
springerlink.com

of this strategy is that walls and rather the ceiling are almost obstacle free enabling robots to move freely and quickly in any direction. An application which addresses the need for a robot to climb on the walls and crawl on the ceiling inside a building is the MATS robot [1]. Another examples include the mobile robot platform described in [7] and the FLORA walking assisting system developed by FATEC Corporation [5]. Both of them are based on a ceiling absorbed mobile cart utilizing permanent magnets to keep and move the cart on the ceiling. The system [7] has a working unit that is positioned by three parallel telescopic arms. The FLORA robot has a specially designed cable suspended sit harness that can be used to compensate for the body weight of elderly or physically impaired people during walking. These platform concepts solve the problem of avoiding obstacles on the floor, while they are able to roam over almost the whole inner space of a room, and compared to gantry cranes, they enable the use of co-operating multiple units.

Similary, the ACROBOTER service robot is a ceiling based platform. Figures 1 and 4 present that the Climbing Unit (CU) can move in the plane of the suspended ceiling equipped with anchor points. In this first concept the CU is a planar robotic arm, which swaps between the anchor points and provides the crawling motion of the suspension of the robot. The windable suspending cable holds the Swinging Unit (SU) to which the carried objects can be connected. The system has a pendulum like structure, but compared to the above described systems [5, 7], here, the positioning of the payload is controlled by the actuators (fluid mass flow generators) of the SU. The proposed concept combine the planar stepping motion of the arm and the thrusted-hoisted pendulum-like motion of the working unit in 3D relative to the arm.

A design of a tethered aerial robot with a swinging actuator is presented in [6]. In this concept the weight of the robot is carried by a tether and the thrusting forces are generated by two fans with parallel axes. The main task of the robot is to carry a camera (and/or tiny robotic agents) used in rescue operations above unstructured

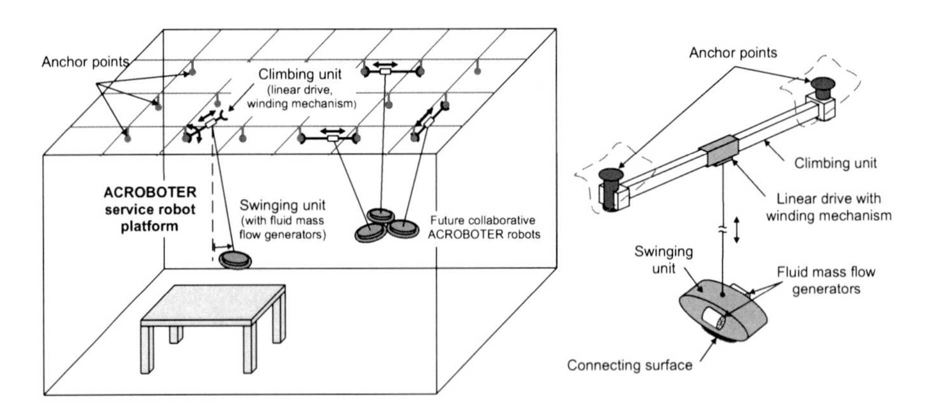

Fig. 1 The ACROBOTER concept: pendulum-like indoor service robot platform (left), main functional components of the system (right)

environments (e.g. at earthquake sites). Despite of the conceptual similarities to tethered aerial systems the different application scenarios of ACROBOTER requires a completely different design. The ACROBOTER platform ensures higher payload capability than aerial robots which may results in high inertial forces and nonlinear oscillations of the payload. In addition the pendulum like behavior of the system may also be utilized to provide better maneuverability and larger workspace in case of proper motion planning of the climbing unit.

2 Conceptual Design

The main task of ACROBOTER is to carry objects in the 3D space of inner environments. This can be accomplished in a set-point manner or the task may require the tracking of some desired trajectories. Hence, the swinging unit (see Fig. 1) needs to be fully actuated and controlled in the 3D environment. The CU has to move the suspension point of the swinging unit smoothly in the plane of the ceiling, while the actuators of the SU have to provide the desired orientation of the unit and its position relative to the CU.

In case of ACROBOTER the ceiling based traction unit, i.e. the climbing unit, is a planar RRT robot. The redundancy of this robot enables smooth planar movements when the CU has to swap between different anchor points (see left in Fig. 4). Different solutions like [5, 7] that provide the smooth movement of the suspension can also be considered.

The present chapter focuses on the design aspects of the swinging unit. The first prototype of the SU is presented in Fig. 2. In this concept two large diameter ducted fans are used to provide the desired nutation of the unit, while three windable orienting cables are used to control the roll and pitch of the platform. Ducted fans are

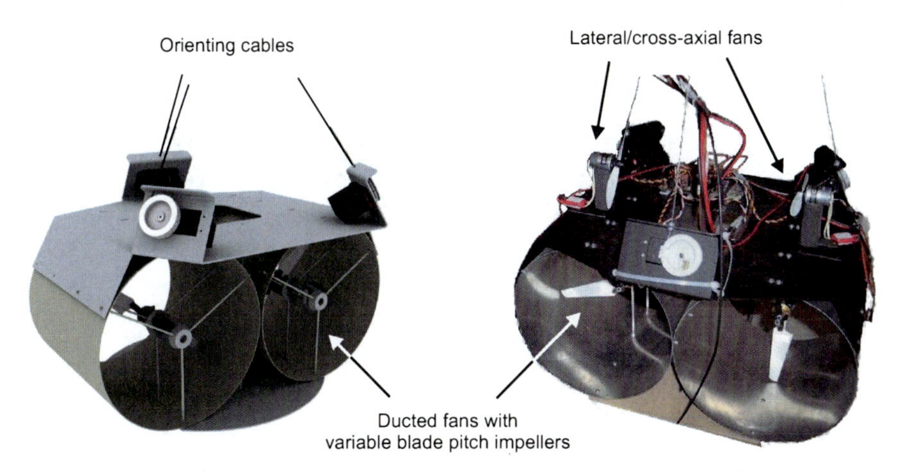

Fig. 2 First prototype of ACROBOTER: CAD design (left), built prototype with complementary lateral/cross-axial fans (right)

lightweight solution for on board thrust generation. The magnitude of the generated thrust is, however, proportional to the diameter of the fans which increases the size of the device when heavy payloads need to be nutated. The moment generated by the two parallel axis fans controls the yaw angle. In addition, the applied ducted fans have variable blade pitch impellers that can adjust the magnitude of the thrust forces or even they can quickly be reversed. This solution can provide large thrust forces, but the maneuverability of the concept is limited. The two ducted fans with parallel axes can only provide a thrust force paralell to the axes of the fans and a resultant moment acting upon the suspended payload as independent control inputs. Thus, the number of actuators is lower than the degrees of freedom of the suspended payload. To resolve the underactuation of the concept a pair of lateral/cross-axial fans were attached to the top of the first prototype of the SU (see right in Fig. 2).

The recognition of the under actuated character of the first design result in a new concept, which was also motivated by the need for decreasing the vertical dimension of the unit. The CAD models of the new concept are shown in Fig. 3 and the corresponding second prototype is depicted at right in Fig. 4. In this concept six identical ducted fans are used as thrusters that are placed around the circumference of a disk. The advantage of this solution is that the swinging unit can be fully actuated by using same ducted fan modules each providing a one-directional thrust force. The price that has to be paid is the lower resultant thrust that the smaller fans can provide. Therefore, developing the second prototype it was assumed that the SU will not provide large nutation for the carried objects. Instead, the fans will more effectively stabilize the motion of the unit around a desired trajectory. The gross motion is generated by the CU while the SU continuously compensates for the tracking errors.

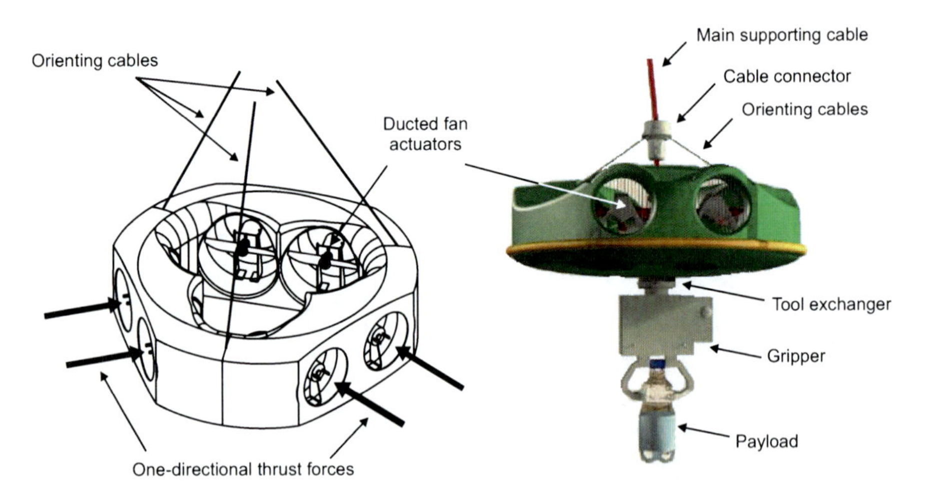

Fig. 3 Second prototype of ACROBOTER with six equally sized one-directional ducted fan actuators

3 Dynamics Modeling

The whole system prototype of ACROBOTER is presented in Fig. 4. This figure shows that the climbing unit is an RRT robot which has 3 degrees-of-freedom, when its upper (anchor) arm is attached to an anchor point. In this respect the CU can be described by conventional robotic approaches using the minimum set of descriptor coordinated, i.e. the generalized coordinates associated with the serial arm structure. The lower (rotation) arm of the CU is a linear axis that carries the winding mechanism hoisting the SU. Thus, including the winding mechanism, the climbing unit has 4DoFs. The main suspending cable and the orienting cables of the SU are connected by the cable connector (CC), which is a relatively small sized component and therefore can be modeled as a point mass with additional 3 DoFs. Then, considering the spatial 6DoFs of the SU, the system has 12DoFs in total, which requires the same number of generalized coordinates to describe its configuration.

Considering the completely different joint structure and actuation of the CU and the SU, the two systems were modeled independently from each other. The CU and the SU have separate motion controllers that are synchronized by the global motion controller of the system. The kinematic description of the CU follows the conventional robotic description and therefore not described here. The SU and the CC forms a cable suspended structure with a closed kinematic chain, where the ducted fan actuators cannot be associated with the real and/or virtual "joints" of the robotic structure. Thus, in case of the kinematic description of the SU the choice of coordinates has a key importance in obtaining a still complex but computationally affordable dynamical model.

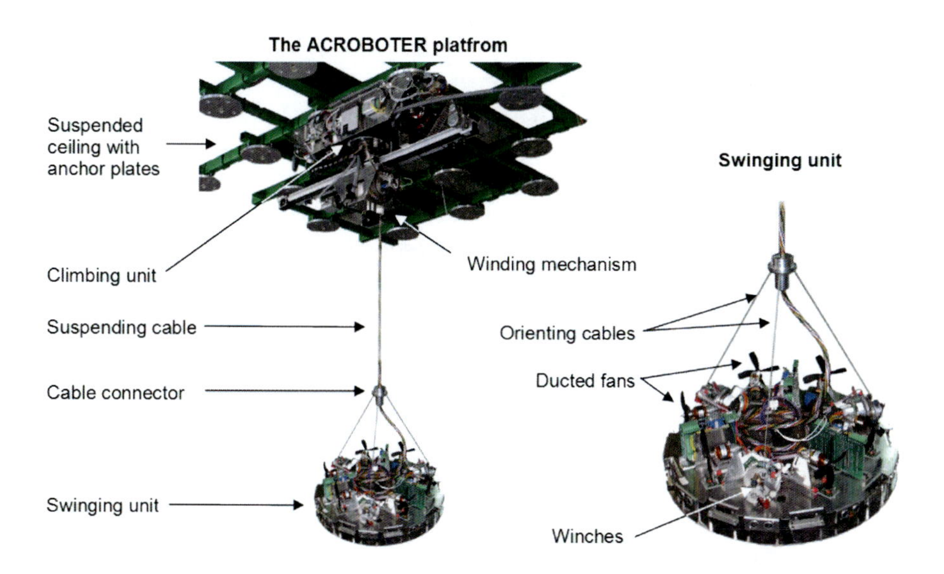

Fig. 4 The ACROBOTER service robot platform

An efficient parameterization of the affine transformation between the global (world) and the local (body) coordinate system of the SU is the use of natural coordinates originally introduced by [2]. This formalism uses a non-minimum set of specially chosen descriptor coordinates for the kinematic description of multibody systems including robotic structures, and the corresponding dynamics modeling is based on the Lagrangian equations of the first kind. This leads to the equations of motion in the well-known set of differential-algebraic equations of index 3

$$\mathbf{M}\ddot{\mathbf{q}} + \boldsymbol{\Phi}_{\mathbf{q}}^{\mathrm{T}}(\mathbf{q})\boldsymbol{\lambda} = \mathbf{Q}_g + \mathbf{H}(\mathbf{q})\mathbf{u}\,, \tag{1}$$

$$\boldsymbol{\phi}(\mathbf{q}) = \mathbf{0}\,, \tag{2}$$

where $\mathbf{q}$ denotes a redundant set of coordinates associated with the CC and the SU. The position of the CC is given by its Cartesian coordinates, while the pose of the SU is described by the coordinates of the co-planar points identified by the outlets of the winches and a unit vector which is perpendicular to the base of the SU. According to [2], the selection of these descriptor coordinates results in a constant mass matrix. The mass matrix $\mathbf{M}$, here, is a block diagonal matrix containing the mass matrices of the CC and SU, respectively. In eq. (1) the matrix $\boldsymbol{\Phi}_{\mathbf{q}} = \partial\boldsymbol{\phi}/\partial\mathbf{q}$ is the constraint Jacobian, $\boldsymbol{\lambda}$ is the vector of Lagrangian multipliers , $\mathbf{Q}_g$ is the constant generalized gravity force and $\mathbf{H}$ is the transmission matrix corresponding to the input vector $\mathbf{u}$. Note that in case of ACROBOTER the constraint equations (2) stand for the squared distances of the basic points (selected as cable outlets of the winches), and the perpendicularity and length of the unit vector the coordinates of which are also used as descriptor coordinates. Consequently, the constraint Jacobian is a linear function of the descriptor coordinates. Although, the number of coordinates are relatively high (15 in case of the model of the SU and the CC), the properties and the special structure of the resulting equations of motion make it possible to derive a real-time dynamic model of ACROBOTER.

Various methods exist for the solution of the equations of motion (1, 2). Simulation techniques involve the classical method of Lagrangian multipliers with Baumgarte stabilization [2, 8], and the projection method [3] which transforms the original set of differential algebraic equations (DAE) of motion to ordinary differential equations (ODE). Another possibility is to substitute the algebraic equations with singularly perturbed differential equations and use available stiff ODE solvers. The direct solution of equations (1) and (2) is possible by the index-reduction of the DAE problem which leads to the full descriptor form of index 1

$$\dot{\mathbf{q}} = \mathbf{p} \tag{3}$$

$$\dot{\mathbf{p}} = \mathbf{a} \tag{4}$$

$$\begin{bmatrix} \mathbf{M} & \boldsymbol{\Phi}_{\mathbf{q}}^{\mathrm{T}} \\ \boldsymbol{\Phi}_{\mathbf{q}} & \mathbf{0} \end{bmatrix} \begin{bmatrix} \mathbf{a} \\ \boldsymbol{\lambda} \end{bmatrix} = \begin{bmatrix} \mathbf{Q}_g + \mathbf{H}(\mathbf{q})\mathbf{u} \\ -\dot{\boldsymbol{\Phi}}_{\mathbf{q}}(\mathbf{q},\mathbf{p})\mathbf{p} \end{bmatrix} \tag{5}$$

$$\boldsymbol{\phi}(\mathbf{q}) = \mathbf{0}\,. \tag{6}$$

The system of equations (4–6) can numerically be solved by backward difference methods (like the Backward-Euler method) via applying the Newton-Raphson iteration scheme to the resulting system of implicit algebraic equations.

4 Control Aspects

Based on the kinematic structure of ACROBOTER described in Section 3 it can be seen that the CU is a redundant manipulator, and moreover the base coordinate system of this arm is changing during swapping between the anchor points. This is an important issue in the control design of the climbing unit. Since the main suspending cable and the orienting cables can equally move the SU in the vertical direction the cable suspension system has a redundant character too. By assuming that the CU regulates the motion of the upper end of the main cable perfectly, the motion of this suspension point can be seen as a constraint on the independent dynamic model of the SU. This way the system formed by the SU and the CC has 9DoFs controlled by 7 actuators only. These include the 4 cable winches and a fictitious compound actuator that provides the two components of the resultant force and the resultant moment generated by the ducted fans. Thus the system is under–actuated, which means that two coordinates out of nine cannot be prescribed arbitrarily because they depend on the internal dynamics of the system. For example, consider that the SU have to move horizontally and the the elevation of the CC above the SU is prescribed. Then the motion of the CC parallel to the base of the SU cannot be actuated. Existing under–actuated robot control techniques (like [9]) are available for the class of systems where the equations of motions without control inputs can easily be identified, which is often the case for serial manipulators. When the control inputs are coupled by a transmission matrix the equation that describes the internal dynamics of the system can be achieved by projecting the equation of motion into the null-space of this matrix. Then, separating the coordinates into controlled and uncontrolled ones, the projected equation can be solved for the uncontrolled coordinates. These calculated coordinates can be seen as prescribed (uncontrolled) coordinates, which enables the generalization of the computed torque control method to under actuated robotic systems [4] modeled by minimum set of generalized coordinates. The computed torque control of ACROBOTER is based on the direct discretization of the differential algebraic equations of motion, which method is described in detail in Part 2 of the present work.

5 Conclusions

The main idea of a novel locomotion technology was presented in this paper and the design concepts of the cable suspended ACROBOTER platform were discussed. The main differences between the presented concepts are their vertical dimension and their maneuverability. Independently of the selected second prototype of the swinging unit it was concluded that the ACROBOTER has a complex spatial structure and it is advantageous to derive its equation of motion in terms of a redundant set of descriptor coordinates. The applied DAE model enables the efficient simulation of the model, while the use of the selected natural coordinates make it straightforward to calculate the forward/inverse kinematics of the system. The actuation scheme of the ACROBOTER robot were also discussed with a view on the possible realization of a computed torque controller.

Acknowledgements. This work was supported in part by the Hungarian National Science Foundation under grant no. OTKA K068910, the ACROBOTER (IST-2006-045530) project, the Hungarian Academy of Sciences under grant no. MTA-NSF/103 and the HAS-BME Research Group on Dynamics of Machines and Vehicles.

References

[1] Balaguer, C., Gimenez, A., Huete, A.J., Sabatini, A.M., Topping, M., Bolmsjö, G.: The MATS robot. IEEE Robotics and Automation Magazine 13(1), 51–58 (2006)
[2] de Jalón, J.G., Bayo, E.: Kinematic and dynamic simulation of multibody systems: the real-time challenge. Springer, Heidelberg (1994)
[3] Kövecses, J., Piedoboeuf, J.-C., Lange, C.: Dynamic modeling and simulation of constrained robotic systems. IEEE/ASME Transactions on mechatronics 8(2), 165–177 (2003)
[4] Lammerts, I.M.M.: Adaptive computed reference computed torque control. PhD thesis, Eindhoven University of Technology (1993)
[5] FATEC Co. Walking assist system by supporting from ceiling, FLORA (2004), `http://www.fa-tec.co.jp/FLORA-C/index.htm`
[6] McKerrow, P.J., Ratner, D.: The design of a tethered aerial robot. In: Proceedings of IEEE International Conference on Robotics and Automation, pp. 355–360 (2007)
[7] Sato, T., Fukui, R., Mofushita, H., Mori, T.: Construction of ceiling adsorbed mobile robots platform utilizing permanent magnet inductive traction method. In: Proceedings of 2004 IEEEiRSJ International Conference on Intelligent Robots and Systems, pp. 552–558 (2004)
[8] von Schwerin, R.: Multibody System Simulation: Numerical Methods, Algorithms, and Software. Springer, Heidelberg (1999)
[9] Spong, M.W.: Partial feedback linearization of underactuated mechanical systems. In: Proceedings of the IEEE International Conference on Intelligent Robots and Systems, vol. 1, pp. 314–321 (1994)

The ACROBOTER Platform - Part 2: Servo-Constraints in Computed Torque Control

Ambrus Zelei and Gábor Stépán

Abstract. The paper presents the motion control of the ceiling based service robot platform ACROBOTER that contains two main subsystems. The climbing unit is a serial robot, which realizes planar motion in the plane of the ceiling. The swinging unit is hoisted by the climbing unit and it is actuated by windable cables and ducted fans. The two subsystems form a serial and subsequent closed-loop kinematic chain segments. Because of the complexity of the system we use natural (Cartesian) coordinates to describe the configuration of the robot, while a set of algebraic equations represents the geometric constraints. Thus the dynamical model of the system is given in the form of differential-algebraic equations (DAE). The system is under-actuated and the the inverse kinematics and dynamics cannot be solved in closed form. The control task is defined by the servo-constraints which are algebraic equations that have to be considered during the calculation of control forces. In this paper the desired control inputs are determined via the numerical solution of the resulting DAE problem using the Backward Euler discretization method.

1 Introduction

Indoor service robots can effectively use the ceiling of the indoor environment to provide obstacle free motion of the base of these robots, while the carried working units can practically move in the whole inner space of the environment [6]. The present paper describes the motion control of a new service robot platform developed within the ACROBOTER (IST-2006-045530) project (see details in Part 1 of

Ambrus Zelei
HAS-BME Research Group on Dynamics of Machines and Vehicles,
H-1521 Budapest, Hungary
e-mail: zelei.ambrus@gmail.com

Gábor Stépán
Department of Applied Mechanics, Budapest University of Technology and Economics,
H-1521 Budapest, Hungary
e-mail: stepan@mm.bme.hu

G. Stépán et al. (Eds.): Dynamics Modeling & Interaction Cont., IUTAM BOOK SERIES 30, pp. 11–18.
springerlink.com

the present paper). The developed robot utilizing the ceiling for the planar motion, and its cable suspended pendulum-like subsystem is the working unit.

A major challenge in ceiling based locomotion is that the ceiling based unit has to hold the total weight of the robot and the payload safely, and has to provide fast motion of the carried objects at the same time. To satisfy these requirements, permanent magnets are applied to develop a ceiling absorbed mobile base in [6]. In case of ACROBOTER, a serial robot based climbing unit was developed that can crawl on an anchor point system installed on the ceiling (see left in Fig. 1).

The other main subsystem of ACROBOTER is the swinging unit. It is connected to the climbing unit via a windable cable that is called the main cable hereafter. The swinging unit has a mechanical interface to connect different tools. This unit can be positioned and oriented with three orineting secondary cables and ducted fan actuators. For more detailed description of the ACROBOTER design the reader is referred to Part 1.The kinematic structure of its planar mechanical model is described in detail in Section 2.

In this paper, first, the dynamical model of the investigated complex robotic structure is described by natural coordinates [2]. The resulting equations of motion are formulated as Lagrangian equations of motion of the first kind. In addition to the introduced geometric constraints, the task of the robot is defined by the so-called servo-constraints which introduce further algebraic equations associated with the original DAE problem of calculating the desired control inputs of the computed torque control (CTC) of ACROBOTER. In the second part of this work the numerical solution of the DAE system of equations is presented by using the Backward Euler discretization method. At the end a real parameter case simulation study is provided to demonstrate the applicability of the proposed controller.

2 Structure of the ACROBOTER Platform

The mechanical structure of the ACROBOTER can be seen left in Fig. 1. The climbing unit is an RRT robot that provides the ceiling based locomotion of the system. Its task is to position the suspension point (cable outlet) in the plane of the ceiling. The climbing unit consists of the anchor arm, the rotation arm and a linear axis moving the winding mechanism. The anchor arm swaps between neighboring anchor points, while the rotation arm and the linear axis provide additional two degrees-of-freedom (DoFs). Thus the climbing unit is a kinematically redundant planar manipulator, except in the case when both ends of the anchor arm is fixed to the ceiling. The winding mechanism hoists the swinging unit via the main cable, which unit contains three additional cable actuators. The main role of these cables to control the orientation of the unit, but they also can regulate its elevation yielding a further redundancy. In addition to these cables, three pairs of ducted fans are employed to orient and position the swinging unit. The orienting cables are assumed to be ideal in the model.

In the planar model shown right in Fig. 1, the climbing unit is considered as a single linear axis. The cable connector modeled as a point mass with 2 DoFs and

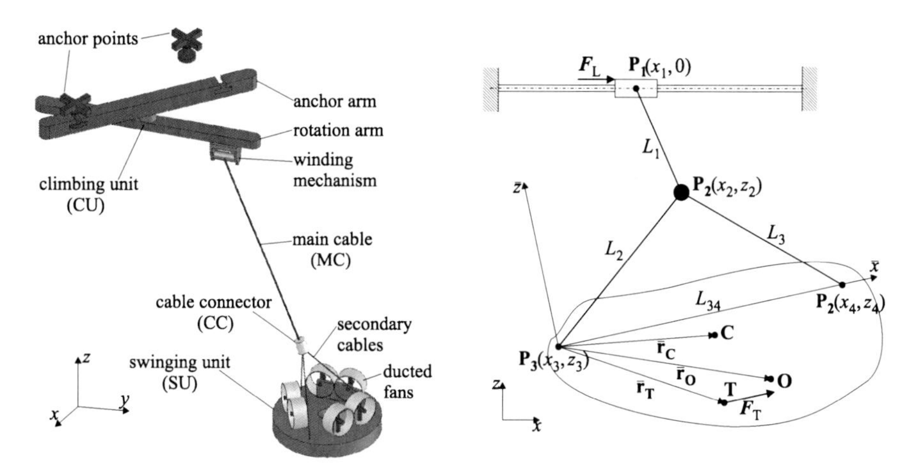

Fig. 1 The ACROBOTER structure (left), planar mechanical model (right)

the swinging unit is a rigid body with 3 DoFs. The kinematics of the swinging unit is described by the redundant set of coordinates associated with points $\mathbf{P}_3$ and $\mathbf{P}_4$.

The total number of DoFs is 6 and we use 7 descriptor coordinates plus one geometric constraint which represents the constant distance L_{34}.

The position of the climbing unit is controlled by the force F_L, while the swinging unit is actuated via the cable forces F_1, F_2 and F_3 and the thrust force F_T. In the model $\mathbf{C}$ denotes the center of gravity of the swinging unit. Point $\mathbf{T}$ determines the line of action of the thrust force F_T being parallel to the local axis $\bar{x}$. And $\mathbf{O}$ is an arbitrarily selected point that has to be controller to move on the desired trajectory.

3 CTC with Backward Euler Discretication

The equations of motion (1) and (2) of the system is derived in the form of the Lagrangian equation of motion of the first kind:

$$\mathbf{M}\ddot{\mathbf{q}} + \mathbf{\Phi}_{\mathbf{q}}^{\mathrm{T}}(\mathbf{q})\boldsymbol{\lambda} = \mathbf{Q}_g + \mathbf{H}(\mathbf{q})\mathbf{u}\,, \tag{1}$$

$$\boldsymbol{\phi}(\mathbf{q}) = \mathbf{0}\,, \tag{2}$$

where $\mathbf{q} \in \mathbb{R}^n$ denotes the descriptor coordinates and $\mathbf{M} \in \mathbb{R}^{n\times n}$ is the constant mass matrix. In eq. (2) the vector $\boldsymbol{\phi}(\mathbf{q}) \in \mathbb{R}^m$ represents geometric constraints. Matrix $\mathbf{\Phi}_{\mathbf{q}}(\mathbf{q}) = \partial\boldsymbol{\phi}(\mathbf{q})/\partial\mathbf{q} \in \mathbb{R}^{m\times n}$ is the constraint Jacobian. Vector $\boldsymbol{\lambda} \in \mathbb{R}^m$ contains the Lagrange multipliers and $\mathbf{Q}_g \in \mathbb{R}^n$ is the constant generalized force vector of the gravitational terms. The control input vector is $\mathbf{u} \in \mathbb{R}^l$ is mapped by the input matrix $\mathbf{H}(\mathbf{q}) \in \mathbb{R}^{n\times l}$. The above formalism can directly be applied to the spatial case yielding the same form of the equations of motion as (1) and (2).

The inverse kinematical and dynamical calculations have unique solution if the number of control inputs and the dimension of the task is equal [1]. Thus the task have to be defined by l number of algebraic equations. This set of additional constraint equations are the so-called servo-constraints (control-constraints) $\boldsymbol{\phi}_s(\mathbf{q},\mathbf{p}(t)) = \mathbf{0}$. We assume that these servo-constraint equations can be written in the form $\boldsymbol{\phi}_s(\mathbf{q},\mathbf{p}(t)) = \mathbf{g}(\mathbf{q}) - \mathbf{p}(t)$ where $\mathbf{g}(\mathbf{q})$ represents, for example, the end-effector position of the robot and $\mathbf{p}(t)$ is the performance goal to be realized [1].

For under-actuated robotic systems modeled by Lagrangian equation of motion of the second kind, the computed torque control method was generalized in [5]. The generalized method is called Computed Desired Computed Torque Control (CDCTC) method. Here the expression "computed desired" refers to the fact that a set of uncontrolled coordinates can be separated from the controlled ones, and the desired values of these uncontrolled coordinates have to be calculated by considering the internal dynamics of the system.

The CDCTC method proposed in [5] can be applied to dynamical systems that are described by ordinary differential equations only. This problem can be resolved by projecting the equations of motion 1 and 2 to the subspace of admissible motions associated with the geometric constraints [4]. The simultaneous application of this projection (including the configuration corrections during the numerical solution) and the CDCTC algorithm is complex and computationally expensive. In addition, it has to be noted that the selection of the controlled and uncontrolled coordinates might be highly intuitive in case of complex (non-convetional) robotic structure like ACROBOTER. The introduction of this kind of distinct coordinates is possible only if the servo-constraint equations can be solved in closed form for the set of controlled coordinates.

Instead of the application of the CDCTC method, we apply the Backward Euler discretization for the DAE system the resulting set of implicit equations are solved by the Newton-Raphson method for the desired control inputs $\mathbf{u}$. Considering a PD controller with gain matrices $\mathbf{K}_P$ and $\mathbf{K}_D$ the control law can be formulated as

$$\mathbf{M}\ddot{\mathbf{q}}^d + \boldsymbol{\Phi}_\mathbf{q}^\mathrm{T}(\mathbf{q}^d)\boldsymbol{\lambda}^d = \mathbf{Q}_g + \mathbf{H}(\mathbf{q}^d)\mathbf{u} - \mathbf{K}_P(\mathbf{q}^d - \mathbf{q}) - \mathbf{K}_D(\dot{\mathbf{q}}^d - \dot{\mathbf{q}}) \,, \tag{3}$$

$$\boldsymbol{\phi}_s(\mathbf{q}^d,\mathbf{p}(t)) = \mathbf{0} \,, \tag{4}$$

$$\boldsymbol{\phi}(\mathbf{q}^d) = \mathbf{0} \,, \tag{5}$$

where superscript d refers to desired quantities. Then, the first order form of equation (3) reads

$$\dot{\mathbf{q}}^d = \mathbf{y}^d \tag{6}$$

$$\dot{\mathbf{y}}^d = \mathbf{M}^{-1}\left[-\boldsymbol{\Phi}_\mathbf{q}^\mathrm{T}(\mathbf{q}^d)\boldsymbol{\lambda}^d + \mathbf{Q}_g + \mathbf{H}(\mathbf{q}^d)\mathbf{u} - \mathbf{K}_P(\mathbf{q}^d - \mathbf{q}) - \mathbf{K}_D(\dot{\mathbf{q}}^d - \dot{\mathbf{q}})\right] \,. \tag{7}$$

Equations (6) and (7) are first order ordinary differential equations, while equations (4) and (5) are algebraic ones. We use the Backward Euler formula with timestep h to discretize the DAE system, that result in the set of nonlinear algebraic equations for the unknowns desired values $\mathbf{z}_{i+1} = \left[\mathbf{q}_{i+1}^d \; \mathbf{y}_{i+1}^d \; \mathbf{u}_{i+1} \; \boldsymbol{\lambda}_{i+1}^d \right]^{\mathrm{T}}$ in the form:

$$\mathbf{F}(\mathbf{z}_{i+1}) = \begin{bmatrix} \mathbf{q}_{i+1}^d - \mathbf{q}_i^d - h\mathbf{y}^d \\ \mathbf{y}_{i+1}^d - \mathbf{y}_i^d - h\mathbf{M}^{-1}\left[-\boldsymbol{\Phi}_{\mathbf{q}}^{\mathrm{T}}(\mathbf{q}_{i+1}^d)\boldsymbol{\lambda}_{i+1}^d + \mathbf{Q}_g + \mathbf{H}(\mathbf{q}_{i+1}^d)\mathbf{u}_{i+1} - \mathbf{K}\mathbf{e}\right] \\ \boldsymbol{\phi}_s(\mathbf{q}_{i+1}^d, \mathbf{p}(t_{i+1})) \\ \boldsymbol{\phi}(\mathbf{q}_{i+1}^d) \end{bmatrix} \tag{8}$$

$$\text{with} \quad \mathbf{K} = \left[\mathbf{K}_P \,\vdots\, \mathbf{K}_D \right] \text{ and } \mathbf{e} = \begin{bmatrix} \mathbf{q}_{i+1}^d - \mathbf{q}_{i+1} \\ \hdashline \dot{\mathbf{q}}_{i+1}^d - \dot{\mathbf{q}}_{i+1} \end{bmatrix}. \tag{9}$$

For the solution the initial values are $\mathbf{q}_0^d$ and $\mathbf{y}_0^d$ at $i = 0$. They should satisfy the servo-constraints and the geometric constraints. During simulation the initial values of the states $\mathbf{q}_0$ and $\mathbf{y}_0$ only have to satisfy the geometric constraints. The numerical solution of 8 and 9 is based on the well-known Newton-Raphson method. The corresponding Jacobian matrix $\mathbf{J}(\mathbf{z}_{i+1}) = \partial\mathbf{F}(\mathbf{z}_{i+1})/\partial\mathbf{z}_{i+1}$ is calculated numerically, however it could also be constructed semi-analytically [3]. Then the iteration

$$\mathbf{z}_{i+1}^{n+1} = \mathbf{z}_{i+1}^n - \mathbf{J}(\mathbf{z}_{i+1}^n)\mathbf{F}(\mathbf{z}_{i+1}^n) \tag{10}$$

provides the solution at each time instants, where $\mathbf{z}_{i+1}^n$ is the nth approximation of $\mathbf{z}_{i+1}$. The initial guess $\mathbf{z}_{i+1}^0$ in each time step comes from the best approximation $\mathbf{z}_i^N$ of the previous time step. Usually the Newtor-Raphson iteration converges in $N = 2 \div 6$ steps depending also on the required tolerance.

4 Real Parameter Case Simulation

This section presents the simulation results obtained for the planar model of ACROBOTER shown right in Fig. 1. The selected descriptor coordinates are the Cartesian coordinates of the points $\mathbf{P}_i$, $i = 1 \ldots 4$ yielding $\mathbf{q} = [x_1 \; x_2 \; z_2 \; x_3 \; z_3 \; x_4 \; z_4]^{\mathrm{T}}$. The control inputs are colleted in vector $\mathbf{u} = [F_L \; F_1 \; F_2 \; F_3 \; F_T]^{\mathrm{T}}$. The single geometric constraint represents the constant distance L_{34} between the points $\mathbf{P}_3$ and $\mathbf{P}_4$ and can be written as:

$$\boldsymbol{\phi}(\mathbf{q}) = \left[(x_3 - x_4)^2 + (z_3 - z_4)^2 - L_{34}^2\right] \tag{11}$$

The mass of the swinging unit is $m_{SU} = 9.3$kg and its moment of inertia with respect to the axis at $\mathbf{P}_3$ is $I_{SU} = 0.4\text{kgm}^2$. The mass of the cable connector is $m_{CC} = 0.5$kg and the mass of the linear drive that represents the climber unit is $m_{CU} = 20$kg. The distance L_{34} is set to be 0.4m. The position of the center of gravity is given by $\bar{\mathbf{r}}_C = [0.2 \; 0.05]^{\mathrm{T}}$, while the point of application of the thrust force and the position of point $\mathbf{O}$ are defined by the vectors $\bar{\mathbf{r}}_T = [0.2 \; -0.05]^{\mathrm{T}}$ and $\bar{\mathbf{r}}_O = [0.2 \; 0]^{\mathrm{T}}$ respectively in the local frame $(\bar{x}, \bar{z})$ measured in meters.

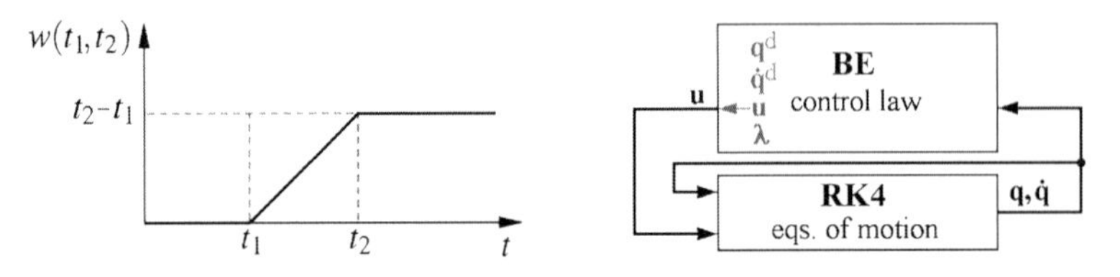

Fig. 2 Modified ramp function (left), block diagram of the simulation (right)

The task of the robot is to track a given trajectory of the point **O**. At the same time the elevation of the cable connector and the horizontality of the swinging unit are also prescribed. The servo-constraint $\boldsymbol{\phi}_s(\mathbf{q},\mathbf{p}(t)) = \mathbf{g}(\mathbf{q}) - \mathbf{p}(t)$ is defined by

$$\mathbf{g}(\mathbf{q}) = \left[x_1 \; z_2 - \frac{z_3+z_4}{2} \; \frac{x_3+x_4}{2} \; \frac{z_3+z_4}{2} \; z_3 - z_4\right]^{\mathrm{T}}, \tag{12}$$

$$\mathbf{p}(t) = \left[x_{CU}^d \; h_{CC}^d \; x_{SU}^d \; z_{SU}^d \; 0\right]^{\mathrm{T}}, \tag{13}$$

where the desired climbing unit position $x_{CU}^d = 0.4w(3,4)$, the desired cable connector elevation $h_{CC}^d = 0.8 - 0.2w(2,4)$, the desired swinging unit horizontal position $x_{SU}^d = 0.2w(2,4)$ and the vertical position $z_{SU}^d = -1.5 + 0.4w(2,4)$ are given in meters and they are used to calculate the reference values of the controller. The corresponding weighting functions $w(2,4)$ and $w(3,4)$ are defined by $w(t_1,t_2)$ as shown in Fig. 2. In the investigated simple case, it is possible to solve the servo-constraint equations and the geometric constraint equation for the intuitively chosen set of controlled coordinates $\mathbf{q}_c = [x_1 \; z_2 \; x_3 \; z_3 \; z_4]^{\mathrm{T}}$. with $\mathbf{q}_u = [x_2 \; x_4]^{\mathrm{T}}$. The corresponding solution for the controlled coordinates reads:

$$x_1 = x_{CU}^d, \quad z_2 = z_{SU}^d + h_{CC}^d, \quad x_3 = x_{SU}^d - \frac{L_{34}}{2}, \quad z_3 = z_{SU}^d \text{ and } z_4 = z_{SU}^d \tag{14}$$

The uncontrolled coordinate x_4 comes directly from the geometric constraint equation. Despite of the available closed form solution, here, we do not separate the controlled and the uncontrolled coordinates. Instead the servo-constrains are directly attached to the control law (3-5) proposed in Section 3.

The equations of motion was solved using the fourth order Runge-Kutta method, and the control input was calculated by the simultaneously applied Backward-Euler algorithm as shown in Fig. 2. The simulation of the DAE system was accomplished by using Baumgarte's method [2] under the assumption that the geometric constraints do not depend on time explicitly:

$$\begin{bmatrix} \mathbf{M} & \boldsymbol{\Phi}_{\mathrm{q}}^{\mathrm{T}} \\ \boldsymbol{\Phi}_{\mathrm{q}} & \mathbf{0} \end{bmatrix} \begin{bmatrix} \ddot{\mathbf{q}} \\ \boldsymbol{\lambda} \end{bmatrix} = \begin{bmatrix} \mathbf{Q}_{\mathrm{g}} + \mathbf{H}\mathbf{u} \\ -\dot{\boldsymbol{\Phi}}_{\mathrm{q}}\dot{\mathbf{q}} - 2\alpha\boldsymbol{\Phi}_{\mathrm{q}}\dot{\mathbf{q}} - \beta^2\boldsymbol{\phi} \end{bmatrix} \tag{15}$$

In equation (15) $\alpha = 40$ and $\beta = 60$ are constant numbers that effects the suppression of the geometric constraint errors. The time step of the simulation was set to $h = 0.01\,\mathrm{s}$.

Using the experimentally tuned gain matrices $\mathbf{K}_P = \mathrm{diag}(150,0,10,1,3,1,3)$ and $\mathbf{K}_D = \mathrm{diag}(1000,0.5,10,15,20,15,20)$, the simulated motion of the system is presented in Fig. 3, where panel (a) shows the stroboscopic motion of the planar ACROBOTER. The realized path of point $\mathbf{O}$ is denoted by the thick curve that slightly oscillates around, but converges to the desired path depicted as a thin straight line. The desired configurations are shown by dashed lines, while the continuous lines presents the realized configurations of the robot. According to the task equations (12) and (13) the robot is commanded to stand still till $t = 2$s, then the reference point $\mathbf{O}$ is commanded to move along a straight line with constant velocity. During the same period of time the desired elevation of the cable connector is decreasing.

The climbing unit is commanded to start moving with constant velocity at $t = 3$s. Then, at $t = 4$s the task is to keep the swinging unit in a certain fixed position. Panel (b) in Fig. 3 shows the constraint violation with the maximum of 4mm. Note that the constraint violation depend on the α and β parameters of eq. (15). Panels (c) and (d) show the servo-constraint errors. The error in the climbing unit's position is $\phi_{s,1}$, the elevation error of the cable connector is $\phi_{s,2}$, the horizontal and vertical position errors corresponding to the coordinates of point $\mathbf{O}$ are $\phi_{s,3}$ and $\phi_{s,4}$ and the orientation error of the swinging unit is $\phi_{s,5}$ (see eqs. (12) and (13)).

The servo-constraint errors show that the large initial errors decreasing in the first 2 second. Then the sudden change of the desired velocities causes further, but settling oscillations. When the system has to stop at $t = 4$s the oscillations are suppressed, too. However, the relatively high frequency oscillation of $\phi_{s,4}$ dies out slowly. This corresponds to the oscillations of the cable connector having relatively small mass compared to the swinging unit. It is hard to suppress the horizontal

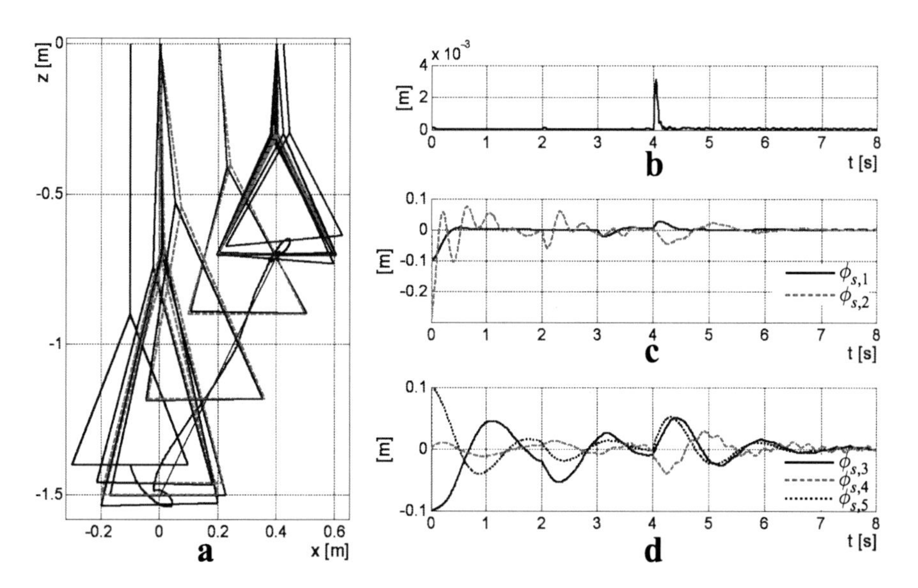

Fig. 3 Simulation results: (a) stroboscopic movement of the system, (b) violation of the geometric constraint, (c) and (d) servo constraint violations

vibration of the cable connector due to the under-actuated character of the robot. These oscillations could be decreased by prescribing smooth trajectories.

5 Conclusions

The computed torque control method was generalized and applied for the tracking control of ACROBOTER using descriptor type system modeling. In contrast to [5], the control inputs were determined via the direct solution of the corresponding DAE problem. The control task was defined by servo-constraint equations incorporated in the algebraic equations. Considering the experimentally tuned control parameters the presented simulation results show the applicability of the proposed simple PD controller for the tracking control of the under-actuated ACROBOTER system.

Acknowledgements. This work was supported in part by the Hungarian National Science Foundation under grant no. OTKA K068910, the ACROBOTER (IST-2006-045530) project and the HAS-BME Research Group on Dynamics of Machines and Vehicles.

References

[1] Blajer, W., Kolodziejczyk, K.: Modeling of underactuated mechanical systems in partly specified motion. Journal of Theoretical and Applied Mechanics 46(2), 383–394 (2008)
[2] de Jalón, J.G., Bayo, E.: Kinematic and dynamic simulation of multibody systems: the real-time challenge. Springer, Heidelberg (1994)
[3] Kovács, L.L., Zelei, A., Bencsik, L., Turi, J., Stépán, G.: Motion control of an under-actuated service robot using natural coordinates. In: RoManSy 18 - Robot Design, Dynamics and Control, Udine, Italy, pp. 331–338 (2010)
[4] Kövecses, J., Piedoboeuf, J.-C., Lange, C.: Dynamic modeling and simulation of constrained robotic systems. IEEE/ASME Transactions on mechatronics 8(2), 165–177 (2003)
[5] Lammerts, I.M.M.: Adaptive computed reference computed torque control. PhD thesis, Eindhoven University of Technology (1993)
[6] Sato, T., Fukui, R., Mofushita, H., Mori, T.: Construction of ceiling adsorbed mobile robots platform utilizing permanent magnet inductive traction method. In: Proceedings of 2004 IEEEiRSJ International Conference on Intelligent Robots and Systems, pp. 552–558 (2004)

The 3D-Pendulum at the World Exhibition 2010 – Control Design and Experimental Results

Thomas Gorius, Robert Seifried, and Peter Eberhard

Abstract. In this paper the control design and experimental results for stabilization and trajectory tracking of a large interactive 3D-pendulum are presented. This pendulum is part of the German Pavilion at the World Exhibition EXPO 2010 in Shanghai.

1 Introduction

During the World Exhibition EXPO 2010 in Shanghai the German Pavilion shows a large interactive pendulum as its main attraction. The most important part of this pendulum is its sphere with 3 m diameter which is equipped with almost 400.000 LEDs and which has a mass of 1.2 t. The sphere is mounted via a filigree bar with a length of 4.2 m and a Cardan joint at an electrically driven cross table used for the excitation of the system. The cross table is integrated in the roof of a theater-like building. Visitors perceive the EXPO-pendulum as a large levitating, shining sphere that features a motion and light show.

To make the sphere to perform large circular and pendular motions a feedback-control must be used. However, to keep up the impression of a non-driven pendulum, the motion of the mounting point, i.e. the motion of the cross table, should only be very small. The practical and widely investigated counterpart of this pendulum is a gantry crane. Because of their practical importance gantry cranes are often-used examples in the control theory literature, [1, 4]. Despite this apparent similarity these approaches cannot be used for the EXPO-pendulum since, roughly speaking, the control objectives of gantry cranes and this pendulum are the direct opposite to each other. In order to provide an impressive event the sphere should perform large displacements while ensuring only a very small motion of the cross table.

Thomas Gorius · Robert Seifried · Peter Eberhard
Institute of Engineering and Computational Mechanics, University of Stuttgart
Pfaffenwaldring 9, 70569 Stuttgart, Germany
e-mail: {gorius, seifried, eberhard}@itm.uni-stuttgart.de

G. Stépán et al. (Eds.): Dynamics Modeling & Interaction Cont., IUTAM BOOK SERIES 30, pp. 19–26.
springerlink.com

The main part of this paper is the presentation of experimental results of the closed loop using the control approaches given in [2, 3]. The paper is organized in the following way: After the description of the mechanical model and the control objectives and constraints in Sect. 2, a brief overview of the controller design for stabilization and tracking of the pendulum is given in Sect. 3. The closed loop performance is demontrated in Sect. 4 by experiments on the real pendulum. A short conclusion of the presented results is given in Sect. 5.

2 System Description

Figure 1 shows the EXPO-pendulum on the test rig in Stuttgart. The important parts of the system are the sphere, the connecting rod, a Cardan joint and the electrically driven cross table. In the case of the test rig the cross table is mounted on a steel structure. The real system can be modeled as the well-known pendulum-cart-system illustrated with four degrees of freedom, e.g. with the two directions of the crosstable x and y and the two Cardan angles α and β. The two acting forces on the carts F_x and F_y are the inputs of the system produced by the electric drives. For a detailed description of the system see [2, 3]. Mainly caused by the big pendulum length of 5.6 m, i.e. the distance from the mounting point to the center point of the sphere, the system's dynamical behaviour is quite slow. In the case of a fixed mounting point the eigenfrequency of the pendulum is approximately 0.21 Hz.

During the operation several hard constraints must be fulfilled at all times. As described in Sect. 1 the maximal displacement of the cross table is constrained to

Fig. 1 The EXPO-pendulum on the test rig in Stuttgart

achieve the impression of a non-driven pendulum. Additionally, the velocities and the accelerations of the carts are highly bounded due to the load on the mechanical parts, especially the joint, and the susceptibility to flexural vibrations of the rod. Finally, because of the design of the Cardan joint the angles are also limited. Thus, the constraints are given by

$$\begin{aligned} |x|, |y| \leq 0.35\,\mathrm{m}\,, \quad |\dot{x}|, |\dot{y}| \leq 0.4\,\frac{\mathrm{m}}{\mathrm{s}}\,, \quad |\ddot{x}|, |\ddot{y}| \leq 0.5\,\frac{\mathrm{m}}{\mathrm{s}^2}\,, \\ |\alpha|, |\beta| \leq 27^\circ\,, \quad |F_x|, |F_y| \leq 15\,\mathrm{kN}\,. \end{aligned} \tag{1}$$

It should be noted that the mechanical system and the control system are designed to achieve even larger Cardan angles. However, due to large safety margins and end switches, the above stated limits must be met.

The control objectives for the EXPO-pendulum are the following:

- stabilizing the equilibrium $(x^*, y^*, \alpha^*, \beta^*)^T = \mathbf{0}$,
- tracking in the sense of making the sphere's center point to track circles in the horizontal plane with slowly variable radius, and
- achieving a swinging of the sphere's center point in a vertical plane rotated along the vertical axis with variable amplitude.

Stabilization mainly means to bring the pendulum from an arbitrary motion into rest. Due to the limitation of the Cardan angles the maximum amplitude of the sphere's center is about 2.55 m. The motion of the pendulum is demanded online so that the controller must be able to switch between these kinds of motion at any time. Of course the transient behaviour of the closed loop should be as fast as possible because time is a leading factor during this event.

3 Controller Design

In this section a brief overview of the implemented controllers is given. It should be pointed out that for the controller design the acting forces on the carts are not considered as the inputs of the system. The reason is that the operation of the EXPO-pendulum has to proceed in due consideration of high safety and reliability guidelines. Therefore, the whole system is only allowed to contain certificated parts, e.g. electronic devices and safety systems. Commercial components like electric drive trains as used for common industry applications can be combined well with such certificated devices which makes it easier to design a safety control and communication. The practical problem is that the drive train used for the EXPO-pendulum cannot be controlled by demanding a drive torque, which is equivalent to an acting force on the cart, because of severe communication and safety control problems. Therefore, the internal position and velocity cascade control of the drives must be used. Hence, the displacements of the carts x and y must be considered as the input of the dynamical system instead of the acting forces on the carts. The use of the internal position controller has the main advantage that the state contraints (1) become now input constraints which can be incorporated much easier. Additionally, model uncertainties of the carts' dynamical behaviour such as friction or elasticity of the drive train are directly compensated by this cascade control.

3.1 Stabilization

After a linearization of the reduced system, i.e. the pendulum considering the position of the carts x and y as the inputs, two SISO-controllers can be designed to stabilize the pendulum where x is used to control β and y is used to control α, see Fig. 2. It can be proven that the closed loop consisting of the linearized reduced system and the controller K being Laplace transformed

$$K(s) = \frac{s + T_{10}}{s^3 + T_{22}\,s^2 + T_{21}\,s + T_{20}} \tag{2}$$

is asymptotically stable *for any* controller gain $V > 0$ and *for all physically meaningful parameters* of the pendulum if

$$T_{10}, T_{21}, T_{20} > 0\,, \quad T_{22} > T_{10}\,, \quad T_{20} < T_{10}\,T_{21} \tag{3}$$

hold. Note that Eq. (3) also implies

$$\lim_{t\to\infty} \beta(t) = 0 \quad \longrightarrow \quad \lim_{t\to\infty} x(t) = 0\,, \quad \lim_{t\to\infty} \alpha(t) = 0 \quad \longrightarrow \quad \lim_{t\to\infty} y(t) = 0\,. \tag{4}$$

Hence, the asymptotic stability of the equilibrium $(x^*, y^*, \alpha^*, \beta^*)^T = \mathbf{0}$ is achieved. Instead of using a constant V, piecewise constant controller gains V_α and V_β are applied chosen dependent on the maximum value of α and β, denoted by ${}_iA_\alpha$ and ${}_iA_\beta$, that are observed in a time intervall $t \in [(i-1)\,T_A, i\,T_A)$, $i \in \mathbb{N}^0$, $T_A > 0$ fixed. For α and analogously for β it is

$$\begin{aligned} {}_iA_\alpha &= \max_{(i-1)T_A \le t < iT_A} |\alpha(t)|\,, \quad {}_0A_\alpha = \alpha_{\max}\,, \quad h(t) = \begin{cases} 1 & \text{if } \; 0 \le t < T_A\,, \\ 0 & \text{else}\,, \end{cases} \\ A_\alpha(t) &= \sum_{i\ge 0} {}_iA_\alpha\, h(t - i\,T_A)\,, \quad V_\alpha(t) = V_\alpha\,(A_\alpha\,(t))\,. \end{aligned} \tag{5}$$

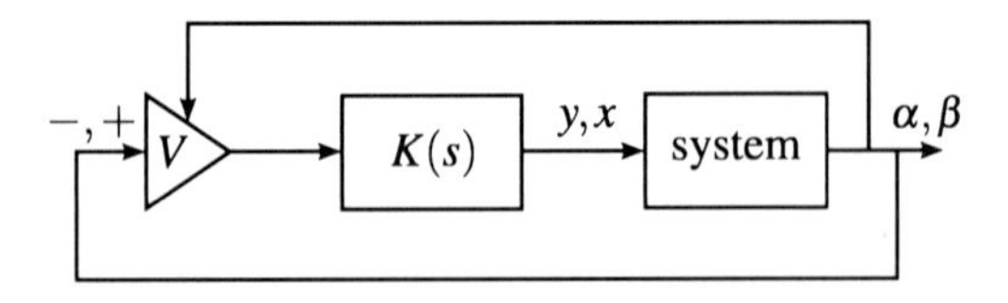

Fig. 2 Control structure for the stabilization of the pendulum using gain adaption

With A_α being a piecewise constant function, V_α is also piecewise constant. The adaption of the controller gains is very useful to control the eigenvalues of the closed loop in such a way that its performance is significantly improved while ensuring all constraints (1) to be fulfilled, see Fig. 3.

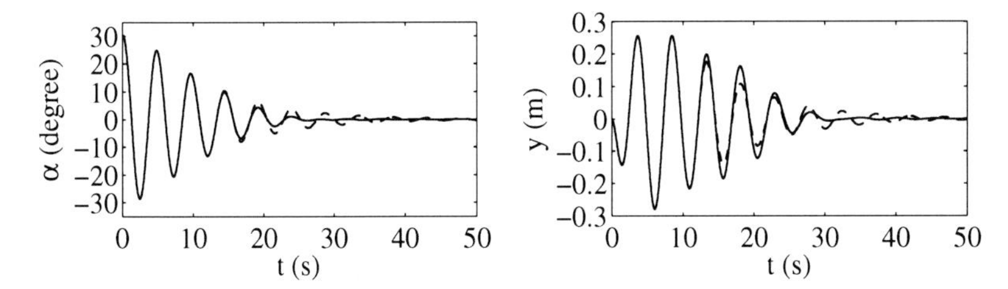

Fig. 3 Stabilization with constant (*dashed lines*) and piecewise constant controller gain (*solid lines*), where the gain increases if A_α is sufficiently small

Exploiting the symmetry of the pendulum, the control approach given in Fig. 2 can be combined with a state transformation to perform pendular motions in vertical planes rotated along the vertical axis, see [2, 3].

3.2 Tracking of Circular Motion

Finally a tracking controller should be designed such that the absolute position of the sphere's center point denoted by w_x and w_y, tracks circles in the horizontal plane with a desired amplitude R and a given frequency ω_0. Using the internal model principle a suitable controller for the control structure of Fig. 4 is

$$K_T(s) = \frac{c_4 s^4 + c_3 s^3 + c_2 s^2 + c_1 s + c_0}{s^4 + d_3 s^3 + d_2 s^2 + d_1 s + d_0} \frac{\frac{s}{T_2}+1}{\frac{s}{T_1}+1} \frac{1}{s^2+\omega_0^2}. \tag{6}$$

The coefficients c_i and d_i are calculated from the desired poles of the error dynamics. By changing the parameters T_1 and T_2 the offset and noise rejection of the closed loop can be influenced.

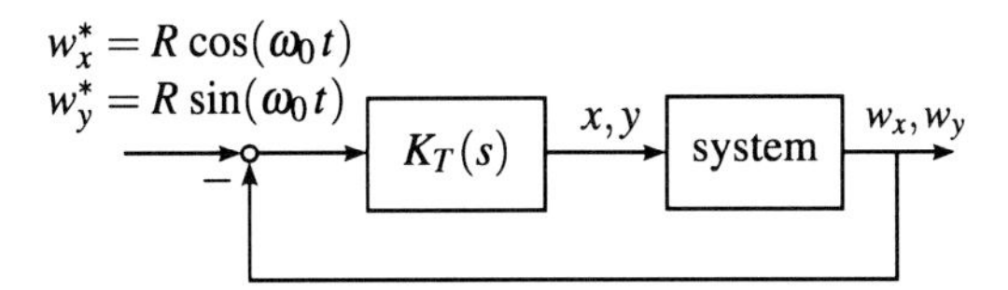

Fig. 4 Control structure for tracking of circular motions

4 Experimental Results

The control approaches of Sect. 3 are integrated into a finite state machine to handle the user inputs and are implemented in a numerical control software. This section gives the results of experiments which are performed on the test rig shown in Fig. 1.

To illustrate the possible combination of different kinds of motion Fig. 5a shows the trajectory of the sphere's center point when switching from circular to pendular motion followed by a rotation of the oscillation plane. First the pendulum is accelerated from rest to circular motion with a radius of 2.3 m within 40 s. Roughly speaking, the subsequent switching process to pendular motion in a vertical oscillation plane rotated with 45° along the vertical axis can be done by stabilizing the motion that is orthogonal to this plane. This results in ellipse-like trajectories of the sphere. By afterwards changing the orientation of this plane Fig. 5a shows the efficiency of this approach.

As described in the sections before the displacements of the carts are really small compared to the motion of the sphere. Figure 5b clearly demonstrates this huge difference which creates the impression of a free pendulum.

The tracking performance during the circular motion in Fig. 5 is shown in detail in Fig. 6. It is seen that the tracking controller yields very good tracking of the desired circular motion in the entire phase. For pure circular motion, i.e. R in Fig. 4 is constant, the tracking errors are less than 1.5 cm.

The last given experiment is the stabilization from circular motion with $R = 2.4$ m to rest. Figure 7a shows the visitors view to the stabilizing process while Fig. 7b gives details of the sphere's motion over time. The breaking starts at approximately 73 s. One can see that due to the high induced damping effect of the applied controller no significant motion is left after 20 s.

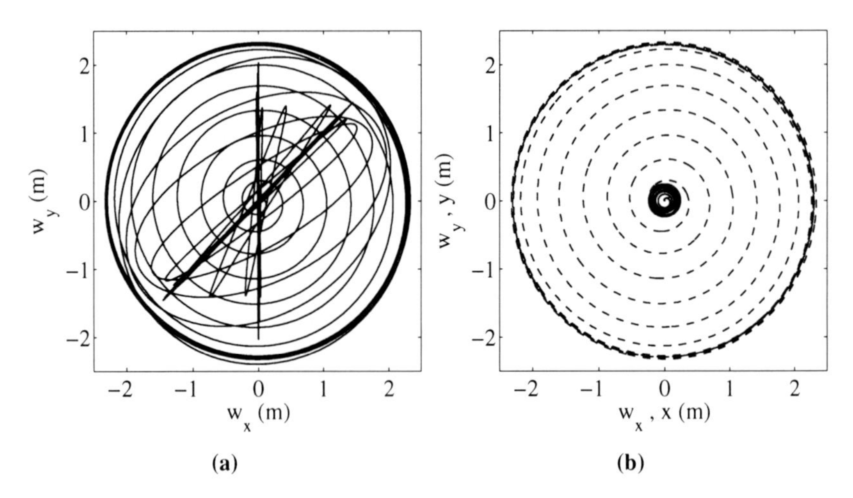

Fig. 5 Trajectories of the pendulum. **a** Trajectory of the sphere's center point accelerating from rest to circular motion, afterwards switching to pendular motion and rotating the plane of motion from 45° to 90°. **b** Comparision between the circular displacements of the sphere's center point (*dashed line*) and the displacements of the cross table produced by the tracking controller (*solid line*)

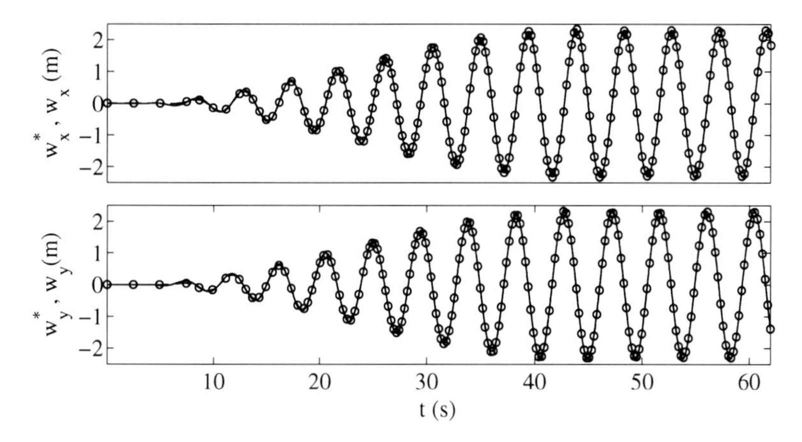

Fig. 6 Performance of the tracking controller. *Solid lines* are the desired trajectories, *dashed-circled lines* are the measured signals

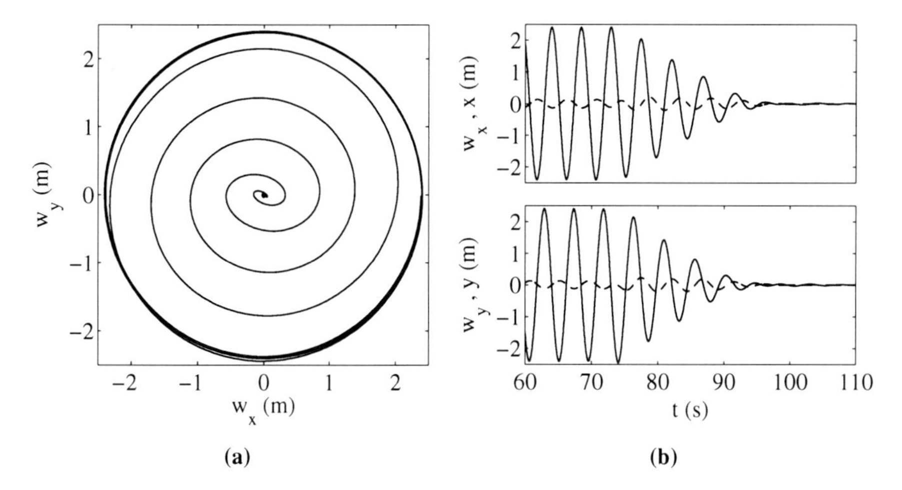

Fig. 7 Stabilization of the pendulum. **a** Trajectory of the sphere's center point when stabilizing the pendulum from circular motion. **b** Performance of the stabilizing controller. *Solid lines* are the position of the sphere's center point, *dashed lines* are the positions of the carts

5 Conclusion

The large 3D-show-pendulum for the German Pavilion at the World Exhibition EXPO 2010 in Shanghai was presented. For the desired shows, stabilization of the equilibrium as well as pendular motion and trajectory tracking of circular motion must be achieved. Therefore, a feedback control approach was designed and implemented on the real-life system. This approach is very well suitable to meet the control goals and the technical limitations of the actual hardware. The efficiency of this approach was shown by experiments on the real-life system.

Acknowledgments. Such a large project is the work of many people. Especially the authors want to thank Milla & Partner and Mr. I. Kaske who are responsible for the idea, concept and the coordination, the Institute for Control Engineering of Machine Tools and Manufacturing Units (ISW) (Dipl.-Ing. A. Hafla, Dipl.-Ing. H.-P. Bock, Dipl.-Ing. P. Sekler) who did the control hardware, drives and implementation, and the Institute of Machine Components (IMA) (Dr.-Ing. A. Krolo, Dipl.-Ing. B. Klein) who contributed to the mechanical design. The actual manufacturing was done by Metron and the video hardware is from ICT. The German Pavilion has been commissioned by the Federal Ministry of Economics and Technology, organization and operation is conducted by Koelnmesse International GmbH.

References

[1] Fang, Y., Dixon, W.E., Dawson, D.M., Zergeroglu, E.: Nonlinear Coupling Control Laws for a 3-DOF Overhead Crane System. In: Proceedings of the 40th IEEE Conference on Decision and Control, vol. 4, pp. 3766–3771 (2001)

[2] Gorius, T., Seifried, R., Eberhard, P.: Control Approaches for a 3D-Pendulum on Display at the EXPO 2010. In: Terze, Z., Lacor, C. (eds.) Proceedings of the International Symposium on Coupled Methods in Numerical Dynamics, Faculty of Mechanical Engineering and Naval Architecture, Zagreb, pp. 281–300 (2009)

[3] Seifried, R., Gorius, T., Eberhard, P.: Control of the Interactive 3D-Pendulum Presented at the World Exhibition EXPO 2010. In: Proceedings of the 9th International Conference on Motion and Vibration Control, MOVIC (2010) (submitted for publication)

[4] Singhal, R., Patayane, R., Banavar, R.N.: Tracking a Trajectory for a Gantry Crane: Comparision Between IDA-PBC and Direct Lyapunov Approach. In: IEEE International Conference on Industrial Technology, pp. 1788–1793 (2006)

Contact to Sample Surface by Self-excited Micro-cantilever Probe in AFM

H. Yabuno, M. Kuroda, and T. Someya

Abstract. AFM (atomic force microscope) has widely used in the fields of surface science, biological science, and so on. The measurement of biological samples requires the detection without contact between the samples and the micro-cantilever probe, because the contact damages the observation objects. The self-excitation technique for the micro-cantilever probe is a powerful tool for the detection in a liquid environment where the micro-cantilever has a very low quality factor Q. In the previous study, we propose an amplitude control method for a van der Pol type self-excited micro-cantilever probe by nonlinear feedback based on bifurcation control. In this presentation, we experimentally investigate the characteristics of detection depending on the magnitude of the response amplitude of the micro-cantilever probe by using our own making AFM. We change the response steady state amplitude of the micro-cantilever probe by setting the nonlinear feedback gain to some values. Based on frequency modulation method, we detect the sample surface and discuss the effect of the magnitude of the amplitude on the performance in air. It is concluded from experimental results that the micro-cantilever probe with small amplitude under high gain nonlinear feedback control enhances the detection performance of AFM.

Hiorshi Yabuno
Keio University, 3-14-1 Hiyoshi,
Kohoku Yokohama, 223-8522, Japan
e-mail: `yabuno@mech.keio.ac.jp`

Masaharu Kuroda
National Institute of Advanced Industrial Science and Technology,
1-2-1, Namiki, Tsukuba, 305-8564, Japan
e-mail: `m-kuroda@aist.go.jp`

Takashi Someya
University of Tsukuba,
1-1-1 Ten-no-dai, Tsukuba, 305-8573, Japan

G. Stépán et al. (Eds.): Dynamics Modeling & Interaction Cont., IUTAM BOOK SERIES 30, pp. 27–33.
springerlink.com

1 Introduction

Atomic force microscopes (AFMs) have been applied widely as a powerful tool for modern nano-science for nanoscale imaging and surface manipulation. It has been difficult to apply AFM to measurement of biomolecules. First, to keep the samples biologically active, the measurement must be performed in a high-viscosity liquid, where the quality factor Q is very low. Secondly, to impart no damage to soft materials with an irregular surface, non-contact measurement is required so that the contact of the micro-cantilever probe to the surface is avoided. As mentioned in the next section, noncontact AFM (NC-AFM) method detects a frequency shift in the mechanical resonant oscillation of the micro-cantilever due to the attractive force between the tip apex atom and the sample surface atom [1]. In this paper, the method to reduce the response amplitude is introduced to realize the noncontact measurement. We perform linear plus nonlinear feedback control so that the micro-cantilever behaves as a van der Pol type self-excited oscillator [2]. Linear feedback causes self-excited oscillation in the micro-cantilever probe even in a high viscous liquid environment and high gain nonlinear feedback decreases the steady state amplitude of the self-excited oscillation. By using our own making AFM, we experimentally confirm that the high gain nonlinear feedback gain realizes the small response amplitude and the self-excited micro-cantilever with small amplitude enhances the detection method of AFM. when the response amplitude is kept smaller.

2 Principle of Measurement by AFM

Toward the discussion on measurement by noncontact AFM (NC-AFM) with van der Pol type self-excited micro-cantilever probe, we show by using a simple spring-mass model instead of micro-cantilever probe how to measure the profile of sample surface without probe contact. Atomic force between the surface and the tip of micro-cantilever probe is changed depending on the distance of them. Based on the fact that the variation of the atomic force corresponds to the profile of sample surface, NC-AFM carries out measuring the profile without contact scanning a micro-cantilever probe on the surface as shown in Fig. 1.

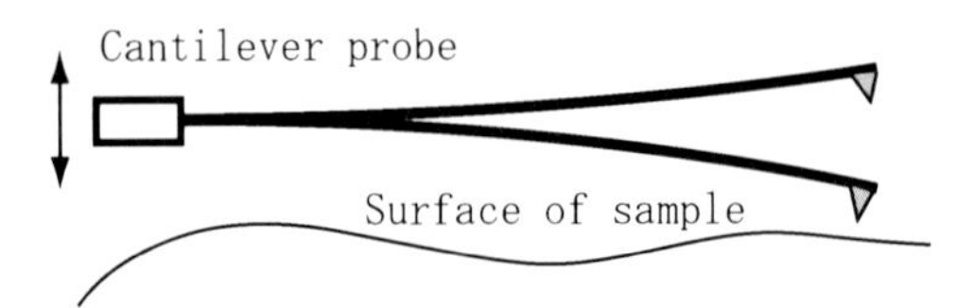

Fig. 1 Measurement by micro-cantilever probe

By considering only the first mode, we introduce a simple model of the micro-cantilever as Fig. 2(a), which consists of mass and spring whose natural length is

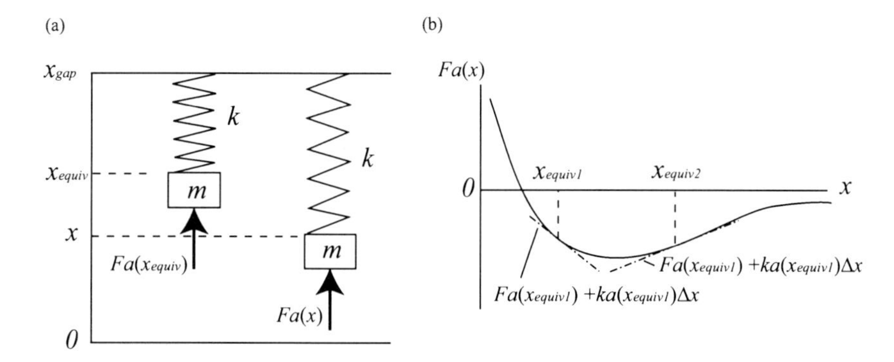

Fig. 2 Spring-mass model for cantilever probe

zero; the restoring force of the spring is assumed to be zero in this model when the position of the mass x is x_{gap}. The bottom line stands for the position of surface of the sample. The static equilibrium points are x_{gap} and x_{equiv} in the cases without and with effect of atomic force, respectively. The upper horizontal line in Fig. 2(a) is the supporting point of the micro-cantilever, i.e., the position of the tip of the micro-cantilever in the state where it is not deflected. The atomic force between the tip apex atom on the cantilever and the sample surface atom depends on the tip-sample distance x_{gap} as Fig. 2(b).

The equation of motion of the spring mass system is expressed as

$$m\ddot{x} = -k(x - x_{gap}) + F_a(x) - c\dot{x}, \tag{1}$$

where k is a equivalent spring constant of micro cantilever, x_{gap} is the distance between the surface, and $F_a(x)$ is atomic force. The equilibrium point of x_{equiv} satisfies the following equilibrium equation:

$$0 = -k(x_{equiv} - x_{gap}) + F_a(x_{equiv}). \tag{2}$$

The equilibrium point x_{equiv} is changed with the gap x_{gap} in the static equilibrium state under no effect of atomic force. Expanding $F_a(x)$ around the equilibrium point with respect to $\Delta x (= x - x_{equiv})$ yields

$$F_a(x) = F_a(x_{equiv}) + \left.\frac{dF_a}{dx}\right|_{x=x_{equiv}} \Delta x \tag{3}$$

By using Eqs. (1), (2), and (3), we obtain the equation governing of the oscillation in the neighborhood of the equilibrium point $x = x_{equiv}$ as

$$m\Delta\ddot{x} + c\Delta\dot{x} + \left\{k - k_a(x_{gap})\right\} \Delta x = 0, \tag{4}$$

where $k_a(x_{gap}) = dF_a/dx|_{x=x_{equiv}(x_{gap})}$. The effect of viscous damping is also assumed. It is noticed again that the natural frequency $\sqrt{\{k - k_a(x_{gap})\}/m}$ of the mass subjected to atomic force is varied by the tip-sample distance x_{gap}. While horizontally scanning, we move vertically the position of the supporting point of the micro-cantilever probe, i. e., x_{gap} in the spring mass model of Fig. 2(a), so that the natural frequency is kept constant. Then, the vertical motion of the supporting point is equivalent to the profile of the sample.

It is important for obtaining atomic high resolution to detect the natural frequency of the cantilever probe within high accuracy. The utilization of self-excited micro-cantilever probe under positive linear velocity feedback is proposed [3]. In general, the response frequency of self-excited system is equal to the natural frequency under small response amplitude; the frequency is related to the response amplitude, but if the response amplitude is very small and the error of the square of the response amplitude is neglected, the response amplitude is equal to the linear natural frequency. Therefore, this frequency detection method based on self-excited oscillation is much easier to detect the natural frequency within high accuracy than the conventional method based on external excitation, even in a liquid environment where Q factor of the micro-cantilever probe is very low; Q factor quantitatively describes by using the frequency response curve how damped an resonator is [4].

In order to avoid the contact of the micro-cantilever to the sample, we apply more nonlinear feedback so that the micro-cantilever behaves as van der Pol oscillator. Then, the equation of motion is as follows:

$$m\Delta\ddot{x} + c\Delta\dot{x} + (k - k_a(x_{gap}))\,\Delta x = -k_{cont}\Delta\dot{x} - k_{non}\Delta x^2\Delta\dot{x}, \tag{5}$$

which is rewritten as

$$m\Delta\ddot{x} + \left\{(c + k_{cont}) - k_{non}\Delta x^2\right\}\Delta\dot{x} + (k - k_a(x_{gap}))\,\Delta x = 0. \tag{6}$$

From this governing equation, we can realize the van der Pol self-excited micro cantilever probe by suitable linear and nonlinear feedback gains: the linear feedback gain k_{cont} satisfying $c + k_{cont} < 0$ causes self-excited oscillation and the high gain nonlinear feedback realizes the small limit cycle, i.e., small steady state amplitude.

3 Experiment of Noncontact Measurement

We apply the control method proposed in section 2 to our own making AFM as shown in Fig. 3. A micro-cantilever probe is equipped in the AFM. The dimensions is $225\mu\text{m}\times 33\mu\text{m}\times 5\mu\text{m}$ and the natural frequency is around 120 kHz. By an optical lever based on a laser diode and a photo detector, the deflection angle of the mico-cantilever probe is measured and from the data, the linear and nonlinear feedback is calculated in real time and the piezo actuator attached to the supporting point of the micro-cantilever probe is actuated to realize the dynamics of van der Pol oscillator [2]. The tube scanner moves the sample in the vertical direction and the vertical position is measured by a piezo-electric displacement sensor.

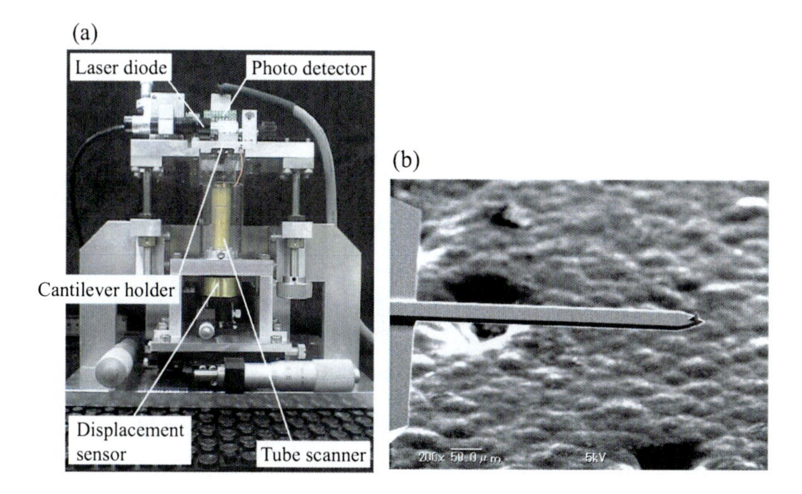

Fig. 3 Picture of micro-cantilever ((a) overview of AFM, (b) micro-cantilever probe (dimension is $225\mu m \times 33\mu m \times 5\mu m$ and natural frequency is 121kHz)

We experimentally examine the shift of the natural frequency of the micro-cantilever depending on the tip-sample distance. By changing the nonlinear feedback gain, we perform experiments about self-excited micro-cantilever probe under two different steady state amplitudes. First, we set the response amplitude to be 22.7nm when the tip-sample distance is 20 nm and we approach the sample stage to the micro-cantilever probe. The experimental results are shown in Fig. 4(a). The ordinates are the response frequency of the self-excited micro-cantilever probe and the response amplitude. The abscissa is the position of the sample stage and the absolute value corresponds to the gap of the tip-sample distance in the equilibrium state; 0 means the contact position of the stage to the tip in the equilibrium state. It is seen from Fig. 4 (a) that in the case when the gap is greater than around 5nm, the natural frequency is not changed. Therefore, the AFM cannot detect the existence of the sample in such a gap. However, at the cap of 3.9nm, the AFM can detect it. In this experiment, the detection criterion while approaching the sample stage to the micro-cantilever probe is set as follows: If the two conditions are satisfied that when the average of the slope of adjacent 5 points on the graph between the natural frequency and the gap is smaller than -3×10^{-9} and that the natural frequency decreases in nearer region to the sample, the first point of the five points is regarded as the point where AFM detects the existence of the sample.

Next, when the tip-sample distance is 20 nm, by high gain nonlinear feedback, we set the response amplitude to be much smaller 5.4 nm compared with in the case of Fig. 4(a). Similar to the experiment for Fig. 4(a), we approach the sample stage to the micro-cantilever probe. The experimental results are shown in Fig. 4(b). Under the same criterion of the detection as the previous experiment, AFM can detect the sample at the tip-sample distance of 8.5nm. Compared with the case of the relatively

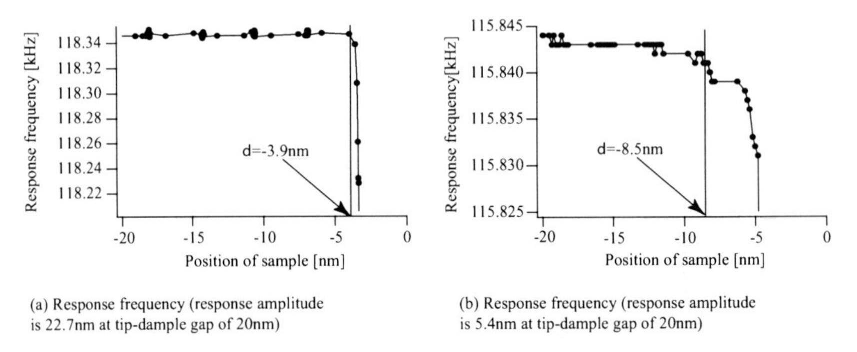

(a) Response frequency (response amplitude is 22.7nm at tip-dample gap of 20nm)

(b) Response frequency (response amplitude is 5.4nm at tip-dample gap of 20nm)

Fig. 4 Relationship between equivalent natural frequency and position of sample in air (Sampling step : 0.2nm, Sample : mica)

large response amplitude of the micro-cantilever in Fig. 4(a), AFM can detect the existence of the sample at the position of the micro-cantilever probe farther from the sample.

4 Conclusions

Performance of NC-AFM by using van der Pol self-excited micro-cantilever probe is experimentally examined. First, we introduce a simple spring-mass system subjected to atomic force and describe the measurement principle of NC-AFM. We show a van der Pol type self-excited micro-cantilever based on linear plus nonlinear feedback control; the response amplitude depends on the magnitude of the nonlinear feedback gain [5]. We use our own making AFM and investigate the dependence of the detection performance on the response amplitude of the van der Pol type self-excited micro-cantilever probe. It is experimentally concluded that the micro-cantilever probe with relatively small response amplitude can detect the existence of the sample in the state where the probe is farther from the sample, compared with the case with larger response amplitude.

Acknowledgements. This work was supported by a Grant-in-Aid for Science Research from the Japanese Ministry of Education, Culture, Sports, Science and Technology, No. 22360096.

References

1. Morita, S., Wiesendanger, R., Meyer, E.: Noncontact Atomic Force Microscopy. Springer, Heidelberg (2002)
2. Yabuno, H., Kaneko, H., Kuroda, M., Kobayashi, T.: Van der Pol Type Self-Excited Micro-cantilever Probe of Atomic Force Microscopy. Nonlinear Dynamics 54, 137–149 (2008)

3. Okajima, T., Sekiguchi, H., Arakawa, H., Ikai, A.: Self-Oscillation Technique for AFM in Liquids. Applied Surface Science 210, 68–72 (2003)
4. Tooley, M.: Electronic circuits: fundamentals and application, Newnes (2006)
5. Yabuno, H.: Stabilization and Utilization of Nonlinear Phenomena Based on Bifurcation Control for Slow Dynamics. Journal of Sound and Vibration 315, 766–780

Dynamics and Trajectory Planning of a Space Robot with Control of the Base Attitude

Fumiya Matsumoto and Hiroaki Yoshimura

Abstract. This paper develops a trajectory planning for a space robot, which enables us to simultaneously control its base attitude as well as the end effector trajectory. First, it is shown how the space robot dynamics can be formulated in the context of regular Lagrangian systems with holonomic constraints. Second, geometry of the space robot motion is explored; namely, it is shown how geometric phases corresponding to deviations of the base attitude are yielded in conjunction with the end effector motion. In our trajectory planning, it is demonstrated how the base attitude of the space robot can be controlled by the end effector in iteratively drawing complementary circles to reduce the geometric phase. Finally, we demonstrate the validity of our approach with numerical simulations.

1 Introduction

For a space robot that is floating in space, the position and attitude of the base may be changed by reaction forces when the end effector moves (see [2, 3]). The motion control of such a space robot is known as a falling cat problem ([1, 4]); namely, a falling cat can change its attitude by twisting her body although she has no angular momentum in the initial state.

In this paper, we study dynamics and trajectory generation of a space robot with control of its base attitude. First, we derive holonomic constraints in the two-dimensional Euclidean space and develop required kinematical equations using

Fumiya Matsumoto
Department of Applied Mechanics and Aerospace Engineering,
Waseda University, 3-4-1 Okubo Shinjuku, Tokyo 169-8555, Japan
e-mail: 11.08.gen@ruri.waseda.jp

Hiroaki Yoshimura
Department of Applied Mechanics and Aerospace Engineering,
Waseda University, 3-4-1 Okubo Shinjuku, Tokyo 169-8555, Japan
e-mail: yoshimura@waseda.jp

G. Stépán et al. (Eds.): Dynamics Modeling & Interaction Cont., IUTAM BOOK SERIES 30, pp. 35–43.
springerlink.com

local coordinates in a configuration manifold. Second, we develop a mathematical model of the space robot in the context of regular Lagrangian systems with holonomic constraints; we show how the set of translational and rotational equations of motion can be formulated. Then, we propose a trajectory planning of the space robot with controlling the base attitude using geometric phases; namely, the base attitude is shown to be controlled by the end effector in iteratively drawing complementary circles to reduce the geometric phase. Finally, we illustrate the validity of our approach with numerical simulations.

2 Space Robots

Consider a model of space robots as shown in Fig. 1 and every parameter is shown in Table 1.

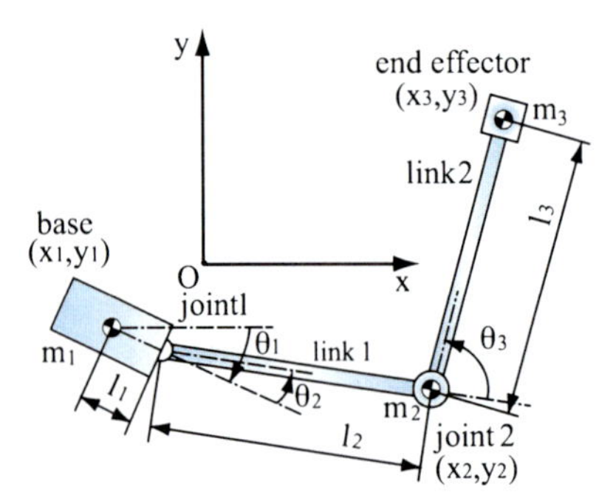

Fig. 1 Space robot model

Table 1 Parameters of the space robot

Symbol	Meaning	Value
m_1	Mass of base	100.0 kg
m_2	Mass of joint 2	5.0 kg
m_3	Mass of hand	7.5 kg
M	Total mass	112.5 kg
I_1	Inertia moment of base	500.0 kg m^2
I_2	Inertia moment of joint 2	25.0 kg m^2
I_3	Inertia moment of hand	37.5 kg m^2
l_1	Distance from the center of base to joint 1	1.0 m
l_2	Length of link 1	5.0 m
l_3	Length of link 2	5.0 m

We assume that two links consist of massless rigid bodies, each of which is connected by an ideal rotational joint and also that the base body consists of a rigid body. We also suppose that no external forces or torques are imposed on the space robot. Furthermore, the joint 2 and the end effector are modeled by point masses.

Let $Q = \mathbb{R}^2 \times \mathbb{R}^2 \times \mathbb{R}^2 \times S^1 \times S^1 \times S^1$ be a configuration space, whose local coordinates are given by $(x_1, y_1, x_2, y_2, x_3, y_3, \theta_1, \theta_2, \theta_3)$. Now, we have holonomic constraints denoted by

$$\begin{aligned}
&x_1 - x_2 + l_1 \cos\theta_1 + l_2 \cos(\theta_1 + \theta_2) = 0,\\
&y_1 - y_2 + l_1 \sin\theta_1 + l_2 \sin(\theta_1 + \theta_2) = 0,\\
&x_2 - x_3 + l_3 \cos(\theta_1 + \theta_2 + \theta_3) = 0,\\
&y_2 - y_3 + l_3 \sin(\theta_1 + \theta_2 + \theta_3) = 0.
\end{aligned} \tag{1}$$

By the time derivative of holonomic constraints in (1), it follows

$$\begin{aligned}
&\dot{x}_1 - \dot{x}_2 - l_1 \dot{\theta}_1 \sin\theta_1 - l_2(\dot{\theta}_1 + \dot{\theta}_2) \sin(\theta_1 + \theta_2) = 0,\\
&\dot{y}_1 - \dot{y}_2 + l_1 \dot{\theta}_1 \cos\theta_1 + l_2(\dot{\theta}_1 + \dot{\theta}_2) \cos(\theta_1 + \theta_2) = 0,\\
&\dot{x}_2 - \dot{x}_3 - l_3(\dot{\theta}_1 + \dot{\theta}_2 + \dot{\theta}_3) \sin(\theta_1 + \theta_2 + \theta_3) = 0,\\
&\dot{y}_2 - \dot{y}_3 + l_3(\dot{\theta}_1 + \dot{\theta}_2 + \dot{\theta}_3) \cos(\theta_1 + \theta_2 + \theta_3) = 0.
\end{aligned} \tag{2}$$

3 Lagrangian Dynamics with Holonomic Constraints

The Lagrangian of the space robot is given by

$$\begin{aligned}
L\ &(x_1, y_1, x_2, y_2, x_3, y_3, \theta_1, \theta_2, \theta_3, v_{x_1}, v_{y_1}, v_{x_2}, v_{y_2}, v_{x_3}, v_{y_3} v_{\theta_1}, v_{\theta_2}, v_{\theta_3})\\
&= \tfrac{1}{2}\{m_1 \left(v_{x_1}^2 + v_{y_1}^2\right) + m_2 \left(v_{x_2}^2 + v_{y_2}^2\right) + m_3 \left(v_{x_3}^2 + v_{y3}^2\right)\\
&\qquad + I_1 v_{\theta_1}^2 + I_2 \left(v_{\theta_1} + v_{\theta_2}\right)^2 + I_3 \left(v_{\theta_1} + v_{\theta_2} + v_{\theta_3}\right)^2\},
\end{aligned} \tag{3}$$

which is *regular* since the determinant of the Hessian of L is nonsingular. In the context of regular Lagrangian systems with constraints (see [5]), dynamics of the space robot can be denoted by using Lagrange's multipliers $\mu_1, \ldots, \mu_4$ as follows:

① For the translational motion,

$$\begin{aligned}
&m_1 \dot{v}_{x_1} = -\mu_1, \quad m_1 \dot{v}_{y_1} = -\mu_2, \quad m_2 \dot{v}_{x_2} = \mu_1 - \mu_3, \quad m_2 \dot{v}_{y_2} = \mu_2 - \mu_4,\\
&m_3 \dot{v}_{x_3} = \mu_3, \quad m_3 \dot{v}_{y_3} = \mu_4.
\end{aligned}$$

② For the rotational motion,

$$\begin{aligned}
&(I_1 + I_2 + I_3)\dot{v}_{\theta_1} + (I_2 + I_3)\dot{v}_{\theta_2} + I_3 \dot{v}_{\theta_3}\\
&\quad = \mu_1 \{l_1 \sin\theta_1 + l_2 \sin(\theta_1 + \theta_2)\}\\
&\qquad - \mu_2 \{l_1 \cos\theta_1 + l_2 \cos(\theta_1 + \theta_2)\}\\
&\qquad + \mu_3 l_3 \sin(\theta_1 + \theta_2 + \theta_3) - \mu_4 l_3 \cos(\theta_1 + \theta_2 + \theta_3)
\end{aligned}$$

and

$$
\begin{aligned}
\tau_1 &= (I_2+I_3)(\dot v_{\theta_1}+\dot v_{\theta_2})+I_3\dot v_{\theta_3}-\mu_1 l_2\sin(\theta_1+\theta_2)\\
&\quad +\mu_2 l_2\cos(\theta_1+\theta_2)-\mu_3 l_3\sin(\theta_1+\theta_2+\theta_3)+\mu_4 l_3\cos(\theta_1+\theta_2+\theta_3),\\
\tau_2 &= I_3(\dot v_{\theta_1}+\dot v_{\theta_2}+\dot v_{\theta_3})-\mu_3 l_3\sin(\theta_1+\theta_2+\theta_3)+\mu_4 l_3\cos(\theta_1+\theta_2+\theta_3).
\end{aligned}
$$

Eliminating the Lagrange multipliers, it follows

$$
\begin{aligned}
& m_1\dot v_{x_1}+m_2\dot v_{x_2}+m_3\dot v_{x_3}=0,\\
& m_1\dot v_{y_1}+m_2\dot v_{y_2}+m_3\dot v_{y_3}=0,\\
& m_2\dot v_{x_2}\{l_1\sin\theta_1+l_2\sin(\theta_1+\theta_2)\}\\
&\quad +m_3\dot v_{x_3}\{l_1\sin\theta_1+l_2\sin(\theta_1+\theta_2)+l_3\sin(\theta_1+\theta_2+\theta_3)\}\\
&\quad -m_2\dot v_{y_2}\{l_1\cos\theta_1+l_2\cos(\theta_1+\theta_2)\}\\
&\quad -m_3\dot v_{y_3}\{l_1\cos\theta_1+l_2\cos(\theta_1+\theta_2)+l_3\cos(\theta_1+\theta_2+\theta_3)\}\\
&\quad -\{(I_1+I_2+I_3)\dot v_{\theta_1}+(I_2+I_3)\dot v_{\theta_2}+I_3\dot v_{\theta_3}\}=0,\\
& \tau_1+m_2\dot v_{x_2}\{l_2\sin(\theta_1+\theta_2)\}\\
&\quad +m_3\dot v_{x_3}\{l_2\sin(\theta_1+\theta_2)+l_3\sin(\theta_1+\theta_2+\theta_3)\}\\
&\quad -m_2\dot v_{y_2}\{l_2\cos(\theta_1+\theta_2)\}\\
&\quad -m_3\dot v_{y_3}\{l_2\cos(\theta_1+\theta_2)+l_3\cos(\theta_1+\theta_2+\theta_3)\}\\
&\quad -\{(I_2+I_3)(\dot v_{\theta_1}+\dot v_{\theta_2})+I_3\dot v_{\theta_3}\}=0,\\
& \tau_2+l_3\sin(\theta_1+\theta_2+\theta_3)m_3\dot v_{x_3}-l_3\cos(\theta_1+\theta_2+\theta_3)m_3\dot v_{y_3}\\
&\quad -I_3(\dot v_{\theta_1}+\dot v_{\theta_2}+\dot v_{\theta_3})=0,
\end{aligned}
\tag{4}
$$

where τ_1 and τ_2 are input torques associated to the joint 1 and joint 2. In the above, the second-order conditions hold:

$$
\begin{aligned}
&\dot x_1=v_{x_1},\quad \dot y_1=v_{y_1},\quad \dot\theta_1=v_{\theta_1},\quad \dot x_2=v_{x_2},\quad \dot y_2=v_{y_2},\quad \dot\theta_2=v_{\theta_2},\\
&\dot x_3=v_{x_3},\quad \dot y_3=v_{y_3},\quad \dot\theta_3=v_{\theta_3}.
\end{aligned}
\tag{5}
$$

Note that equation (5) are essential to consider the space robot dynamics in the *velocity phase space*, although we can also eliminate the velocity variables.

4 Geometric Phases in Space Robot Motion

Since there is no external force, the linear momentum is to be conserved as

$$
m_1\dot x_1+m_2\dot x_2+m_3\dot x_3=0,\quad m_1\dot y_1+m_2\dot y_2+m_3\dot y_3=0. \tag{6}
$$

Equation (6) can be directly obtained by integrating the first and second equations associated to the translational motion in (4), which can be integrated again to obtain

$$
m_1x_1+m_2x_2+m_3x_3=0,\quad m_1y_1+m_2y_2+m_3y_3=0. \tag{7}
$$

It follows from the third equation in (4) that the conservation of angular momentum are given by

$$m_1(x_1\dot{y}_1 - y_1\dot{x}_1) + m_2(x_2\dot{y}_2 - y_2\dot{x}_2) + m_3(x_3\dot{y}_3 - y_3\dot{x}_3) \\ + I_1\dot{\theta}_1 + I_2(\dot{\theta}_1 + \dot{\theta}_2) + I_3(\dot{\theta}_1 + \dot{\theta}_2 + \dot{\theta}_3) = 0,$$

which can be transformed, by using equations (1), (2), (6) and (7), into

$$(I_1 + I_2 + I_3 - \frac{m_2 m_3}{m_1} l_3^2)\dot{\theta}_1 + (I_2 + I_3 - \frac{m_2 m_3}{m_1} l_3^2)\dot{\theta}_2 + (I_3 - \frac{m_2 m_3}{m_1} l_3^2)\dot{\theta}_3 \\ + \frac{m_2(m_1 + m_2 + m_3)}{m_1}(x_2\dot{y}_2 - y_2\dot{x}_2) + \frac{m_3(m_1 + m_2 + m_3)}{m_1}(x_3\dot{y}_3 - y_3\dot{x}_3) = 0.$$

Then, we can develop the one-form $d\theta_1$ as

$$d\theta_1 = -a_1 d\theta_2 - a_2 d\theta_3 - a_3(x_2 dy_2 - y_2 dx_2) - a_4(x_3 dy_3 - y_3 dx_3),$$

where

$$a_1 = -\frac{m_2 m_3 l_3^2 - m_1(I_2 + I_3)}{m_1(I_1 + I_2 + I_3) - m_2 m_3 l_3^2}, \quad a_2 = -\frac{m_2 m_3 l_3^2 - m_1 I_3}{m_1(I_1 + I_2 + I_3) - m_2 m_3 l_3^2},$$
$$a_3 = \frac{m_2(m_1 + m_2 + m_3)}{m_1(I_1 + I_2 + I_3) - m_2 m_3 l_3^2}, \quad a_4 = \frac{m_3(m_1 + m_2 + m_3)}{m_1(I_1 + I_2 + I_3) - m_2 m_3 l_3^2}.$$

Now, introduce the one-form ω as

$$\omega = d\theta_1 + a_1 d\theta_2 + a_2 d\theta_3 + a_3(x_2 dy_2 - y_2 dx_2) + a_4(x_3 dy_3 - y_3 dx_3),$$

which is not *closed* since $d\omega = 2a_3(dx_2 \wedge dy_2) + 2a_4(dx_3 \wedge dy_3) \neq 0$.

Let A be the initial position and B be the final position in Q respectively. Let c be a path from A to B and let c_1 be a path from $B \to A$ via the origin O in Q. Under the conditions $x_2 dy_2 - y_2 dx_2 = 0$ and $x_3 dy_3 - y_3 dx_3 = 0$ imposed on the path c_1, we may regard the one-form ω on the path as an exact one-form and hence its integral is only dependent on the initial and final positions. Namely, one has

$$\int_{c_1} \omega = [\theta_1 + a_1\theta_2 + a_2\theta_3]_B^A \\ = (\theta_{1A} - \theta_{1B}) + a_1(\theta_{2A} - \theta_{2B}) + a_3(\theta_{3A} - \theta_{3B}) = \Delta\theta_1 + a_1\Delta\theta_2 + a_2\Delta\theta_3.$$

Further, the integral on c can be computed by

$$\int_c \omega = \int_\gamma \omega - \int_{c_1} \omega,$$

where $\gamma = c + c_1$ is a closed curve in Q. Let D be a compact surface in Q with the boundary γ, namely, $\partial D = \gamma = c + c_1$. Using Stokes' theorem, it follows

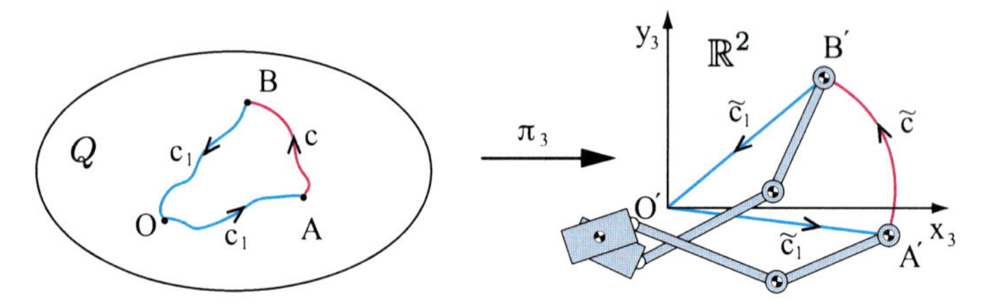

Fig. 2 The path of the end effector

$$
\begin{aligned}
\int_{\partial D} \omega = \int_D d\omega = \int_D (2a_3 dx_2 \wedge dy_2 + 2a_4 dx_3 \wedge dy_3) \\
= 2a_3 S_2(D_2) + 2a_4 S_3(D_3).
\end{aligned}
$$

Define the projection $\pi_3 : Q \to \mathbb{R}^2$ by $(x_1, y_1, x_2, y_2, x_3, y_3, \theta_1, \theta_2, \theta_3) \mapsto (x_3, y_3)$. So, the point O in Q is projected to $O' = \pi_3(O)$ in $\mathbb{R}^2$. Similarly, the points A and B are projected to $A' = \pi_3(A)$ and $B' = \pi_3(B)$ in $\mathbb{R}^2$. Then, we introduce $\tilde{\gamma} := \pi_3 \circ \gamma$ in $\mathbb{R}^2$ from the closed loop $\gamma = c + c_1$ in Q as in Fig. 2 and hence, $\tilde{c} = \pi_3 \circ c$, $\tilde{c}_1 = \pi_3 \circ c_1$ and $\tilde{\gamma} = \tilde{c} + \tilde{c}_1$. Moreover, $S_3(D_3) = \int_{D_3} dx_3 \wedge dy_3$ is an area of D_3 in $\mathbb{R}^2$, which is obtained by $D_3 = \pi_3(D)$. Similarly, $S_2(D_2) = \int_{D_2} dx_2 \wedge dy_2$ is an area of D_2 in $\mathbb{R}^2$ with boundary $\partial D_2 := \pi_2 \circ \gamma = \pi_2(c + c_1)$, where $D_2 = \pi_2(D)$ using the projection $\pi_2 : Q \to \mathbb{R}^2$; $(x_1, y_1, x_2, y_2, x_3, y_3, \theta_1, \theta_2, \theta_3) \mapsto (x_2, y_2)$. Since $\omega = 0$, one has

$$
\begin{aligned}
\int_c \omega = \int_{\partial D} \omega - \int_{c_1} \omega \\
= 2a_3 S_2(D_2) + 2a_4 S_3(D_3) - \Delta\theta_1 - a_1 \Delta\theta_2 - a_2 \Delta\theta_3 = 0.
\end{aligned}
$$

Hence, the geometric phase $\Delta\theta_1$ can be computed by

$$
\Delta\theta_1 = -a_1 \Delta\theta_2 - a_2 \Delta\theta_3 + 2a_3 S_2(D_2) + 2a_4 S_3(D_3). \tag{8}
$$

The deviation of the base attitude associated to the end effector of the space robot is equal to $\Delta\theta_1$, and it follows from (8) that $\Delta\theta_1$ is not dependent on time t but only on $\Delta\theta_2, \Delta\theta_3$ as well as $S_2(D_2)$ and $S_3(D_3)$, which can be obtained from the projection of the area of the closed surface D with the boundary $\partial D = \gamma = c + c_1$.

Similar derivations of geometric phases were done by [2] by using Stokes' theorem. From more mathematical points of view, see [4].

5 Trajectory Planning

Let us consider the trajectory planning of the end effector from the initial point A to the target point B in Q with the base body attitude control by computing geometric phases. Set the points A and B to the values in the below:

$$\begin{aligned}A &= (x_{1A}, y_{1A}, x_{2A}, y_{2A}, x_{3A}, y_{3A}, \theta_{1A}, \theta_{2A}, \theta_{3A})\\ &= (-1, 0, 5, 0, 10, 0, 0, 0, 0),\\ B &= (x_{1B}, y_{1B}, x_{2B}, y_{2B}, x_{3B}, y_{3B}, \theta_{1B}, \theta_{2B}, \theta_{3B})\\ &= (-0.225, -0.516, 4.499, 2.818, 0, 5, 0, 0.730, 1.960),\end{aligned}$$

each of which may be obtained consistently with equations (1) and (7). In the subspace $\mathbb{R}^2$, given an *underlying* trajectory of the end effector $(x_3(t), y_3(t))$ from $A' = \pi_3(A) = (10,0)$ to $B' = \pi_3(B) = (0,5)$ as in Fig. 3, it inevitably causes some deviation $\Delta\theta_1$ of the base attitude angle θ_1 due to the geometric phase. So, *the problem that we shall consider here is to control the deviation of the base body attitude as well as the trajectory of the end effector.*

It follows from the initial and target points A and B that we can calculate the deviations $\Delta\theta_1 = \theta_{1A} - \theta_{1B}, \Delta\theta_2 = \theta_{2A} - \theta_{2B}, \Delta\theta_3 = \theta_{3A} - \theta_{3B}$, and hence we can simply calculate $S_3(D_3)$ under the condition without the term $S_2(D_2)$ in (8). However, there exists the term $S_2(D_2)$ in fact, and the main stumbling block lies in the fact that it is very difficult to calculate $S_2(D_2)$ to suppress $\Delta\theta_1 = 0$. So, we shall consider a control algorithm to achieve $\Delta\theta_1 = 0$ of calculating the geometric phase by iteratively drawing the complementary circle trajectory and with its area as

$$S_3(D_{(3,k)}) = \frac{1}{2a_4}\Delta\theta_{(1,k)} + \frac{a_1}{2a_4}\Delta\theta_{(2,k)} + \frac{a_2}{2a_4}\Delta\theta_{(3,k)}. \tag{9}$$

In the above, $\Delta\theta_{(i,k)} (i = 1,2,3)$ are deviations from the target associated to the k-th value of the angles $\theta_{(i,k)}$ when the end effector reaches to point B.

The iterated control algorithm is given as follows; first, we set the initial values of the angles $\theta_{(i,0)}$; second, the complementary circle trajectory is iteratively given at Point B to cancel the geometric phase until the deviation $\Delta\theta_{(1,k)}$ of the base attitude is to be zero. The radius of the k-th complementary circle trajectory is given by

$$r_k = \sqrt{\frac{|S_3(D_{(3,k)})|}{\pi}} \quad (k \geq 1).$$

Finally, the end effector trajectory is obtained as in Fig. 4.

6 Numerical Analysis of Dynamics

In order to verify the proposed trajectory planning with the attitude control of the base as shown in Fig. 4, let us make numerical analysis of dynamics of the space robot. Given the trajectories θ_k, $k = 1,2,3$ as in Fig. 4 and Fig. 5 obtained from the algorithm in §5, we can simply calculate required torques τ_1 and τ_2 as in Fig. 6 by the *inverse dynamics* using (4) and (5). Conversely, given the torques τ_1 and τ_2 in Fig. 6, it follows from *direct dynamics* that we can obtain the trajectories of the space robot as in Fig. 4 and Fig. 5, from which we can verify the attitude control of

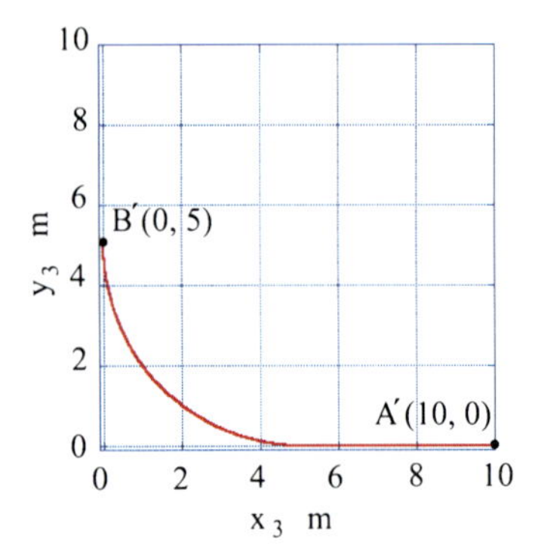

Fig. 3 Underlying trajectory

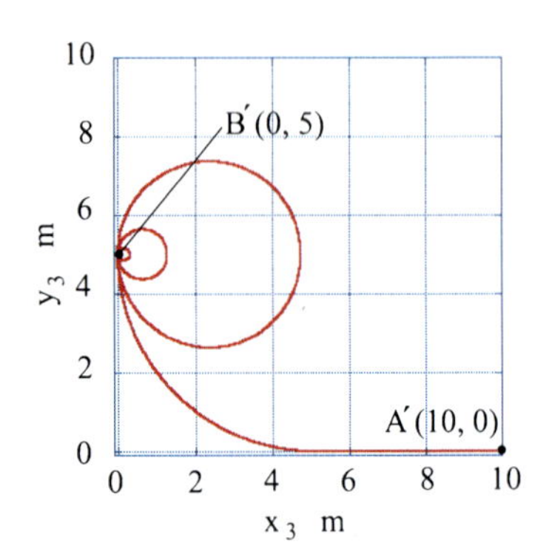

Fig. 4 Trajectory for the base attitude control

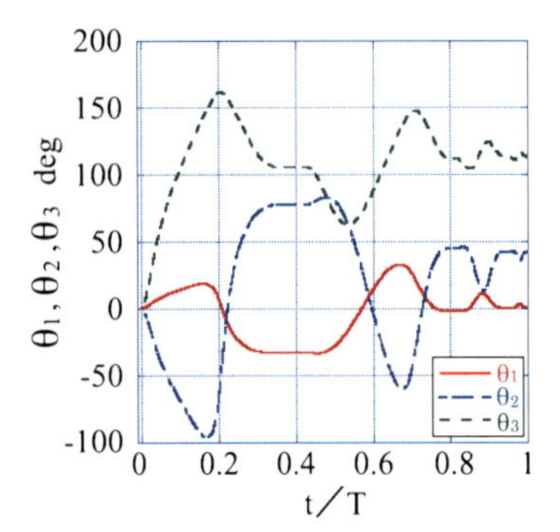

Fig. 5 The base attitude and joint angles

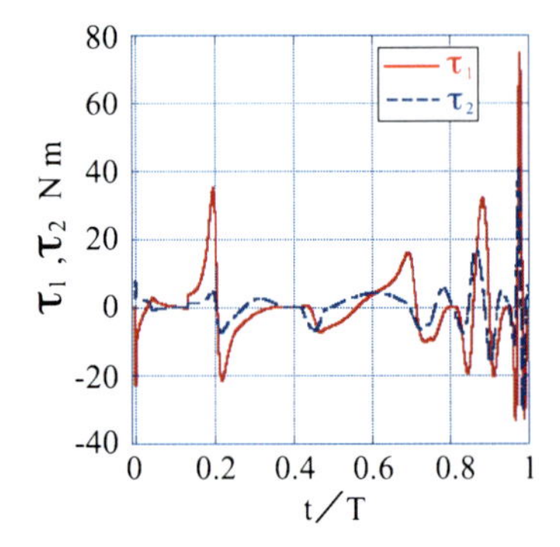

Fig. 6 Joint torques

the base. Thus, we obtain the control input for the proposed trajectory planning of the end effector of the space robot and with the base attitude.

7 Conclusions

We have developed a trajectory planning for space robots, which enables us to simultaneously control the base attitude and the end effector trajectory. We have shown that dynamics can be formulated in the context of regular Lagrangian systems. The trajectory planning with control of the base attitude of the space robot is based on calculating geometric phases. Finally, we have demonstrated the validity of our approach with numerical simulations.

Acknowledgements. We thank Jerrold E. Marsden for helpful discussions on geometric mechanics and control. This research is partially supported by Grant-in-Aid for Scientific Research in JSPS (C)1650216, JST-CREST and Waseda University Grant for Special Research Projects 2010A-606.

References

1. Kane, T.R., Scher, M.P.: A dynamical explanation of the falling cat phenomenon. Int. J. Solid Structures 5, 663–670 (1969)
2. Yamada, K., Tsuchiya, K.: On the relation of a hand trajectory of a space robot and its attitude variation. Transactions of the Society of Instrument and Control Engineers 28(3), 374–382 (1992)
3. Nakamura, Y., Mukherjee, R.: Nonholonomic path planning of space robots via a bidirectional approach. IEEE Transaction on Robotics and Automation 7(4), 500–514 (1991)
4. Marsden, J.E.: Lectures on Mechanics. London Mathematical Society Lecture Note Series, vol. 174. Cambridge University Press, Cambridge (1992)
5. Yoshimura, H.: On the Lagrangian formalism of nonholonomic mechanical systems, DETC2005-84273. In: Proceedings of ASME International Design Engineering Technical Conferences and Computers and Information in Engineering Conference, IDETC 2005, Long Beach, California, USA, September 24-28, pp. 1–7 (2005)

Applications and Control of Bio-Inspired Robots

The four papers in this section discuss bio-inspired robotic applications. The theoretical work of F.L. Chernousko investigates in detail the progressive locomotion of a two body system in a fluid, the results of which are of interest for biomechanics of swimming. Papers by C. Behn provide a rigorous mathematical framework to describe the kinematics and dynamics of worm-like locomotion systems and contribute to their adaptive control. The experimental results of the paper by Q. Bombled are connected to the ground detection of a six legged walking robot.

Optimal Control of a Two-Link System Moving in a Fluid

Felix L. Chernousko

Abstract. Progressive locomotion of a two-link system in a fluid is considered. The system consists of two rigid bodies, the main body and the tail, connected by a cylindrical joint. The actuator installed at the joint controls the angle between the axes of the body and the tail. The progressive motion of the system is caused by high-frequency angular oscillations of the tail relative to the main body. The fluid acts upon each body with the resistance force proportional to the squared velocity of the body. It is shown that, under certain assumptions, the system can move progressively, if the tail performs the retrieval phase of its oscillations faster than the deflection phase. This result correlates well with the observations of swimming. The optimal periodic motion of the tail is obtained that corresponds to the maximal, under certain constraints, average velocity of the system.

1 Introduction

It is well-known that multibody systems can move inside a resistive medium, if their parts perform certain oscillations relative to each other. This principle of locomotion typical to fish, insects, and some animals [1, 2] is applied in various mobile robots [3-5]. Dynamics of multilink snake-like systems in the presence of dry friction is studied in [6-8].

In the paper, a progressive rectilinear motion of a two-link system is considered. It is assumed that the fluid acts upon each body with a resistance force proportional to the squared velocity of the body. The system consists of two bodies, the main body and the tail, connected by a cylindrical joint where the actuator is installed. The progressive motion occurs as a result of high-frequency oscillations of the tail relative to the main body. The mechanical model of the system and the assumptions made are described in Section 2.

The equation of motion is simplified by means of the method of averaging in Section 3. The case of a piecewise constant angular velocity of the tail is considered in Section 4.

Felix L. Chernousko
Institute for Problems in Mechanics of the Russian Academy of Sciences,
pr. Vernadskogo 101-1, 115926, Moscow, Russia

G. Stépán et al. (Eds.): Dynamics Modeling & Interaction Cont., IUTAM BOOK SERIES 30, pp. 47–55.
springerlink.com

The optimal control problem for the motion of the tail is formulated in Section 5. The solution of the problem is presented in Section 6. As a result, the motion is determined that corresponds to the maximum, under the constraints imposed, average velocity of the system.

In Section 7, the obtained results are extended to the case of two links attached symmetrically to the main body. A numerical example is presented in Section 8.

2 Mechanical Model

Consider a two-body mechanical system that consists of a main body of mass m and a link OA attached to the main body by a cylindrical joint O (Fig.1). The length of the tail OA is denoted by a, and its mass is assumed to be negligible compared to mass m.

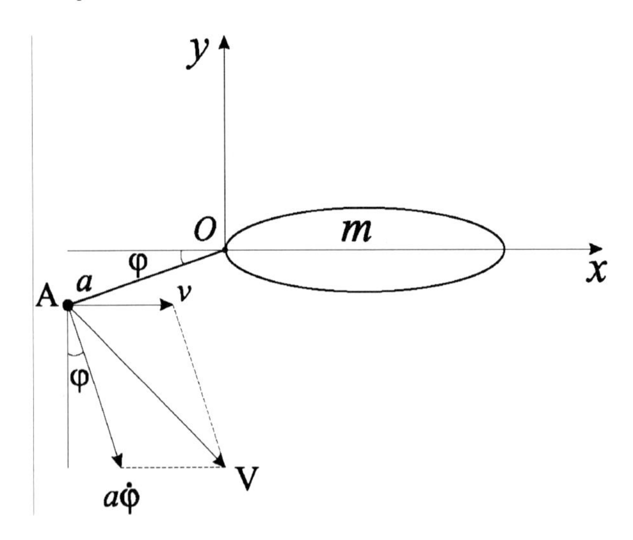

Fig. 1 Mechanical model

The tail performs high-frequency oscillations relative to the main body. Denote by φ the angle between the axes of the main body and the tail, and by T the period of oscillations. Suppose that the oscillations of the tail are periodic and symmetric, so that the following conditions

$$\varphi(t+T)=\varphi(t),\quad \varphi(t+T/2)=-\varphi(t) \tag{1}$$

hold for each t.

Suppose that the fluid acts upon each body with a resistance force proportional to the squared velocity of the body. Under certain assumptions (the main body is symmetric with respect to its axis Ox, its resistance along the axis Oy is much higher than along the axis Ox, its moment of inertia is large, etc.), the angular and lateral motions of the main body can be neglected, and it will move mostly along axis Ox.

Denote by v the velocity of the body (we assume that $v \geq 0$) and by $c_0 v^2$ its resistance force, where c_0 is a constant. Note that the adjoint mass of the body can be included into m. Suppose that the resistance forces acting upon the tail are reduced to the force applied at the point A and equal to $k_0 V^2$, where V is the velocity of the point A, and k_0 is a constant.

The equation of motion of the body can be written as follows

$$\dot{v} = -cv^2 - k(v + a\dot{\varphi}\sin\varphi)\sqrt{v^2 + a^2\dot{\varphi}^2 + 2va\dot{\varphi}\cos\varphi}. \tag{2}$$

Here, $c = c_0 / m$ and $k = k_0 / m$.

3 Averaging

Suppose that $k \ll c$, so that the ratio $\mu = k / c$ is a small parameter. Let us consider the normalized (dimensionless) variables with:

$$t = T\tau,\ T = \mu T_0,\ \varphi = \mu\psi,\ v = \mu(a / T_0)u,\ \mu \ll 1, \tag{3}$$

and simplify equation (2) omitting terms of higher order of μ. We obtain

$$\frac{du}{d\tau} = -\varepsilon\left[u^2 + (u + \psi\frac{d\psi}{d\tau})\left|\frac{d\psi}{d\tau}\right|\right],\ \varepsilon = ac\mu^2. \tag{4}$$

Here, $\psi(\tau)$ is a periodic function of period 1 with properties following from (1)

$$\psi(\tau + 1) = \psi(\tau),\ \psi(\tau + 1/2) = -\psi(\tau). \tag{5}$$

Using the asymptotic method of averaging [9], we obtain from (4) the averaged equation of the first approximation as follows:

$$\frac{du}{d\tau} = -\varepsilon(u^2 + I_1 u + I_0),\ u \geq 0,$$

$$I_1 = \int_0^1 \left|\frac{d\psi}{d\tau}\right| d\tau,\ I_0 = \int_0^1 \psi(\frac{d\psi}{d\tau})\left|\frac{d\psi}{d\tau}\right| d\tau. \tag{6}$$

The solution of the average equation (6) differs from the exact solution of equation (4) by the terms of order of ε on the large time interval $\Delta\tau \sim \varepsilon^{-1}$ [9].

If $I_0 \geq 0$, then, by virtue of (6), u always decreases, and the progressive motion with a positive velocity is impossible. In a more interesting case where $I_0 < 0$, there exists a unique positive stationary velocity u given by

$$u^* = \sqrt{(I_1^2/4) - I_0} - I_1/2. \tag{7}$$

This velocity is asymptotically stable. To calculate u^*, we are to specify the time-history of the angle $\psi(\tau)$.

4 Piecewise Constant Angular Velocity

Suppose that the dimensionless angular velocity $d\psi/d\tau$ is piecewise constant, so that the time-history of $\psi(\tau)$ is given by

$$\psi(\tau) = \begin{cases} b_1\tau, \tau \in (0,\theta), & 0<\theta<1/2, \\ b_2(1/2-\tau), & \tau \in (\theta, 1/2), \\ b_1(1/2-\tau), & \tau \in (1/2, 1/2+\theta), \\ b_2(\tau-1), & \tau \in (1/2+\theta, 1). \end{cases} \tag{8}$$

Here, $b_1 > 0$, $b_2 > 0$, and $\theta \in (0,1/2)$ are constant parameters. They must satisfy the condition

$$b_1\theta = b_2(1/2-\theta) \tag{9}$$

that guarantees the continuity of $\psi(\tau)$. The piecewise linear time-history $\psi(\tau)$ from equation (8) is shown in Fig.2.

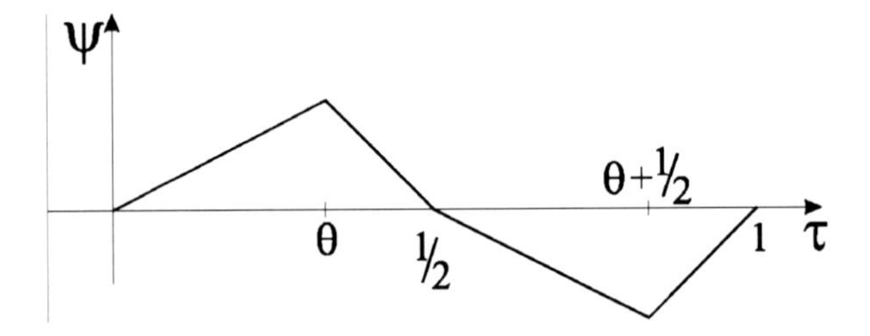

Fig. 2 Piecewise linear $\psi(\tau)$

Inserting $\psi(\tau)$ from (8) into equations (6) and (7), we obtain

$$u^* = \frac{b_1 b_2}{b_1+b_2}\left(\sqrt{1+\frac{b_2-b_1}{4}} - 1\right). \tag{10}$$

The constants b_1 and b_2 are expressed through the constant angular velocities of deflection (ω_1) and retrieval (ω_2) of the tail:

$$b_i = T_0\omega_i\,,\ i = 1,2\,. \tag{11}$$

Using equations (3) and (11), we obtain the value v^* of the dimensional stationary velocity as follows:

$$v^* = \frac{\mu a u^*}{T_0} = \frac{\mu a \omega_1\omega_2}{\omega_1 + \omega_2}\left[\sqrt{1 + \frac{(\omega_2 - \omega_1)T_0}{4}} - 1\right]. \tag{12}$$

Hence, if and only if the retrieval of the tail is performed faster than its deflection ($\omega_2 > \omega_1$), the stationary velocity v^* of the system is positive: $v^* > 0$.

5 Optimal Control Problem

Consider now the optimal control problem for the angular motion of the tail. Let $\Omega = d\psi / d\tau$ be the control subject to the constraints

$$-\Omega_- \le \Omega \le \Omega_+\,, \tag{13}$$

where Ω_+ and Ω_- are given positive constants. Suppose that the normalized angle ψ changes over the interval $\tau \in (0,1/2)$ as follows: it grows from $\psi(0) = 0$ to $\psi(\theta) = \psi_0 > 0$ and then decreases to $\psi(1/2) = 0$. Here, $\theta \in (0,1/2)$ and $\psi_0 > 0$ are constant parameters. The behavior of $\psi(\tau)$ for $\tau \in (1/2,1)$ is determined by equation (5). We look for functions $\Omega(\tau)$ and $\psi(\tau)$ that maximize the stationary velocity u^* defined by (7). It can be shown that, under the assumptions made, $I_1 = 4\psi_0$, so that the maximization of u^* in (7) is reduced to the minimization of the integral functional I_0 defined by equation (6).

The solution of the optimal control problem stated above is obtained using the Pontryagin maximum principle [10]. This solution requires a rather lengthy analysis, and we present here only the final results.

6 Optimal Motion

The solution of the optimal control problem stated in Section 5 is described by the following equations. The optimal control $\Omega(\tau)$ and the corresponding optimal time-history of the normalized angle $\psi(\tau)$ are given by equations

$$\Omega = \Omega_+\,,\ \psi = \Omega_+\tau \ \text{for}\ \tau \in (0,\tau_*)\,,$$

$$\Omega = \Omega_+ \left[1 + \frac{3(\tau - \tau_*)}{2\tau_*}\right]^{-1/3}, \quad \psi = \Omega_+ \tau_* \left[1 + \frac{3(\tau - \tau_*)}{2\tau_*}\right]^{2/3} \quad \text{for } \tau \in (\tau_*, \theta) \tag{14}$$

on the interval $\tau \in (0, \theta)$ and by equations

$$\Omega = -\Omega_-, \quad \psi = \Omega_-(1/2 - \tau) \tag{15}$$

for $\tau \in (\theta, \ 1/2)$. Here, the parameters τ_* and θ are determined by equations

$$\tau_* = s\theta, \quad \theta = 1/2 - \psi_0 / \Omega_-, \tag{16}$$

where s is the only root of the cubic equation

$$s(3-s)^2 = 4\left[\psi_0 / (\Omega_+ \theta)\right]^3 \tag{17}$$

lying in the interval $s \in (0,1)$. Equations (14) and (15) determine the optimal solution on the time interval $\tau \in (0, 1/2)$; for the rest of the period $\tau \in (1/2, 1)$, this solution is defined according to equation (5).

The maximum value of the average stationary velocity of the system that corresponds to the solution obtained is

$$u^* = \left[\psi_0^2 \Omega_- - 4(\psi_0^3 / \theta)(2-s)(3-s)^{-2} + 4\psi_0^2\right]^{1/2} - 2\psi_0, \tag{18}$$

where $s \in (0,1)$ is a root of equation (17).

Let us consider the cases where there are no upper or lower bounds (13) on the angular velocity. If the upper bound is absent $(\Omega_+ \to \infty)$, we have, by virtue of equations (17) and (15), $s = 0$ and $\tau_* = 0$. In this case, we obtain from equations (14), (15), and (18):

$$\Omega = (2/3)\psi_0 \theta^{-2/3} \tau^{-1/3}, \quad \psi = \psi_0 (\tau / \theta)^{2/3} \quad \text{for } \tau \in (0, \theta),$$

$$\Omega = -\Omega_-, \quad \psi = \Omega_-(1/2 - \tau) \quad \text{for } \tau \in (\theta, 1/2),$$

$$u^* = \left[\psi_0^2 \Omega_- - (8/9)(\psi_0^3 / \theta) + 4\psi_0^2\right]^{1/2} - 2\psi_0.$$

If the lower bound on the control is absent $(\Omega_- \to \infty)$, we have, on the strength of equations (16) and (18), $\theta = 1/2$, $u^* \to \infty$. Here, the interval $(\theta, \ 1/2)$ in equations (15) becomes zero, the retrieval phase is instantaneous, and the velocity of the progressive motion tends to infinity.

Thus, the optimal solution is completely described in terms of normalized variables. To return to the original dimensional ones, we are to use equations (3).

7 Three-Body System

Let us consider now a system that consists of a main body of mass m and two links, OA and OA', attached to it by cylindrical joints (Fig.3). Suppose that these links are always symmetric with respect to the axis of the body, their lengths are equal to a, and their masses are negligible compared to m.

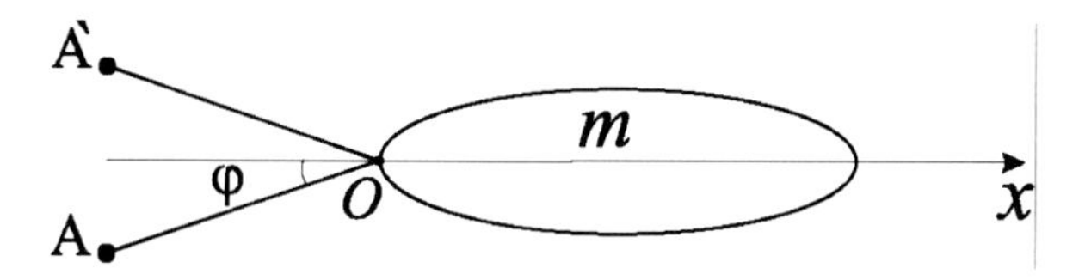

Fig. 3 System with two symmetric lines

The links OA and OA' perform high-frequency oscillations relative to the body so that the respective angles φ and φ' (see Fig.3) change synchronously within the interval $(0,\varphi_0)$. Thus, we have $0 \le \varphi(t) = \varphi'(t) \le \varphi_0$.

While the case of one link attached to the main body models swimming by means of a tail, the case of two links attached corresponds to swimming with the help of two extremities.

The analysis of this case is quite similar to the case of a body with a tail. In equation (2), we are to replace k by $2k$. For a piecewise linear function $\psi(\tau)$ as well as for the optimal motion, the stationary average velocity is two times higher as u^* given by respective equations (10) and (18).

8 Example

Let us consider a numerical example for the case of one link (tail) attached to the main body. We take $\psi_0 = 1$, $\Omega_+ = 4$, $\Omega_- = 10$ and obtain from equations (16), (17), and (18): $\theta = 0.4$, $s = 0.117$, $\tau_* = 0.047$, $u = 1.425$. The optimal time-histories of functions $\Omega(\tau)$ and $\psi(\tau)$ determined by equations (14) and (15) are shown in Figs.4 and 5, respectively.

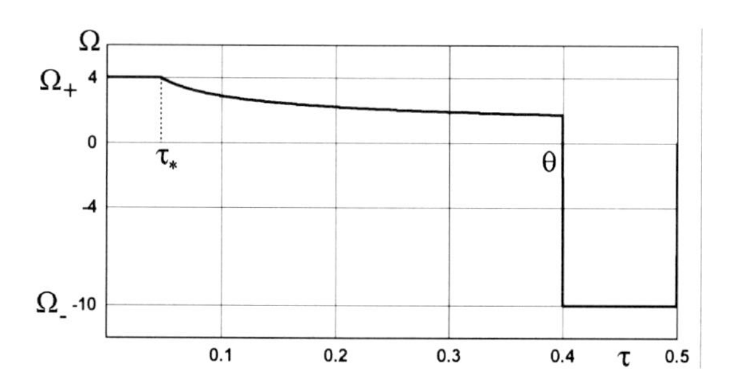

Fig. 4 Optimal time-history of $\Omega(\tau)$

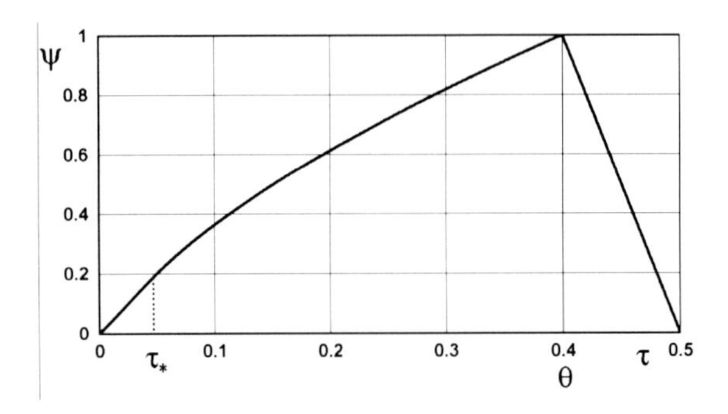

Fig. 5 Optimal time-history of $\psi(\tau)$

Let us compare the optimal solution with the piecewise linear one defined by equation (8). We choose parameters b_1, b_2, and θ so that the function $\psi(\tau)$ coincides with the optimal solution shown in Fig. 5 at $\tau = 0$, $\tau = \theta$, and $\tau = 1/2$. Hence, we have $\theta = 0.4$, $b_1 = 2.5$, and $b_2 = 10$. The corresponding value of u^* defined by equation (10) is $u^* = 1.391$. It differs from the optimal value less than by 2,4 %.

9 Conclusions

It is shown that a mechanical system consisting of a main body and one or two links attached to it by cylindrical joints can move progressively in a resistive fluid, if the links perform certain high-frequency oscillations, and resistant forces are proportional to the squared velocity of the moving body. For the case of a piecewise constant angular velocity of the links, the explicit formula for the average velocity of the body is obtained. The optimal control problem for the angular motion of the links is formulated and solved. The maximum possible velocity of the body under the constraints imposed is evaluated. This optimal solution is close to the case of a piecewise constant angular velocity. The progressive motion is possible, if the angular velocity of the links in the retrieval phase is higher than in the deflection phase. This result correlates well with the observation of swimming.

Acknowledgments. The work was supported by the Russian Foundation for Basic Research (Project 08-01-00411).

References

1. Gray, J.: Animal Locomotion. Norton, New York (1968)
2. Ligthill, M.J.: Large-amplitude elongated-body theory of fish locomotion. Proc. R. Soc. B. 179, 125–138 (1971)
3. Hirose, S.: Biologically Inspired Robots: Snake-like Locomotors and Manipulators. Oxford University Press, Oxford (1993)

4. Terada, Y., Yamamoto, I.: Development of oscillating fin propulsion system and its application to ships and artificial fish. Mitsubishi Heavy Industries Technical Review 36, 84–88 (1999)
5. RoboTuna, http://web.mit.edu/towtank/www/Tuna/tuna.html
6. Chernousko, F.L.: Controllable motions of a two-link mechanism along a horizontal plane. J. Appl. Math. Mech. 65(4), 565–577 (2001)
7. Chernousko, F.L.: Snake-like locomotions of multilink mechanisms. J. Vib. Contr. 9(1-2), 237–256 (2003)
8. Chernousko, F.L.: Modelling of snake-like locomotion. Appl. Math. Compu. 164(2), 415–434 (2005)
9. Bogoliubov, N.N., Mitropolsky, Y.A.: Asymptotic Methods in the Theory of Nonlinear Oscillations. Gordon and Breach Sci. Publ., New York (1961)
10. Pontryagin, L.S., Boltyanskii, V.G., Gamkrelidze, R.V., Mishchenko, E.F.: The Mathematical Theory of Optimal Processes. Gordon and Breach, New York (1986)

Straight Worms under Adaptive Control and Friction - Part 1: Modeling

Carsten Behn and Klaus Zimmermann

Abstract. This paper is a contribution to the adaptive control of worm-systems, which are inspired by biological ideas, in two parts. We introduce a certain type of mathematical models of finite DOF worm-like locomotion systems: modeled as a chain of k interconnected (linked) point masses in a common straight line (a discrete straight worm). We assume that these systems contact the ground via 1) spikes in Part 1 and then 2) stiction combined with Coulomb sliding friction (modification of a Karnopp friction model) in Part 2. We sketch the corresponding theory. In general, one cannot expect to have complete information about a sophisticated mechanical or biological system, only structural properties (known type of actuator with unknown parameters) are known. Additionally, in a rough terrain, unknown or changing friction coefficients lead to uncertain systems, too. The consideration of uncertain systems leads to the use of adaptive control in Part 2 to control such systems. Gaits from the kinematical theory (preferred motion patterns to achieve movement) in Part 1 can be tracked by means of adaptive controllers (λ-trackers) in Part 2. Simulations are aimed at the justification of theoretical results.

1 Introduction

The following is taken as the basis of our theory, [6]:

(i) A worm is a mainly terrestrial (or subterrestrial, possibly also aquatic) locomotion system characterized by one dominant linear dimension with no active (driving) legs or wheels.
(ii) Global displacement is achieved by (periodic) change of shape (in particular local strain: peristalsis) and interaction with the environment (undulatory locomotion).
(iii) The model body of a worm is a 1-dimensional continuum that serves as the support of various physical fields.

Carsten Behn · Klaus Zimmermann
Ilmenau University of Technology, Department of Technical Mechanics,
Max-Planck-Ring 12, 98693 Ilmenau, Germany
e-mail: `{carsten.behn,klaus.zimmermann}@tu-ilmenau.de`

G. Stépán et al. (Eds.): Dynamics Modeling & Interaction Cont., IUTAM BOOK SERIES 30, pp. 57–64.
springerlink.com

The continuum in (iii) is just an interval of a body-fixed coordinate. Most important fields are: *mass*, continuously distributed (with a density function) or in discrete distribution (chain of point masses), *actuators*, i.e., devices which produce internal displacements or forces thus mimicking muscles, *surface structure* causing the interaction with the environment.

It is well known, that, if there is contact between two bodies (worm and ground), there is some kind of "friction", which depends on the physical properties of the surfaces of the bodies. In particular, the friction may be anisotropic (depends on the orientation of the relative displacement). This interaction (mentioned in (ii)) could emerge from a surface texture or from a surface endowed with scales or bristles (we shall speak of *spikes* for short) which suppress or prevent backward displacements. It is responsible for the conversion of (mostly periodic) internal and internally driven motions into a change of external position (undulatory locomotion [3]), see [4].

In this paper only *discrete straight worms* shall be considered: chains of mass-points moving along a straight line, Fig. 1.

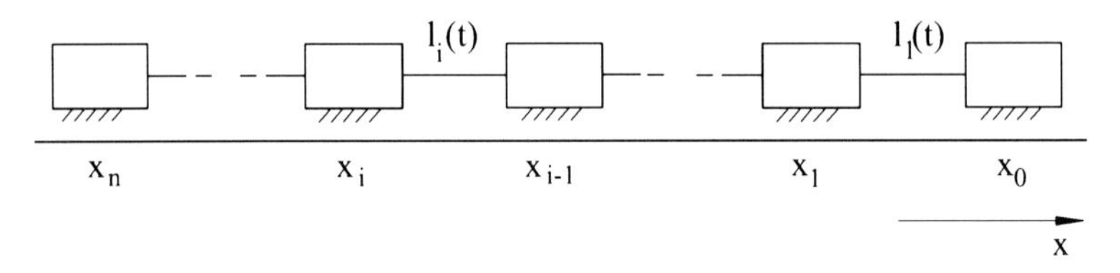

Fig. 1 Chain of point masses with spikes

First we focus on interaction via *spikes* since in this case a thorough kinematic theory is available. Later on we introduce Coulomb friction and try to analyze artificial worms as dynamical (control) systems.

2 Kinematics

Following [5], the motions of the worm system, $t \longmapsto x_i(t)$, are investigated under the **general assumption** to be of differentiability class $D^2(\mathbb{R})$, i.e.,

$$x_i(\cdot) \text{ and } \dot{x}_i(\cdot) \text{ continuous, } \ddot{x}_i(\cdot) \text{ piecewise continuous } (\ddot{x}_i \in D^0(\mathbb{R})). \qquad (1)$$

The spikes (attached to the points $\kappa \in \mathbf{K} \subset \{0, 1, \ldots, n\}$) restrict the velocities of the contact points,

$$\dot{x}_\kappa \geq 0, \; \kappa \in \mathbf{K}. \qquad (2)$$

This is a system of **constraints** in form of differential inequalities which the system's motions are subject to.

Let the distances of consecutive masspoints (= actual lengths of the links) be denoted l_j, $j = 1, \ldots, n$, and the actual distance of the masspoint i from the head S_i,

$$l_j := x_{j-1} - x_j\,, \qquad S_i := x_0 - x_i = \sum_{j=1}^{i} l_j\,. \tag{3}$$

Then there holds for the velocities $\dot{x}_i = \dot{x}_0 - \dot{S}_i$, $i = 0,\ldots,n$, and the constraint (2) yields $\dot{x}_0 - \dot{S}_\kappa \geq 0$, i.e., $\dot{x}_0 \geq \dot{S}_\kappa$ for all $\kappa \in \mathbf{K}$. This necessarily entails

$$\dot{x}_0 \geq V_0 := \max\{\dot{S}_\kappa \mid \kappa \in \mathbf{K}\}. \tag{4}$$

Consequently, the head velocity is

$$\dot{x}_0 = V_0 + w\,, \quad w \geq 0\,, \tag{5}$$

and for the others it follows

$$\dot{x}_i = V_0 - \dot{S}_i + w\,, \quad i = 0,\ldots,n\,. \tag{6}$$

Since w is a common additive term to *all* velocities $\dot{x}_i$, it describes a *rigid part of the motion* of the total system (motion at "frozen" $\dot{l}_j$).

If $0 \in \mathbf{K}$ (head equipped with spike) then because of $S_0 = 0$ the head velocity is non-negative.

The coordinate and velocity of the **center of mass** are obtained by averaging the x_i and $\dot{x}_i$ (remind equal masses m for all i),

$$\begin{aligned} x^* &= x_0 - S\,, \qquad S := \tfrac{1}{n+1}\sum_{i=0}^{n} S_i\,, \\ v^* &:= \dot{x}^* = W_0 + w\,, \quad W_0 := V_0 - \dot{S}\,. \end{aligned} \tag{7}$$

Now there are two representations of the velocities,

$$\dot{x}_i = \dot{x}_0 - \dot{S}_i = w + V_0 - \dot{S}_i\,. \tag{8}$$

They show that, alternatively, the head velocity $\dot{x}_0$ *together with* $\dot{S}_i$ (mind $\dot{S}_0 = 0$), or the rigid velocity part w *together with* $\dot{S}_i$ may serve as **generalized velocities** of the system (DOF = $n+1$).

When considering locomotion under external load it might be of interest to know which and how many of the masspoints κ with ground contact, $\kappa \in \mathbf{K}$, are at rest during the motion of the system, these are the **active spikes** which transmit the propulsive forces from the ground to the system. Now $\dot{x}_\kappa = V_0 - \dot{S}_\kappa + w$ together with $w \geq 0$ and $V_0 = \max\{\dot{S}_i \mid i \in \mathbf{K}\} \geq \dot{S}_\kappa$ imply

$$\dot{x}_\kappa = 0 \Longleftrightarrow w = 0 \wedge V_0 = \dot{S}_\kappa\,, \quad \kappa \in \mathbf{K}. \tag{9}$$

If the head is equipped with a spike, $0 \in \mathbf{K}$, then in view of $\dot{S}_0 = 0$ and the definition of V_0 in (4) it follows

$$\text{If } 0 \in \mathbf{K} \text{ then } \dot{x}_0 = 0 \Longleftrightarrow w = 0 \wedge \dot{S}_\kappa \leq 0 \text{ for all } \kappa \in \mathbf{K}. \tag{10}$$

The worm system is called to move under **kinematic drive** if by means of the actuators *all* distances l_j, $j = 1, \dots, n$, or, equivalently, the relative velocities $\dot{l}_j$ are prescribed as functions of t. Then $\dot{S}_i = \sum_{j=1}^{i} \dot{l}_j$ and $V_0 = \max\{\dot{S}_\kappa \mid \kappa \in \mathbf{K}\}$ become known functions of t, and in the velocities (6)

$$\dot{x}_i = V_0(t) - \dot{S}_i(t) + w, \quad i = 0, \dots, n,$$

the rigid part w is now the only free variable. This corresponds to the fact that for the system of $n+1$ masspoints on the x–axis ($DOF = n+1$) the distance relations (3),

$$x_{j-1} - x_j - l_j(t) = 0, \quad j = 1, \dots, n, \tag{11}$$

now represent n independent *rheonomic holonomic constraints* which shrink the degree of freedom to 1. w is the remaining generalized velocity of the worm system. The differential constraints (2) are, as before, satisfied by definition of V_0.

Once more, the rigid part w of the velocities keeps arbitrary in kinematics. So it seems promising to put it equal to zero, then all velocities of the masspoints are known functions of t. Putting $w = 0$ locks the single degree of freedom, the system has become a *compulsive mechanism with ground contact* (the latter causing locomotion).

There remains a nicely simple **Kinematical theory:** (worm with kinematic drive and $w(t) = 0$)

Prescribe: $l_j(.) \in D^2(\mathbb{R}) : t \mapsto l_j(t) > 0, \; j = 1, \dots, n.$
Determine: $S_i := \sum\limits_{j=1}^{i} l_j, \; V_0 := \max\{\dot{S}_\kappa \mid \kappa \in \mathbf{K}\} \in D^1(\mathbb{R}).$
Result: $x_0(t) = \int\limits_0^t V_0(s)ds, \; x_j(t) = x_0(t) - S_j(t), \; j = 1, \dots, n.$

(12)

The equivalence (9) entails that the kinematical theory is valid iff at any time at least one spike is active. In applications it might be necessary to use a kinematic drive that ensures a prescribed number of spikes to be active at every time.

Example 1
A worm system with $n = 2$ and $\mathbf{K} = \mathbf{N} = \{0, 1, 2\}$ is considered. We present (construction suppressed here) a kinematic drive such that at every time exactly one of the three spikes is active.

Using the Heaviside function

$$h(t_0, t_1, \cdot) : \tau \mapsto h(t_0, t_1, \tau) := \begin{cases} 1, \text{ if } t_0 < \tau \leq t_1 \\ 0, \text{ else} \end{cases}$$

we define

$$\begin{aligned} l_1(t) &:= l_0 + a\, l_0 \left(1 - \cos(\pi t)\right) h(0, 2, t) \\ l_2(t) &:= l_1(t + \tfrac{1}{3}) - l_0 \end{aligned} \tag{13}$$

on the primitive time interval $[0,T]$, $T := 3$, and then take their T-periodic continuation to $\mathbb{R}^+$. Here $l_0 = 2$ is the original length of the links, and $l_0 a$, $a = 0.2$, is the amplitude of the length variation in time.

Applying (12) we obtain the results which are sketched in Fig. 2. As one can see, the cycle of active spikes is $\mathbf{1} \to \mathbf{0} \to \mathbf{2}$. Average speed of center of mass is $\bar{v} \approx 0.5333$.

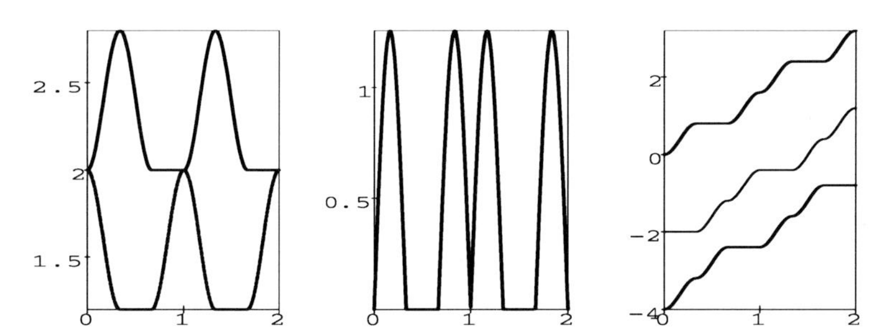

Fig. 2 Left: Gait $l_{1,2}$ vs. t/T; middle: V_0 vs. t/T, right: $x_{0,1,2}$ vs. t/T

3 Dynamics

The dynamics of the worm system are formulated by means of Newton's law for each of the masspoints. The following forces are applied to masspoint i, all acting in $x-$direction, see Fig. 3:

- g_i, the *external impressed* (physically given) *force* (e.g., resultant of viscous friction and weight component backward: $g_i = -k_0 \dot{x}_i - \Gamma_i$).
- μ_i, the *stress resultant* (inner force) of the links (let, formally, $\mu_0 = \mu_{n+1} := 0$).
- z_i, $i \in \mathbf{K}$, the *external reaction force* caused by the constraint (2), acting on the spiked masspoints.

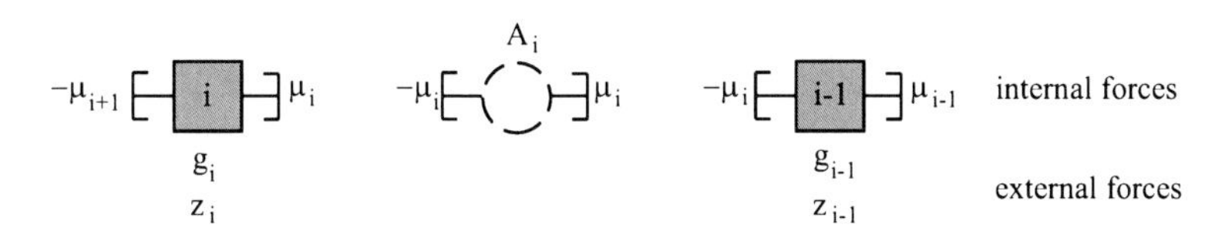

Fig. 3 Masspoints with forces (A_i: actuator)

As the constraint (2) describes a **one-sided** restriction of $\dot{x}_\kappa$, velocity and reaction force are connected by a **complementary-slackness condition**:

$$\dot{x}_\kappa \geq 0, \;\; z_\kappa \geq 0, \;\; \dot{x}_\kappa z_\kappa = 0, \;\; \kappa \in \mathbf{K}. \tag{14}$$

This means that $z_\kappa(t)$ is zero if at time t the masspoint κ is moving forward, i.e., if the velocity inequality is *strict*, $\dot{x}_\kappa(t) > 0$ ("$\dot{x}_\kappa$ has *slack*"), whereas $z_\kappa(t)$ may have arbitrary non-negative values as long as $\dot{x}_\kappa(t) = 0$ (reaction force at resting spike). Positive $z_\kappa(t)$ implies $\dot{x}_\kappa(t) = 0$, simultaneous vanishing, $\dot{x}_\kappa(t) = \lambda_\kappa(t) = 0$ is possible by chance.

Newton's laws for the $n+1$ masspoints

$$\begin{aligned} m\ddot{x}_\kappa &= g_\kappa + \mu_\kappa - \mu_{\kappa+1} + z_\kappa, \quad \kappa \in \mathbf{K}, \\ m\ddot{x}_i &= g_i + \mu_i - \mu_{i+1}, \qquad\quad i \in \mathbf{N} \setminus \mathbf{K} \end{aligned} \tag{15}$$

now appear as a system of equations that together with the slackness conditions (14) and initial conditions for x_i and $\dot{x}_i$, $i = 0, \ldots, n$, governs the motions of the worm system.

An *actuator* is, first, a multipole with input activation signal and energy (immanent energy source - e.g., electrical battery, chemical agents - possible as well), see Fig. 4.

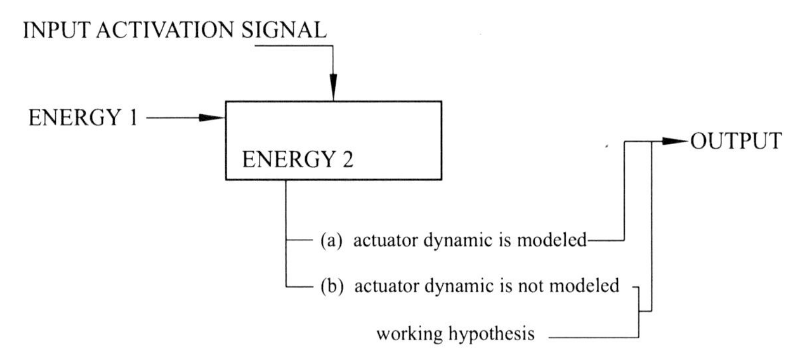

Fig. 4 Actuator, schematically

Its output are forces, torques, displacements, twists, respectively, which depend on time and may be connected with the actual state or the state history of the system or with the system's rheological constitution. In most cases an actuator is treated massless but otherwise it could be handled as one of the masspoints. Often the internal dynamics of an actuator are not modeled, rather the output is connected with the input by means of working hypotheses: the multipole remains a black box. It becomes a (almost) white box if for its internal dynamics a (more or less crude) model is established which yields an output law.

We hint at four physical models of actuators from literature: a) in [2] with output force, b) in [5] with output displacement, c) in [4] a mixed case, and d) in [1] with output torque or rotation.

Here, our starting - point for introducing an actuator is a given output law. Let μ_i be qualified as impressed forces obeying the following law:

$$\mu_i(t,x,\dot{x}) := c_i\big(x_{i-1} - x_i - l_i^0\big) + k_i(\dot{x}_{i-1} - \dot{x}_i) + u_i(t). \tag{16}$$

This mathematical relation describes the parallel arrangement of a linear-elastic spring with a constant stiffness c_i and original length l_i^0, a Stokes damping element with constant coefficient k_i, and a time-dependent force $u_i(t)$. The following figure (Fig. 5) shows the corresponding physical model of this actuator (cf. Fig. 3), where now the small circular box represents a non-modeled device generating the force $u_i(t)$.

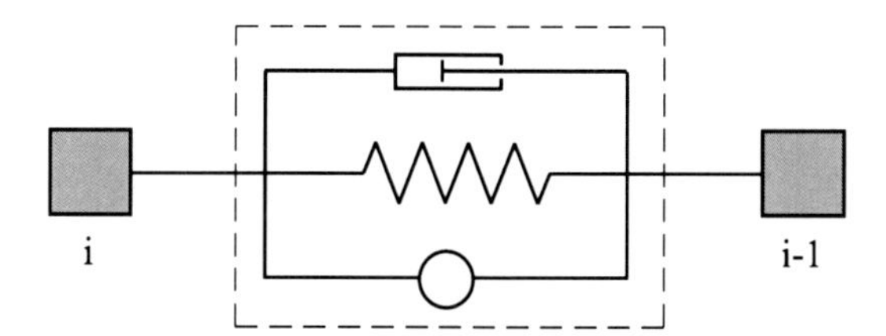

Fig. 5 Actuator, general physical model

Under the assumption that all actuators have the same data (stiffness c, original spring-length l^0, damping coefficient k_{00}) and with $\mathbf{K} = \mathbf{N}$ the equations of motions follow from (15) in the actual form

$$\begin{aligned} m\ddot{x}_0 &= -c(x_0 - x_1 - l^0) - k_{00}(\dot{x}_0 - \dot{x}_1) - k_0\dot{x}_0 - u_1(t) - \Gamma_0 + z_0, \\ m\ddot{x}_j &= -c(2x_j - x_{j+1} - x_{j-1}) - k_{00}(2\dot{x}_j - \dot{x}_{j+1} - \dot{x}_{j-1}) + \\ &\qquad + u_j(t) - u_{j+1}(t) - k_0\dot{x}_j - \Gamma_j + z_j, \\ m\ddot{x}_n &= c(x_{n-1} - x_n - l^0) + k_{00}(\dot{x}_{n-1} - \dot{x}_n) - k_0\dot{x}_n + u_n(t) - \Gamma_n + z_n. \end{aligned} \tag{17}$$

The accompanying complementary slackness conditions can be satisfied through expressing the λ_i by means of the "controller" (see [5])

$$z_i(f_i, \dot{x}_i) = -\frac{1}{2}\big(1 - \mathrm{sign}(\dot{x}_i)\big)\big(1 - \mathrm{sign}(f_i)\big) f_i, \quad i \in \mathbf{N}, \tag{18}$$

where f_i is the resultant of all further forces acting on the masspoint i.

At this stage, the u_i are to be seen as prescribed functions of t - *offline-controls*, later on they will also be handled as depending on the state $(x, \dot{x})$ - *feedback, online-controls*.

If the actuator data are known (l^0, c and k_{00}), and n is small ($n \leq 2$), then an actuator input $u_i(t)$ can be calculated which controls the system in such a way as to track a preferred motion-pattern constructed in kinematical theory (see Example 1 above).

But, as a rule, the actuator data are not known exactly. Then an **adaptive control scheme** is required that, despite of this drawback, achieves tracking at least approximately. This is addressed in Part 2.

References

1. Hirose, S.: Biologically Inspired Robots: Snake-Like Locomotors and Manipulators. Oxford University Press, Oxford (1993)
2. Huang, J.: Modellierung, Simulation und Entwurf biomimetischer Roboter basierend auf apedaler undulatorischer Lokomotion. ISLE, Ilmenau (2003)
3. Ostrowski, J.P., Burdick, J.W., Lewis, A.D., Murray, R.M.: The Mechanics of Undulatory Locomotion: The mixed Kinemtic and Dynamic Case. In: Proceedings IEEE International Conference on Robotics and Automation, Nagoya, Japan (1995)
4. Steigenberger, J.: On a class of biomorphic motion systems. Preprint No. M12/99, Faculty of Mathematics and Natural Sciences, TU Ilmenau (1999)
5. Steigenberger, J.: Modeling Artificial Worms. Preprint No. M02/04. Faculty of Mathematics and Natural Sciences, TU Ilmenau (2004)
6. Zimmermann, K., Zeidis, I., Behn, C.: Mechanics of Terrestrial Locomotion - With a Focus on Non-pedal Motion Systems. Springer, Berlin (2009)

Straight Worms under Adaptive Control and Friction - Part 2: Adaptive Control

Carsten Behn and Klaus Zimmermann

Abstract. This is the second part of the contribution to the adaptive control of worm-systems, which are inspired by biological ideas. Part 1 is the basis for this part. We focus now on the adaptive control since one cannot expect to have complete information about a sophisticated mechanical or biological system in general. Only structural properties (known type of actuator with unknown parameters) are known. Additionally, in a rough terrain, unknown or changing friction coefficients lead to uncertain systems, too. The consideration of uncertain systems leads us now to the use of adaptive control. We still assume that the worm-system contacts the ground via spikes and track gaits from the kinematical theory (preferred motion patterns to achieve movement) by means of adaptive controllers (λ-trackers). Then we replace the worm-ground interaction by stiction combined with Coulomb sliding friction (modification of a Karnopp friction model) and point out the main differences for the worm-like locomotion.

1 Introduction

The starting point is the end of Part 1 (modeling): the dynamics of a worm system, see Fig. 1, with an actuator whose corresponding physical model is presented in Fig. 2 (parallel arrangement of a linear-elastic spring with a constant stiffness c_i and original length l_i^0, a Stokes damping element with constant coefficient k_i, and a time-dependent force $u_i(t)$ - small circular box).

Carsten Behn · Klaus Zimmermann
Ilmenau University of Technology, Department of Technical Mechanics,
Max-Planck-Ring 12, 98693 Ilmenau, Germany
e-mail: {carsten.behn, klaus.zimmermann}@tu-ilmenau.de

G. Stépán et al. (Eds.): Dynamics Modeling & Interaction Cont., IUTAM BOOK SERIES 30, pp. 65–72.
springerlink.com

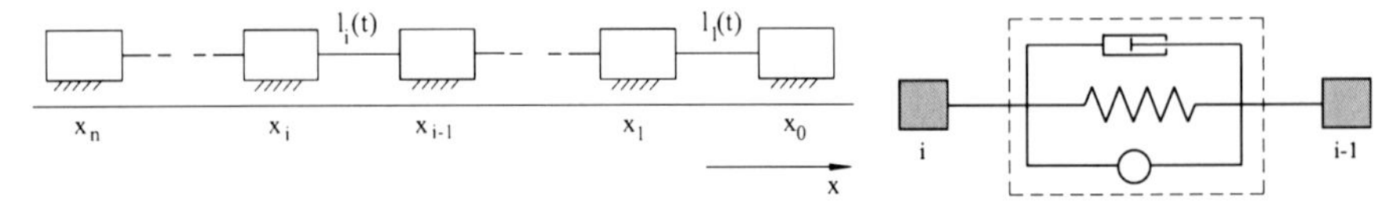

Fig. 1 Chain of point masses with spikes

Fig. 2 Actuator, general physical model

The equations of motion take the form

$$
\begin{aligned}
m\ddot{x}_0 &= -c(x_0 - x_1 - l^0) - k_{00}(\dot{x}_0 - \dot{x}_1) - k_0\dot{x}_0 - u_1(t) - \Gamma_0 + z_0, \\
m\ddot{x}_j &= -c(2x_j - x_{j+1} - x_{j-1}) - k_{00}(2\dot{x}_j - \dot{x}_{j+1} - \dot{x}_{j-1}) + \\
&\qquad + u_j(t) - u_{j+1}(t) - k_0\dot{x}_j - \Gamma_j + z_j, \\
m\ddot{x}_n &= c(x_{n-1} - x_n - l^0) + k_{00}(\dot{x}_{n-1} - \dot{x}_n) - k_0\dot{x}_n + u_n(t) - \Gamma_n + z_n .
\end{aligned}
\tag{1}
$$

The accompanying complementary slackness conditions were

$$\dot{x}_i \geq 0, \ z_i \geq 0, \ \dot{x}_i z_i = 0, \ i \in \{0,1,\ldots,n\}. \tag{2}$$

and could be satisfied through expressing the λ_i by means of the "controller" (see [10])

$$z_i(f_i,\dot{x}_i) = -\frac{1}{2}\big(1 - \mathrm{sign}(\dot{x}_i)\big)\big(1 - \mathrm{sign}(f_i)\big)f_i, \ \ i \in \{0,1,\ldots,n\}, \tag{3}$$

where f_i is the resultant of all further forces acting on the masspoint i.

Now the system of equations that together with the slackness conditions (2) and initial conditions for x_i and $\dot{x}_i$, $i = 0,\ldots,n$, governs the motions of the worm system.

In Part 1, the u_i are to be seen as prescribed functions of t - *offline-controls*. If the actuator data are known (l^0, c and k_{00}), and n is small ($n \leq 2$), then an actuator input $u_i(t)$ can be calculated which controls the system in such a way as to track a preferred motion-pattern constructed in kinematical theory.

But, as a rule, the actuator data are not known exactly. Then an **adaptive control scheme** is required that, despite of this drawback, achieves tracking at least approximately. We have to consider actuator inputs $u_i(t)$ which to be handled as depending on the state $(x,\dot{x})$ - *feedback, online-controls*. This feedback has to generated adaptively. This topic is addressed in the next section.

2 Adaptive Control

One problem in practice is the lack of precise knowledge of the actuator data, moreover, various worm system parameters, specifying environmental contact, may be not exactly known as well: we have to deal with uncertain systems.

Hence, it is impossible to calculate force inputs u to achieve a prescribed movement. We have to design a controller which on its own generates the necessary forces to track a prescribed kinematic gait. This leads us to an adaptive high-gain output feedback controller (learning controller). The aim is not to identify the actuator data or worm system parameter, but to simply control this system in order to track a given reference trajectory (kinematic gait), i.e., to ensure a desired movement of the system. We do not focus on exact tracking, rather we focus on the **λ-tracking control objective** tolerating a pre-specified tracking error of size λ.

Considering the worm system (1) we choose as outputs the actual lengths of the links, i.e., $y_i := x_{i-1} - x_i$, for $i = 1,\dots,n$. λ-tracking now means, given $\lambda > 0$, a control strategy

$$y = (y_1,\dots,y_n) \mapsto u = (u_1,\dots,u_n)$$

is sought which, when applied to this system (1), realizes tracking of any reference signal $y_{\text{ref}} = (y_{\text{ref}\,1},\dots,y_{\text{ref}\,\text{n}}$ in the following way:

(i) every solution of the closed-loop system is defined and bounded on $\mathbb{R}_{\geq 0}$, and
(ii) the output $y(\cdot)$ tracks $y_{\text{ref}}(\cdot)$ with asymptotic accuracy quantified by $\lambda > 0$ in the sense that $\max\left\{0, \|y(t) - y_{\text{ref}}(t)\| - \lambda\right\} \to 0$ as $t \to \infty$.

The last condition means that the error $e(t) := y(t) - y_{\text{ref}}(t)$ is forced, via the adaptive feedback mechanism (controller (4)), towards a ball around zero of arbitrary small pre-specified radius $\lambda > 0$, see Fig. 3.

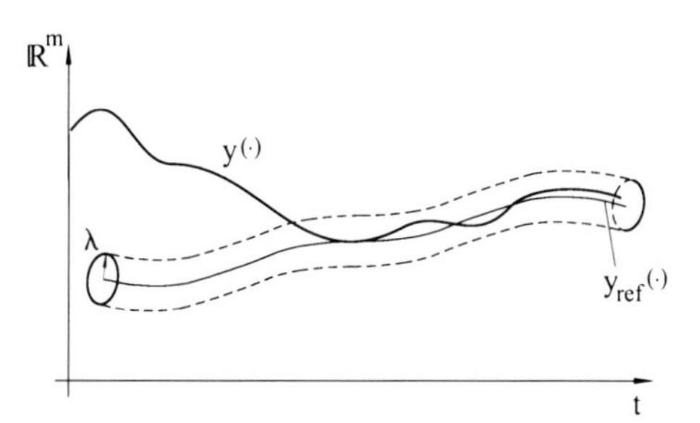

Fig. 3 The λ-tube along the reference signal

The following controller realizes our goal, for a mathematical proof (and a general system class where it works) see [3] and [4]:

$$\left.\begin{aligned} e(t) &:= y(t) - y_{\text{ref}}(t)\,, \\ u(t) &= \left(k(t)e(t) + \tfrac{d}{dt}\big(k(t)e(t)\big)\right), \\ \dot{k}(t) &= \gamma \max\left\{0, \|e(t)\| - \lambda\right\}^2, \end{aligned}\right\} \tag{4}$$

with any $k(0) \in \mathbb{R}$, $\lambda > 0$, $\gamma > 1$, $y_{\text{ref}}(\cdot) \in W^{2,\infty}$ (a Sobolev-Space), $u(t), e(t) \in \mathbb{R}^{\text{m}}$ and $k(t) \in \mathbb{R}$.

For simulations we consider a worm with $n = 2$ links. So the control $u_{1,2}$ is sought such that $y_{1,2}$ gets close to $l_{1,2}$ in the above precise sense. We choose the following data for all simulations in this paper. Additional data needed for simulations shall be given on the spot:

- worm system: $m_0 = m_1 = m_2 = 1$, $c = 10$, $k_{00} = 5$;
- environment: $k_0 = 0$, $\Gamma_{1,2,3} = 2.7$ (ensures kinematical theory to be dynamically feasible);
- reference gait (taken from [10], see also equation (13) of Example 1 in Part 1):

$$\begin{aligned} l_1(t) &:= l_0 + a\,l_0\,\big(1 - \cos(\pi t)\big)\,h(0,2,t) \\ l_2(t) &:= l_1(t + \tfrac{1}{3}) - l_0 \end{aligned} \tag{5}$$

  on the primitive time interval $[0,T]$, $T := 3$, and then take their T-periodic continuation to $\mathbb{R}^+$, $l_0 = 2$ is the original length of the links, and $l_0\,a$ with $a = 0.2$ is the amplitude of the length variation in time;
- controller: $k(0) = 0$, $\lambda = 0.2$, $\gamma = 300$.

The actuator and environmental data $(c\,, k_{00}\,, k_0\,, \Gamma)$ are fixed just for doing the simulations. In general they could be understood as estimates the unknown values may vary about.

For numerical reasons we use the smooth approximation $\mathrm{sign}(x) \approx \tanh(10000x)$.

Simulation 1: Worm with ideal spikes, i.e., (3) with controller (4).

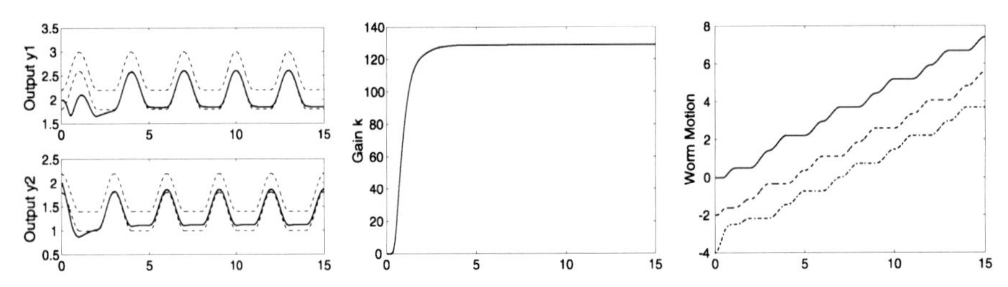

Fig. 4 (a) Outputs y_1 and y_2 with λ-tubes; (b) gain parameter $k(\cdot)$; (c) worm motion; all vs. time t

The figures - to be compared with ones in Example 1 in Part 1 - show a good tracking behavior after a transient phase until $t \approx 3$. Average speed of center of mass is $\overline{v} \approx 0.4938$ (after transient process). Fig. 4 (b) shows a monotonic increase of $k(t)$ towards a limit k_∞. But if some perturbation repeatedly caused the output to leave the λ-strip then $k(t)$ would take larger values again and again.

For practical reasons $k(t)$ must not exceed a feasible upper bound. That is why we introduce an improved adaptation law, see [5], that makes $k(t)$ decrease as long as further growth is not necessary. We distinguish three cases: 1. increasing $k(\cdot)$

while e is outside the tube, 2. constant $k(\cdot)$ after e entered the tube - no longer than a pre-specified duration t_d of stay, and 3. decreasing $k(\cdot)$ after this duration has been exceeded. For instance:

$$\dot{k}(t) = \begin{cases} \gamma\Big(\|e(t)\| - \lambda\Big)^2, \ \|e(t)\| \geq \lambda, \\ 0, \qquad\qquad\qquad\qquad \Big(\|e(t)\| < \lambda\Big) \wedge (t - t_E < t_d), \\ -\sigma k(t), \qquad\qquad\quad\ \Big(\|e(t)\| < \lambda\Big) \wedge (t - t_E \geq t_d), \end{cases} \tag{6}$$

with given $\sigma > 0$, $\gamma \gg 1$, and $t_d > 0$, whereas the entry time t_E is an internal time variable.

Simulation 2: Same system as before, but now controller from (4) with adaptor (6) ($\sigma = 0.2, t_d = 1$).

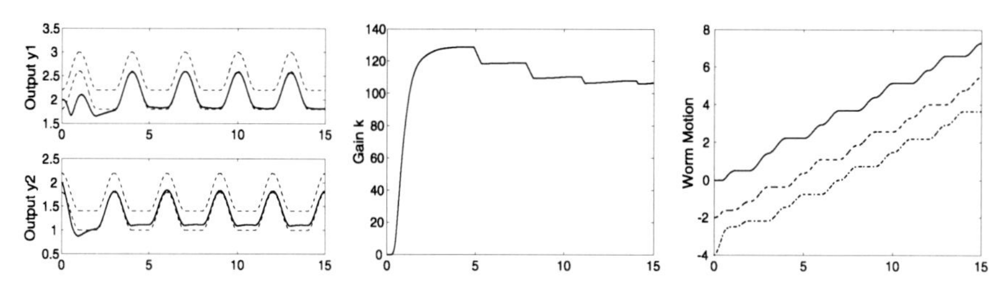

Fig. 5 (a) Outputs y_1 and y_2 with λ-tubes; (b) gain parameter $k(\cdot)$; (c) worm motion; all vs. time t

Fig. 5 (b) indicates that the maximal value $k \approx 130$ is only due to the transient behavior (until e enters the λ-tube) whereas the minimum high gain is obviously $k^* \approx 100$. Average speed of center of mass is $\overline{v} \approx 0.4859$ (after transient process).

3 Friction

We replace the worm-ground interaction via spikes by stiction combined with Coulomb sliding friction. This might be seen as a more realistic description of the interaction or as a model of practical failing of the spikes (being of finite strength).

First, we present a mathematical model for the Coulomb laws that is both theoretically transparent and handy in computing. It is by far simpler than various sophisticated laws in literature (e.g. [1, 2, 7, 9]) but similar to the model given in [8] and it well captures stick-slip effects in application to worm dynamics, see [6] and [11]. We note that this modeling makes friction a real-valued function of two arguments, and not a set-valued one depending solely on the velocity as preferred by most authors. Let

$$F(f,v) := \begin{cases} F^-, & v < -\varepsilon \vee \left(|v| \le \varepsilon \wedge f < -F_0^-\right), \\ -f, & |v| \le \varepsilon \wedge f \in \left[-F_0^-, F_0^+\right], \\ -F^+, & v > \varepsilon \vee \left(|v| \le \varepsilon \wedge f > F_0^+\right). \end{cases} \tag{7}$$

Different $F^{\pm}$ or $F_0^{\pm}$ values characterize a friction anisotropy. A suitable $\varepsilon > 0$ replaces the computer accuracy and mimics the vague processes at small velocities as well.

Using the Heaviside function h from Example 1 in Part 1, F can be given a closed form (disregarding its values at $v = \pm\varepsilon$). In order to avoid difficulties in computing caused by jumps of the h-function we turn to a smooth mathematical model (in the sense of an approximation). Basically, we use a tanh-approximation of the sign-function $\operatorname{sign}(x) \approx \tanh(Ax)$ with some sufficiently large $A \gg 1$, see above.

The smooth mathematical model then is

$$\begin{aligned} F(f,v) = &-f H(-\varepsilon,\varepsilon,v) H(-F_0^-,F_0^+,f) \\ &+F^- \left\{H(-\infty,-\varepsilon,v) + H(-\varepsilon,0,v) H(-\infty,-F_0^-,f)\right\} \\ &-F^+ \left\{H(\varepsilon,+\infty,v) + H(0,\varepsilon,v) H(F_0^+,+\infty,f)\right\}, \end{aligned} \tag{8}$$

where F is now a $\mathscr{C}^\infty$-function in closed analytical form by means of

$$H(a,b,x) := \frac{1}{2}\left\{\tanh\left(A\,(x-a)\right) + \tanh\left(A\,(b-x)\right)\right\}, \tag{9}$$

the smooth approximation of $h(a,b,x)$. We use $A = 10^5$ and $\varepsilon = 0.005$ in the sequel.

Mind that $F_0^- \gg 1$ essentially leads to the theory of ideal spikes, whereas a small F_0^- corresponds with a breakable or broken spike.

Again, adaptive control has to be used when considering uncertain or randomly changing friction data (rough terrain). Successful application is shown by the following simulation results: 1. stiction only, 2. sliding friction only, and 3. both.

Simulation 3: First, we consider only stiction, i.e., $F^+ = F^- = 0$ for (8). We choose $F_0^- = 16$ (guided by spikes-worm theory — not outlined here) and $F_0^+ = 3$. Applying controller (4) with adaptor (6) ($\sigma = 0.2$, $t_d = 1$) yields:

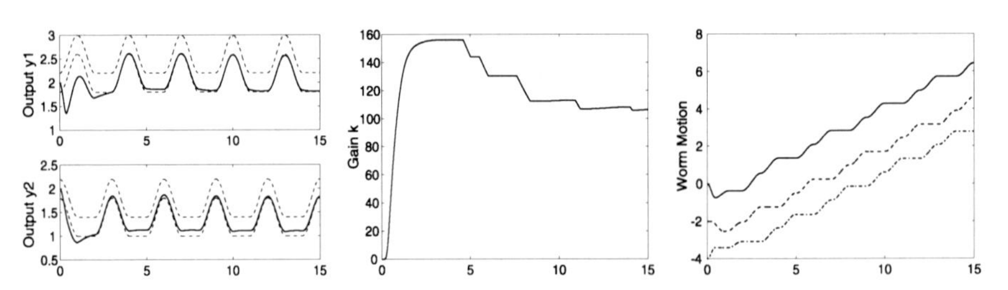

Fig. 6 (a) Outputs y_1 and y_2 with λ-tubes; (b) gain parameter $k(\cdot)$; (c) worm motion; all vs. time t

There are some short backward motions at the beginning, afterwards the motion coincides with that of Simulations 1 and 2. Average speed of center of mass is $\overline{v} \approx 0.4857$ (after transient process).

Simulation 4: Now, we replace stiction by sliding friction. We put $F_0^+ = F_0^- = 0$ and choose instead $F^- = 16$ and $F^+ = 3$. Again applying controller (4) with adaptor (6) ($\sigma = 0.2$, $t_d = 1$) yields:

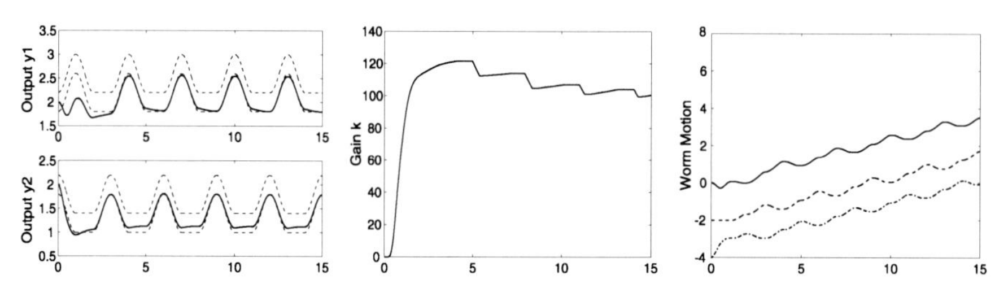

Fig. 7 (a) Outputs y_1 and y_2 with λ-tubes; (b) gain parameter $k(\cdot)$; (c) worm motion; all vs. time t

Though there is again a good tracking of the desired gait from kinematical theory, we observe an unsatisfactory external behavior of the worm (recurring negative velocities), obviously owing to the cancelation of stiction. Average speed of center of mass is $\overline{v} \approx 0.2399$ (after transient process).

Simulation 5: At last, applying the friction model (8) with $F_0^- = 18$, $F_0^+ = 3$ (remind $F_0^- = 16$ in Simulation 3), $F^- = 8$, $F^+ = 1$ (additional sliding friction) and using the same control data as before we obtain the following results:

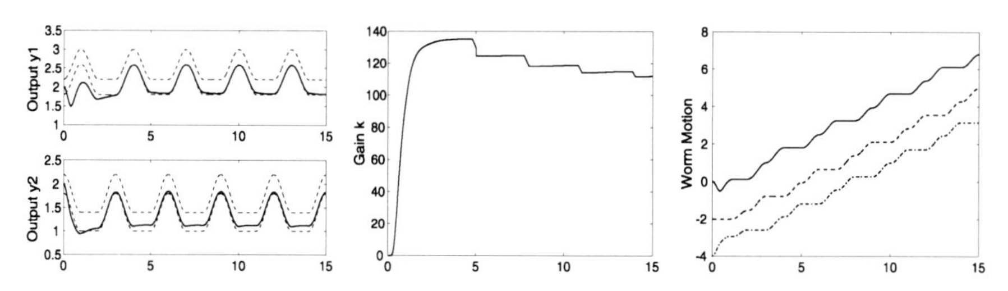

Fig. 8 (a) Outputs y_1 and y_2 with λ-tubes; (b) gain parameter $k(\cdot)$; (c) worm motion; all vs. time t

Good behavior, comparable with that in Simulation 3, but now a bit smaller average speed $\overline{v} \approx 0.4793$ due to the sliding friction.

If $F_0^- = 16$ was used then the worm would run backwards. The need of $F_0^- = 16 + 2$ is caused by the forward sliding friction $F^+ = 1$. This is essentially the same effect as it would be caused by an increase of Γ from 2.7 to 4.7 since two masspoints are sliding at every moment during the motion. So F^+ leads to an additional backward force of magnitude 2 that has to be compensated by stiction.

4 Conclusion and Outlook

In the foregoing examples the (adaptive) control has been directed to ensure a prescribed gait (i.e., a temporal pattern of shape - something internal!). It is intelligible that a changing environment or changing type of interaction influences the global movement and the driving forces u_i despite a good tracking of the gait.

A comparison of Simulations 3 to 5 points at stiction as the essential part of Coulomb interaction with the ground and gives a warning of a careless reducing of the interaction to pure sliding friction.

Finally we sketch some current and future tasks:

- to track a prescribed global movement of the worm (first step: to track a reference head speed — a pure tracking of the prescribed gait is not sufficient to do this),
- to validate the theory by experiments,
- to investigate tracking under friction which randomly changes online, possibly coupled with appropriate change of gaits ('gear shift'),
- to investigate tracking under randomly changing internal data (failing actuator).

References

1. Armstrong-Hélouvry, B., Dupont, P., Canudas de Wit, C.: A Survey of Models, Analysis Tools and Compensation Methods for the Control of Machines with Friction. Automatica 30(7), 1083–1138 (1994)
2. Awrejcewicz, J., Olejnik, P.: Analysis of Dynamic Systems with various Friction Laws. Applied Mechanics Reviews 58, 389–411 (2005)
3. Behn, C.: Ein Beitrag zur adaptiven Regelung technischer Systeme nach biologischem Vorbild. Cuvillier, Göttingen (2005)
4. Behn, C., Zimmermann, K.: Adaptive λ-Tracking For Locomotion Systems. Robotics and Autonomous Systems 54, 529–545 (2006)
5. Behn, C., Steigenberger, J.: Improved Adaptive Controllers For Sensory Systems - First Attempts. In: Awrejcewicz, J. (ed.) Modeling, Simulation and Control of Nonlinear Engineering Dynamical Systems, pp. 161–178. Springer, Heidelberg (2009)
6. Behn, C., Steigenberger, J., Zimmermann, K.: Biologically Inspired Locomotion Systems - Improved Models for Friction and Adaptive Control. In: Bottasso, C.L., Masarati, P., Trainelli, L. (eds.) Proceedings ECCOMAS Thematic Conference in Multibody Dynamics, 20 p. Electronical publication, Milano, Italy (2007)
7. Canudas de Wit, C., Olsson, H., Åström, K.J., Lischinsky, P.: A new Model for Control of Systems with Friction. IEEE Transactions on Automatic Control 40(3), 419–425 (1995)
8. Karnopp, D.: Computer simulation of stick-slip friction in mechanical dynamic systems. ASME J. of Dynamic Systems, Measurement and Control 107(1), 100–103 (1985)
9. Olsson, H., Åström, K.J., Canudas de Wit, C., Gräfert, M., Lischinsky, P.: Friction Models and Friction Compensation. European Journal of Control 4, 176–195 (1998)
10. Steigenberger, J.: Modeling Artificial Worms. Preprint No. M02/04. Faculty of Mathematics and Natural Sciences, TU Ilmenau (2004)
11. Steigenberger, J.: Mathematical representations of dry friction. Faculty of Mathematics and Natural Sciences, TU Ilmenau (2006) (unpublished Paper)
12. Zimmermann, K., Zeidis, I., Behn, C.: Mechanics of Terrestrial Locomotion - With a Focus on Non-pedal Motion Systems. Springer, Berlin (2009)

Current Sensing in a Six-Legged Robot

Q. Bombled and O. Verlinden

Abstract. Basically, the control problem of legged robot could be summarized in one point: all the legs are working together in such a way that the vehicle remains stable and moves according to the specified path. Legs are either in support phase, where they need to be coordinated to provide the desired body profile; or in swing phase, whose aim is to reach a convenient position to begin a new step. In a general way, the ground detection at the end of the swing phase is a keypoint in the walking algorithms. This paper focused on a DC actuator current measurement to detect interaction of a robot leg with the soil. It shows that careful sensing coupled with a comprehensive model of the system can give a quite accurate estimation of the contact force at the leg foot.

1 Usefulness of Current Information in Legged Robot

Ground detection is a keypoint for legged robot walking algorithms. The common solution to achieve this task is to use mechanical switches. They are of limited interest due to their single direction operating mode and their difficulty of treshold adjustment. To obtain a better representation of the interaction forces in two or three directions, force sensors based on strain gauges ([6], [1]) or piezoelectric effect ([5]) are usually embedded in the legs extremities. Other robots have such sensors inside their joints [4].

In this work, the feasibility of the motor current information to interpret accurately the interaction with the environment, especially the leg–ground contact, is investigated. The main idea is to compare current estimated without foot contact with the one really measured. Unlike foot sensors, such measurements have the capability to inform when the leg encounters an obstacle, wherever the impact is located. They are also readily embeddable, which is of full interest in an application like a walking robot. Moreover, such devices are widely spread and easy to implement in a DC motor hardware driver.

Q. Bombled · O. Verlinden
University of Mons, Place du Parc 20, 7000 Mons, Belgium
e-mail: {bombledq, verlindeno}@umons.ac.be

G. Stépán et al. (Eds.): Dynamics Modeling & Interaction Cont., IUTAM BOOK SERIES 30, pp. 73–80.
springerlink.com

A comprehensive model of the leg, developed in Sect. 2, is useful to have a first estimation of the motors currents. Friction measurements setup and results in steady–state conditions are detailed in Sect. 3. Sect. 4 illustrates the use of current information and its model–based estimation to obtain the ground detection ability. Conclusions and potential interest of current sensing are finally discussed in Sect. 5

2 The Leg of AMRU5

The legged robot AMRU5 was developed for demining purposes. It has been built by the Royal Military Academy of Belgium, within the framework of the Autonomous and Mobile Robot in Unstructured environment project. Its weight is about 30 kg and its outer diameter can reach 1.6m.

2.1 *Leg Description*

Fig. 1 depicts the leg CAD model. The foot point is able to move in a 3D space. The degrees of freedom (joints) are:

- q_r giving the tangential motion to the leg extremity;
- q_v for the up and down motion;
- q_h for the back and forth motion.

A pantograph mechanism is used to obtain a decoupling between the vertical and the horizontal motion. Transmission components are the same for the q_v and q_h joints : each motor is followed by a gearbox, whose output shaft is connected to a chain transmission (not represented in Fig. 1). The second sprocket wheel of this transmission is assembled with a ball screw. The displacement of the nut along the ball screw is the degree of freedom of the model.

Transmission for joint q_r is exactly the same, apart from the ball screw. The motor relative to q_r is not represented in Fig. 1.

2.2 *Multibody Model*

Equations of a multibody system are written in the joint space as (1):

$$\mathbf{M}(\mathbf{q})\ddot{\mathbf{q}} + \mathbf{C}(\mathbf{q},\dot{\mathbf{q}})\dot{\mathbf{q}} + \mathbf{g}(\mathbf{q}) + \tau_f = \tau \qquad (1)$$

where $\mathbf{q}$ is the 3x1 vector of joint space variables (q_r, q_v, q_h), $\dot{\mathbf{q}}, \ddot{\mathbf{q}}$ their first and second time derivatives, $\mathbf{M}(\mathbf{q})\ddot{\mathbf{q}}$ gathers the inertia forces, $\mathbf{C}(\mathbf{q},\dot{\mathbf{q}})\dot{\mathbf{q}}$ accounts for centrifugal and Coriolis forces, $\mathbf{g}(\mathbf{q})$ represents the gravity forces, τ is the joint forces vector and τ_f the friction torques that will be studied in detail in Sect. 3.

The free and open source EasyDyn framework [3] library has been used to build the equations of motions, according to a generalized coordinates approach. The user has to specify inertia properties and gravity forces in the global frame, and to

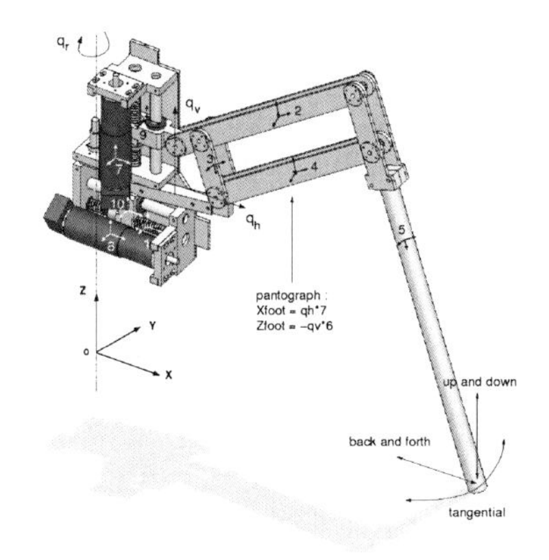

Fig. 1 Real robot and CAD model of the leg. Each modeled body is represented by a frame and numbered. Body 1 — chassis frame; bodies 2 to 5 — pantograph parts; bodies 6 to 8 — rotor shaft; bodies 9 and 10 — the balls nut.

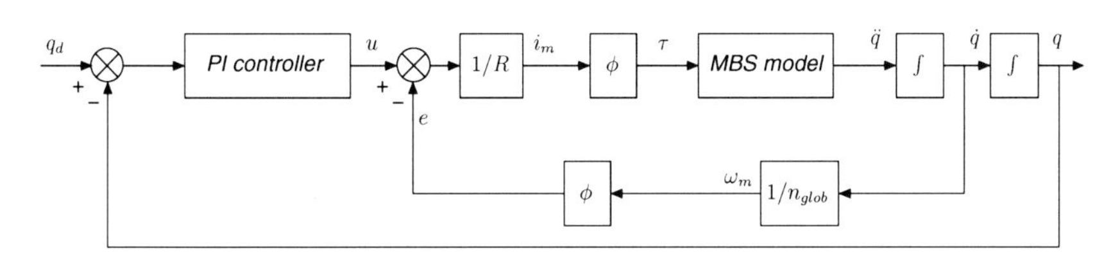

Fig. 2 AMRU5 joint control is decentralized. Each actuator is position–controlled independently of the others. This figure shows the general scheme used in simulation, with ϕ the motor torque constant, R the armature resistance of the motor and n_{glob} the global reduction ratio of the transmission between the rotor and the joint position.

describe the kinematics at position level only, in terms of joint variables $\dot{\mathbf{q}}$. For this, combinations of simple homogeneous transformation matrices are made using a computer algebra system. The complete kinematics (velocity and acceleration in the global frame) is finally deduced by symbolic differentiation. Further details can be found in [7].

2.3 *Actuator and Controller Model*

To complete the multibody model, the actuator and controller dynamics are added. Each joint has its own PI position controller (see Fig. 2). Position error $q - q_d$ (q_d being the desired position) is the controller input. Output u is sent to the motor model, which gives motor current i_m as an output. After some simulations, its has been proved that motor inductance could be neglected. Finally, the rotor torque τ is used as an input for the multibody model. An overview of this closed loop scheme is depicted in Fig. 2. The global multiphysic system is solved using the EasyDyn Newmark integration scheme.

3 Friction Identification

First simulations and comparisons with real measurements have shown that friction effect can not be neglected. The aim of this section is to determine a relationship between the joints velocities and the friction forces applied on the system. Assumptions are the following:

- Friction of each joint is considered separately;
- Model describes the steady–state effects, not friction dynamics;
- Coulomb – viscous approximation is used (Stribeck's effect at low velocities not considered);
- The model is global and relates joint velocity $\dot{q}$ with the friction torque $\tau_{f,rotor}$ acting on the motor shaft. It is put under the form (2) :

$$\tau_{f,rotor} = (\tau_{c,rotor} + \beta\dot{q})sgn(\dot{q}) \quad (2)$$

where $\tau_{c,rotor}$ is the Coulomb friction torque, and $\beta\dot{q}$ the viscous term.

As in [2], constant velocity profiles are imposed to each joint to identify friction parameters. It has the advantage of emphasize the steady–state friction characteristics. Amplitude of the motions are such that the complete stroke of each joint is covered.

3.1 Measurement Chain

The Hall effect current transducer is a LEM LA 25-NP/SP8. Its measuring range is from -3 to +3A. Because of the low armature resistance of the motor (lower than 0.4 Ω), classical current sensing with a shunt resistor is not appropriate.

A/D conversion is made with 10–bit resolution over 5V. Due to the 100Hz control loop frequency, a first low pass analog filter with a cut–off frequency of 30Hz is inserted in the measurement loop. To smooth the signal again, measurements are digitally filtered with an order two Butterworth filter. The delays introduced by the filtering have been designed in such a way that it is negligible with respect to the mechanical effects. The measurement chain is designed to get a resolution of 4 mA.

3.2 Friction Parameters

Identification of friction parameters requires the knowledge of the other effects responsible for current consumption, amongst them:

- the gravity effects;
- the inertia, Coriolis and centrifugal forces : with a constant velocity profile, they are very low excepted at velocity changes. Therefore, friction parameters are estimated outside of this area;
- the chordal velocity variation : mainly due to the small number of teeth in sprocket chain transmissions, velocity variation during the gear mesh induce

motor torque variations and consequently current fluctuations. By averaging the measured current, this effect is strongly reduced.
- the flexibility of the joints : because the parts of the mechanisms are not infinitely rigid, the joints deformation causes energy, and thus current, fluctuations. Although it exists, this effect will be neglected in our approach.

Motors torques τ are computed from current measurements. Friction torques τ_f are then identified by using the multibody model (1). Parameters are gathered in Table 1.

Table 1 Friction parameters estimation. τ_c represents the Coulomb friction level and β the viscous coefficient. The absolute error ε_A between the averaged measurements and the model is given, as well as the corresponding force error ε_F acting at leg tip, along the corresponding direction.

Joint	$sgn(\dot{q})$	$\tau_c(mN.m)$	$\beta(mN.s)$	$\varepsilon_A(mN.m)$	$\varepsilon_F(N)$
q_r	>0	3.9	6.6	0.2	0.9
	<0	-4.0	10.4	0.07	0.3
q_v	>0	3.7	188.0	0.03	0.4
	<0	-4.7	402.0	0.2	2.3
q_h	>0	7.1	66.0	0.08	0.7
	<0	-7.6	109.3	0.2	1.7

4 Environment Detection

The objective of the environment detection is to stop the leg whenever it meets an obstacle, and to adapt the trajectory to bypass this one. We focus only on the ground detection with the vertical joint, which occurs when the difference between estimated and actual current is higher than a predefined level.

4.1 On–Line Estimation of Current

Estimation of currents $\mathbf{i}_{est}$ is provided in real-time by using the global scheme of Fig. 3. Joint velocity $\dot{\mathbf{q}}$ and acceleration $\ddot{\mathbf{q}}$ are evaluated with a numerical derivation of position measurements which have been previously filtered to reduce noise measurement(z^{-1} is the delay operator). The complete kinematics of the 10 bodies (position $\mathbf{e}_b$, velocities $\mathbf{v}_b$ and acceleration $\mathbf{a}_b$ in the global frame) is then computed through the function $\mathbf{k}(.)$ coming from the kinematic description of Sect. 2.2. Motions of the bodies are used as inputs to estimate the current contribution of inertia and gravity forces. Friction contribution is computed with parameters of Sect. 3.2.

Fig. 4 shows sine and sawteeth motions for the vertical joint. Measured and estimated current agree quite well, excepted at sharp velocity changes. Explanation of this discordance is that for extreme position of the q_v joint, there is an important bending of the horizontal ball-screw. Therefore, it stores energy which is given

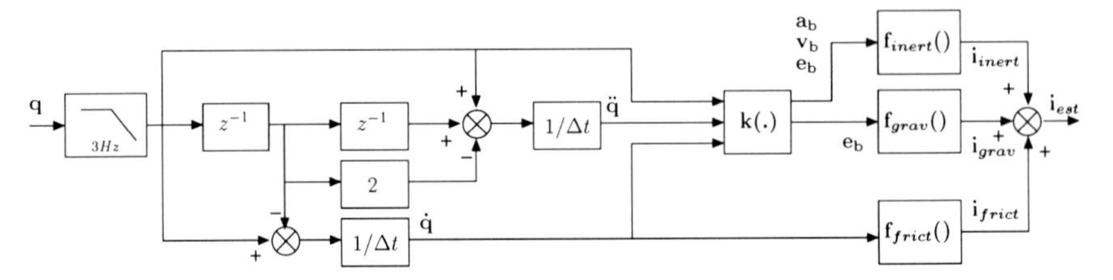

Fig. 3 Current estimation mechanism

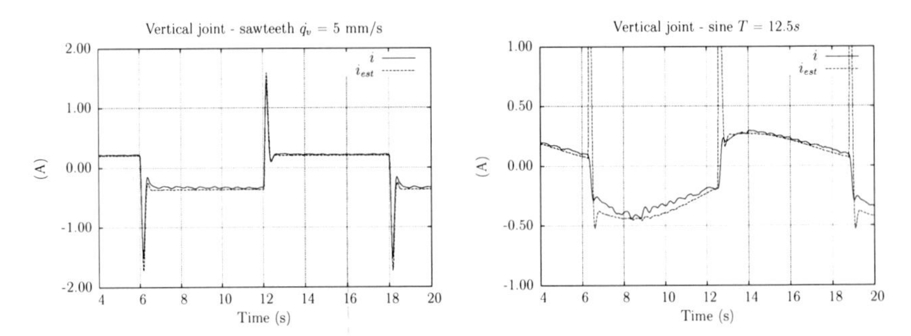

Fig. 4 Comparison of measured and estimated currents, for a sawteeth and a sine motion of the vertical joint. Besides friction forces, effect of inertia and gravity is not a negligible part of the current consumption, especially for motions with variable velocity.

back to the system when the velocity change occurs. Because we do not model the joint compliance, the estimated current at velocity reversals is always higher than the estimated one.

4.2 Contact Force Estimation

Experiments have been made by applying a sawteeth motion to the vertical joint (subscript v). The force acting vertically at the foot is proportional to Δi_v computed via (3):

$$\Delta i_v = i_v - i_{v,est} \tag{3}$$

with $i_{v,est}$ the estimated current that should be flowing through the motor if there wasn't any obstacle, and i_v the measured current. When contact with the ground occurs ($t \sim 13.5s$ in Fig. 5), Δi_v becomes higher than the predefined level, and ground is considered as detected. Motor supply is then stopped. Even with the ground reaction force, the joint position stays locked because of its irreversibility. The controller re-takes the control when the desired joint trajectory decreases again as depicted at $t \sim 16s$ in Fig. 5. During contact phase, the current estimation is not computed: indeed force estimation is lost because the motor is not supplied anymore.

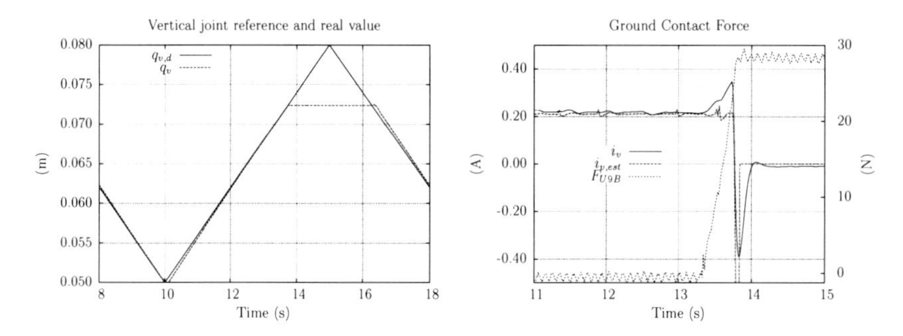

Fig. 5 Ground detection via current measurements. As soon as Δi_v is too important, leg vertical motion and current estimation are stopped.

The measurement of the contact force is provided by a monoaxial force sensor HBM-UB9 fixed on the ground. Table 2 summarizes the Δi_v desired for the ground interaction, the equivalent force F_{eq} and F_{U9B}, the force measured with the external sensor.

Table 2 Predefined current level, equivalent contact force F_{eq} and measured forces F_{U9B}

Δi_v (A)	F_{eq} (N)	F_{U9B} (N)	error (%)
0.12	27.6	27.4	< 1
0.15	34.6	33.1	4
0.18	41.5	37.2	10
0.21	48.4	41.1	15

5 Conclusions

This paper has investigated the feasibility of the ground contact force estimation with current sensing, in the case of a legged robot. It has been shown that an accurate measurement coupled with a comprehensive model of the system can lead to a quite good estimation of this force. The effect of joint compliance should be added to avoid large estimation errors, especially at velocity reversals, or for positions or foot high loads responsible for the mechanism deformation.

But, despite of the numerous advantages of the method like the easy–embeddable feature and the low cost of the electronics, the current sensing is not always relevant. Indeed, in the case of high frictions elements leading to irreversibility, the current measurement is only available during motion.

Nevertheless, our algorithm should be sufficient for ground detection in the case of a walking robot. Identification of friction for the three joints and currents estimation mechanism presented in Sect. 4.1 give the opportunity of a generalization in a 3D space, providing extra abilities to the leg as, for instance, obstacles detection. Furthermore, the proposed method could be applicable for any other mechanism actuated by DC motors.

Acknowledgements. The authors would like to acknowledge Prof. Y. Baudoin and the Royal Military Academy of Belgium, for support in research and the lending of the AMRU5 hexapod.

References

1. Berns, K., Ilg, W., Deck, M., Albiez, J., Dillman, R.: Mechanical construction and computer architecture of the four-legged walking machine BISAM. IEEE/ASME Transactions on Mechatronics 4, 32–38 (1999)
2. Bona, B., Indri, M., Smaldone, N.: Friction identification and model-based digital control of a direct-drive manipulator. In: Menini, L., Zaccarian, L., Abdallah, C.T. (eds.) Current Trends in Nonlinear Systems and Control, pp. 231–251. Birkhäuser, Boston (2006)
3. Easy simulation of dynamic problems, `http://mecara.fpms.ac.be/EasyDyn/index.html`
4. Espenschied, K.S., Quinn, R.D., Beer, R.D., Chiel, H.J.: Biologically based distributed control and local reflexes improve rough terrain locomotion in a hexapod robot. Robotics and Autonomous Systems 18, 59–64 (1996)
5. Galvez, J.A., Estremera, J., Gonzalez de Santos, P.: A new legged-robot configuration for research in force distribution. Mechatronics 13, 907–932 (2003)
6. Gassmann, B., Scholl, K.-U., Berns, K.: Locomotion of LAURON III in rough terrain. In: Proc. 2001 IEEE/ASME International Conference on Advanced Intelligent Mechatronics, pp. 959–964 (2001)
7. Verlinden, O., Habumuremyi, J.C., Kouroussis, G.: Open source model of the AMRU5 hexapod robot. In: International Symposium on Measurement and Control in Robotics (2005)

Vehicle Dynamics and Control

Dynamics and questions of modeling and simulation of complex multibody systems such as vehicles are addressed in this section. The work by W.V. Wedig investigates dynamics of road-vehicle systems on rough roads that generates vertical car vibrations. It is presented that, for critical car speeds, the stationary car vibrations may bifurcate into non-stationary chaos. In the paper by L. Mikelsons, equation based reduction techniques are adopted and extended to generate vehicle models with adjustable accuracy enabling the simulation to run in real-time.

Resonances of Road-Vehicle Systems with Nonlinear Wheel Suspensions

Walter V. Wedig

Abstract. The paper investigates dynamics of road-vehicle systems. The ride on rough roads generates vertical car vibrations whose root-mean-squares become resonant for critical speeds. These investigations are extended to nonlinear wheel suspensions with cubic-progressive spring characteristics and piecewise quadratic damping mechanism. For weak but still positive damping, the vibrations become unstable in the overcritical range of car speeds. This nonlinear behavior of road-vehicle systems is detected by perturbation equations and associated Lyapunov exponents. For critical car speeds, the top Lyapunow exponents become positive indicating that the stationary car vibrations bifurcate into non-stationary chaos.

1 Introduction to the Problem

To introduce road-vehicle systems of interest, consider the quarter car model with one degree of freedom, shown in Fig. (1). The car is riding on randomly shaped road surfaces [1] generating vertical car vibrations described by

$$\ddot{X}_t + 2D\omega_1(1+\delta|\dot{R}_t|)\dot{R}_t + \omega_1^2(1+\gamma R_t^2)R_t = 0, \tag{1}$$

$$dZ_s = \Omega U_s ds, \quad dU_s = -(\sqrt{2}\,U_s + Z_s)\Omega ds + \sigma dW_s. \tag{2}$$

In Eq. (1), ω_1 is the natural frequency of the car model and $D > 0$ denotes its linear damping mechanism. The parameters γ and δ determine the cubic-progressive spring, respectively the quadratic damping of the wheel suspension. X_t is the absolute car vibration in dependence on time t and $R_t = X_t - Z_t$ is its relative motion with respect to the excitation Z_t. The road excitation Z_s of the car is defined by the stochastic differential equations (2) in dependence on the way coordinate s. Herein, $\Omega > 0$ is a fixed road frequency and σ denotes the intensity of the Wiener increments dW_s.

Walter V. Wedig
Institut für Technische Mechanik, Universität Karlsruhe
e-mail: wwedig@t-online.de

G. Stépán et al. (Eds.): Dynamics Modeling & Interaction Cont., IUTAM BOOK SERIES 30, pp. 83–90.
springerlink.com

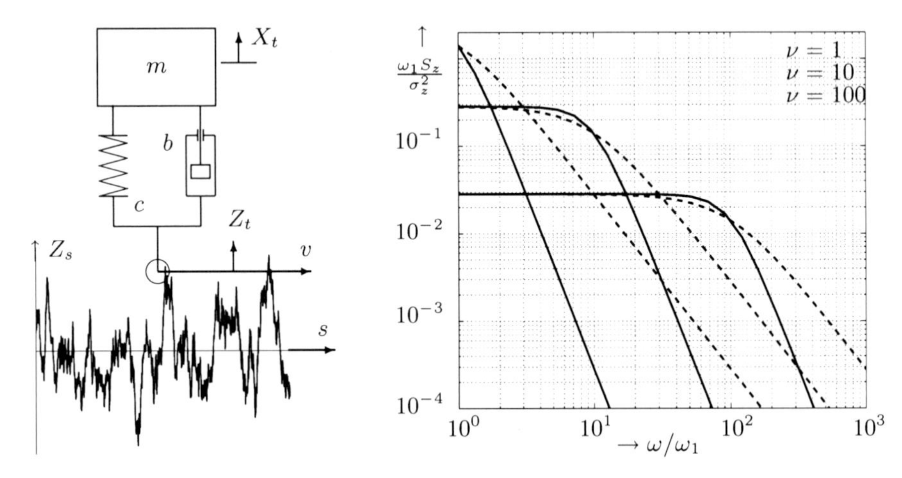

Fig. 1 Car riding on random roads **Fig. 2** Spectra of 1st and 2nd order filter

The transformation from way to time is performed by the increments

$$ds = vdt, \qquad dW_s = \sqrt{v}\, dW_t. \tag{3}$$

The Wiener increments in way and time have the expectations $E(dW_s^2) = ds$ and $E(dW_t^2) = dt$. The insertion of the increments (3) into Eq. (2) gives

$$dZ_t = \Omega v U_t dt, \qquad dU_t = -(\sqrt{2}\, U_t + Z_t)\Omega v dt + \sigma\sqrt{v}\, dW_t, \tag{4}$$

$$S_z(\omega) = \frac{(v\Omega)^2\sigma^2 v}{(v\Omega)^4 + \omega^4}\,, \qquad \sigma_z^2 = \frac{1}{2\pi}\int_{-\infty}^{+\infty} S_z(\omega)d\omega = \frac{\sigma^2}{\sqrt{8}\,\Omega}\,. \tag{5}$$

In the stationary case, the system (4) possesses the power spectrum $S_z(\omega)$ that is integrated over all spectral frequencies $|\omega| < \infty$ to obtain the variance σ_z^2 of the road excitation. Note, that the road spectrum $S_z(\omega)$ depends on speed and vanishes for sufficiently slow or fast speeds. Its variance, however, is independent on speed. Fig. (2) shows the related spectrum $S_z\omega_1/\sigma_z^2$ plotted by solid lines versus the frequency ratio ω/ω_1 in a double-logarithm scaling. In comparison to this, the first order filter

$$d\hat{Z}_t = -v\Omega\hat{Z}_t dt + \sigma\sqrt{v}\, dW_t, \qquad \frac{S_{\hat{z}}\omega_1}{\sigma_z^2} = \frac{2\nu\sqrt{2}}{\nu^2 + (\omega/\omega_1)^2}\,. \tag{6}$$

possesses the spectrum $S_{\hat{z}}(\omega)$ noted in (6) and plotted by interrupted lines for three values of the related car speed $\nu = v\Omega/\omega_1$. Spectra of the first order form (6) are already introduced in [2].

2 Digital Simulation of Stochastic Systems

For digital simulations of the second order filter equations (4) it is convenient to introduce the dimensionless time and noise increments

$$d\tau = \omega_1 dt, \qquad dW_\tau = \sqrt{\omega_1}\, dW_t, \qquad E(dW_\tau^2) = d\tau, \tag{7}$$
$$Z_\tau = \sigma_z \bar{Z}_\tau, \qquad U_\tau = \sigma_z \bar{U}_\tau, \qquad \sigma_z^2 = \sigma^2/(\Omega\sqrt{8}), \tag{8}$$

as well as the related road excitations $\bar{Z}_\tau$ and $\bar{U}_\tau$ as noted in Eq. (8). The insertion of the relations above into the Eq. (4) leads to

$$d\bar{Z}_\tau = \nu\bar{U}_\tau d\tau, \qquad d\bar{U}_\tau = -(\sqrt{2}\,\bar{U}_\tau + \bar{Z}_\tau)\nu d\tau + \sqrt{\nu\sqrt{8}}\, dW_\tau, \tag{9}$$
$$d\mathbf{V}_\tau = A\mathbf{V}_\tau d\tau + \mathbf{g} dW_\tau\,, \qquad \mathbf{V}_\tau = (\bar{Z}_\tau, \bar{U}_\tau)^T, \qquad (\nu = v\Omega/\omega_1) \tag{10}$$

and therewith to the two-dimensional vector form (10). Note, that the related processes possess the stationary rms-values $\sigma_{\bar{z}} = \sigma_{\bar{u}} = 1$. Following Ref. [3], Eq. (10) is written into the integral equation

$$V_{\tau+\Delta\tau}^{(n+1)} = \mathbf{V}_\tau + A\int_\tau^{\tau+\Delta\tau} \mathbf{V}_\sigma^{(n)} d\sigma + \mathbf{g}\int_\tau^{\tau+\Delta\tau} dW_\sigma\,. \tag{11}$$

It is valid for time increments $\Delta\tau \geq 0$, assumed to be small, and for the initial $\mathbf{V}_\tau$ at $\Delta\tau = 0$. Starting with the zeroth approximation $\mathbf{V}_\sigma^{(0)} = \mathbf{V}_\tau$, Eq. (11) is stepwise solved for $n = 0, 1, \ldots$ giving the second order solution

$$\mathbf{V}_{\tau+\Delta\tau}^{(2)} = \mathbf{V}_\tau + A\mathbf{V}_\tau \int_\tau^{\tau+\Delta\tau} d\sigma + \mathbf{g}\int_\tau^{\tau+\Delta\tau} dW_\sigma + $$
$$+ A^2\mathbf{V}_\tau \int_\tau^{\tau+\Delta\tau}\int_\tau^{\sigma} d\sigma_1 d\sigma + A\mathbf{g}\int_\tau^{\tau+\Delta\tau}\int_\tau^{\sigma} dW_{\sigma_1} d\sigma. \tag{12}$$

The iterated integrals obtained in Eq. (12) are evaluated to

$$\mathbf{V}_{\tau+\Delta\tau}^{(2)} = [I + A\Delta\tau + A^2\tfrac{1}{2}(\Delta\tau)^2]\mathbf{V}_\tau + (I + A\tfrac{1}{\sqrt{3}}\Delta\tau)\mathbf{g}\Delta W_\tau, \tag{13}$$

where I denotes the unit matrix. A and $\mathbf{g}$ are given by Eq. (9) and (10). Fig. (3) shows numerical results of the recurrence formula (13). Its evaluation for the system of interest gives the explicit formula

$$\bar{Z}_{\tau+\Delta\tau}^{(2)} = \bar{Z}_\tau + \nu\Delta\tau\bar{U}_\tau - \tfrac{1}{2}(\nu\Delta\tau)^2(\bar{Z}_\tau + \sqrt{2}\,\bar{U}_\tau) + \tfrac{1}{\sqrt{3}}\nu\Delta\tau\sqrt{\nu\sqrt{8}}\Delta W_\tau\,,$$
$$\bar{U}_{\tau+\Delta\tau}^{(2)} = \bar{U}_\tau - \nu\Delta\tau(\bar{Z}_\tau + \sqrt{2}\,\bar{U}_\tau) + \tfrac{1}{2}(\nu\Delta\tau)^2(\sqrt{2}\,\bar{Z}_\tau + \bar{U}_\tau) +$$
$$+ (1 - \sqrt{\tfrac{2}{3}}\,\nu\Delta\tau)\sqrt{\nu\sqrt{8}}\,\Delta W_\tau\,, \qquad (\Delta W_\tau = \sqrt{\Delta\tau}\,N_n). \tag{14}$$

Herein, ΔW_τ is approximated by normally distributed numbers N_n with zero mean and $E(N_n^2) = 1$. The recursion (14) is applied for the scan rate $\Delta\tau = 0.01$ and 10^6 sample points. Solid lines with circles and squares mark the rms-values $\sigma_{\bar{z}}$ and $\sigma_{\bar{u}}$, respectively. Interrupted lines represent first order approximations marked by crosses and diamonds, correspondingly. Both results are plotted versus the car speed in the first half part $0 \leq \mu \leq 1$ by the linear scaling $\nu = \mu$ and a rational scaling with $\nu = 1/(2-\mu)$ in $1 \leq \mu \leq 2$. Obviously, larger measurement deviations are only observable on the left or right end of the diagram. For fast car speeds, the recursion becomes unstable. For slow speeds, the measuring time is too short.

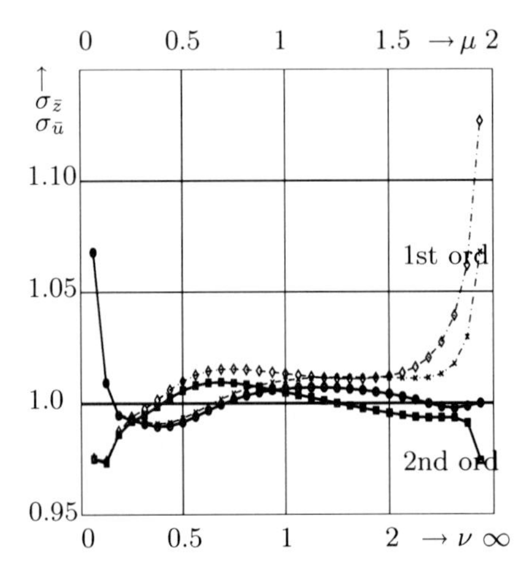

Fig. 3 RMS-values vs car speed

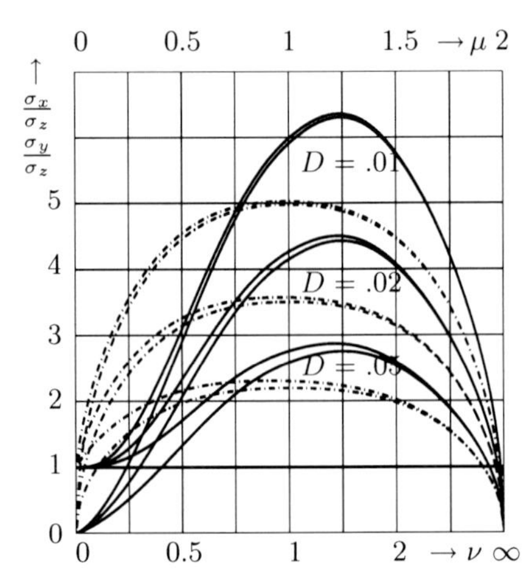

Fig. 4 Calculated values vs speed

Note, that the mixed integral in Eq. (12) is normally distributed. For $\Delta\tau, \Delta W_\tau \to 0$ they are proportional to the integration area $\Delta\tau\Delta W_\tau$, i.e.

$$I_S = \int_\tau^{\tau+\Delta\tau} \int_\tau^\sigma dW_{\sigma_1} d\sigma = \int_\tau^{\tau+\Delta\tau} (W_\sigma - W_\tau) d\sigma = \alpha \Delta\tau \Delta W_\tau . \tag{15}$$

To calculate α, the mean square of the mixed integral is taken by

$$E(I_S^2) = \int_\tau^{\tau+\Delta\tau} \int_\tau^{\tau+\Delta\tau} E[(W_{\sigma_1} - W_\tau)(W_{\sigma_2} - W_\tau)] d\sigma_1 d\sigma_2 = \tfrac{1}{3}\Delta\tau^3 . \tag{16}$$

The double integral in Eq. (16) is evaluated using $E(W_{t_1} W_{t_2}) = \min(t_1, t_2)$. Taking into account that $E[\Delta W_\tau^2] = \Delta\tau$ and $E(I_S^2) = \alpha^2 \Delta\tau^2 E[\Delta W_\tau^2]$ the α- value is finally determined to $\alpha = 1/\sqrt{3}$.

3 MC-Simulation of Road-Vehicle Systems

In the linear case $\gamma, \delta = 0$, the analysis of complex transfer functions can be applied to Eq. (1) and (4) to obtain the rms-values plotted in Fig. (4). Herein, solid lines are related rms-values of the absolute car vibrations and their time derivatives. The first one is calculated by the formula

$$\frac{\sigma_x^2}{\sigma_z^2} = \frac{1}{2D} \frac{2D[1 + 2\nu^2(1 + 2D^2)] + \sqrt{2}\nu[\nu^2(1 + 4D^2) + 4D^2]}{\nu^4 + (2D\nu)^2 + 1 + 2\sqrt{2}D\nu(1 + \nu^2)} . \tag{17}$$

It is plotted for three different damping values starting with $\sigma_x/\sigma_z = 1$ and ending in $\sigma_x/\sigma_z = 0$. In opposite to this, the related rms-velocity and the relative rms-value are starting with zero and end with the finite value $\sigma_y/\sigma_z = 2D$ and $\sigma_r/\sigma_z = 1$, respectively. Both rms-values are calculated to

$$\frac{\sigma_y^2}{\sigma_z^2} = \frac{\nu^2}{2D} \frac{2D[1 + (2D\nu)^2] + \sqrt{2}\nu(1 + 4D^2)}{\nu^4 + (2D\nu)^2 + 1 + 2\sqrt{2}D\nu(1 + \nu^2)} , \tag{18}$$

$$\frac{\sigma_r^2}{\sigma_z^2} = \frac{\nu^3}{2D} \frac{\sqrt{2} + 2D\nu}{\nu^4 + (2D\nu)^2 + 1 + 2\sqrt{2}D\nu(1 + \nu^2)} . \tag{19}$$

Between the extreme slow and fast speeds there is a critical speed range near $\nu = 1.\bar{3}$, where the rms-values of the car vibrations become maximal. Obviously, this critical range is shifted to lower speeds, when first order filter equations are applied. In Fig. (4) the associated results are plotted by interrupted lines. According to [4] they are calculated to

$$\frac{\sigma_{\hat{x}}^2}{\sigma_{\hat{z}}^2} = \frac{1}{2D} \frac{2D + \nu + 4D^2\nu}{1 + \nu^2 + 2D\nu} , \qquad \frac{\sigma_{\hat{y}}^2}{\sigma_{\hat{z}}^2} = \frac{\nu}{2D} \frac{1 + 4D^2 + 8D^3\nu}{1 + \nu^2 + 2D\nu} . \tag{20}$$

For slow and fast speeds they have exactly the same limiting values. Fig. (5) shows numerical simulation results performed by first order approximations applied to the equations (1) and (4). Herein, the rms-values (17) and (18) of the stationary car vibration and their vertical velocity are plotted by solid and interrupted lines, respectively. The associated simulation results are marked by circles and squares for the car vibrations as well as by crosses and diamonds for the road excitations. The Euler scheme was applied for 10^6 sample points with the scan rate $\Delta\tau = 0.001$. For the three damping measures $D = .01, .02$ and $.05$ there is a good coincidence between the results obtained by numerical simulation and the theoretical values calculated by Eq. (17) and (18).

To extend the numerics above to the nonlinear road-vehicle system of interest, the vertical velocity process $Y_t = \dot{X}_t/\omega_1$, the dimensionless time $dt = d\tau/\omega_1$ and the related system processes $Y_\tau = \sigma_z\bar{Y}_\tau$, and $X_\tau = \sigma_z\bar{X}_\tau$ are introduced into the nonlinear equation (1) of motion. Together with Eq. (9), the entire road-vehicle system is described by the four equations

$$\begin{aligned} d\bar{Z}_\tau &= \nu\bar{U}_\tau d\tau, \qquad d\bar{U}_\tau = -(\sqrt{2}\,\bar{U}_\tau + \bar{Z}_\tau)\nu d\tau + \sqrt{\nu\sqrt{8}}\, dW_\tau, \\ d\bar{Y}_\tau &= -2D(\bar{Y}_\tau - \nu\bar{U}_\tau)[1 + \bar{\delta}|\bar{Y}_\tau - \nu\bar{U}_\tau|]d\tau - (\bar{X}_\tau - \bar{Z}_\tau)\times \\ d\bar{X}_\tau &= \bar{Y}_\tau d\tau, \qquad \times[1 + \bar{\gamma}(\bar{X}_\tau - \bar{Z}_\tau)^2]d\tau. \end{aligned} \tag{21}$$

Herein, new coefficients of the cubic spring and the quadratic damping are introduced by $\bar{\gamma} = \gamma \sigma_z^2$ and $\bar{\delta} = \delta \omega_1 \sigma_z$, respectively. D denotes the linear damping and $\nu = v\Omega/\omega_1$ is the related car speed, as already noted.

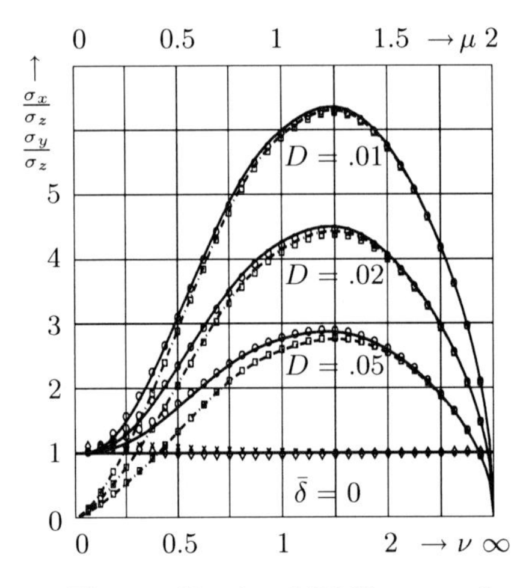

Fig. 5 Simulated RMS vs speed

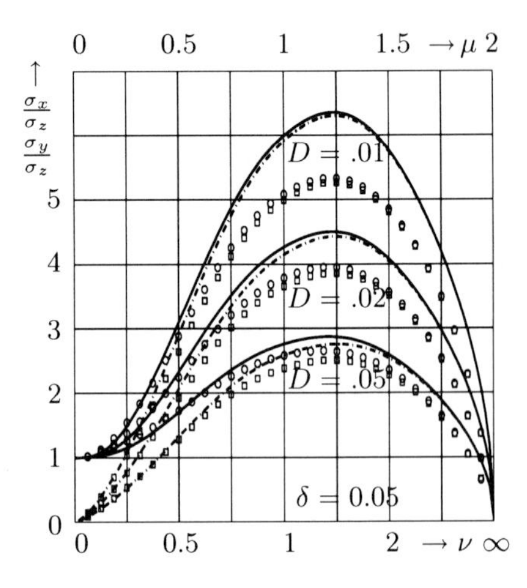

Fig. 6 RMS-reduction by damping

Fig. (6) shows numerical simulation results of the road-vehicle system (21) obtained by Euler schemes applying the scan rate $\Delta\tau = 0.001$ for 10^6 samples. The damping measures selected were $D = .01, .02$ and $.05$. The nonlinearity parameters were $\bar{\delta} = 0.05$ and $\bar{\gamma} = 0$. The obtained simulation results are marked by circles and squares for the car vibrations and their velocities, respectively. In comparison to these numerical results, solid and interrupted lines represent rms-values of the car vibrations for the linear case $\bar{\gamma}, \bar{\delta} = 0$. Obviously, the quadratic damping mechanism reduces the associated rms-values of the linear car vibrations and their velocities. This situation is completely changed in case of nonlinear springs. For $\bar{\delta} = 0$ and $\bar{\gamma} = 0.05$, Fig. (7) shows corresponding numerical results obtained applying the scan rate $\Delta\tau = 0.001$ for 10^7 samples and the two damping values $D = .1$ and $D = .05$. The associated linear rms-values are marked by interrupted lines. According to [4], cubic-progressive springs reduce the relative rms-values of the car vibrations in the under-critical speed range, only. For over-critical speeds, however, there are amplifications due to instabilities of the nonlinear car vibrations.

4 Stability of the Vertical Car Vibrations

To investigate the stability of the nonlinear car vibrations (see e.g. Ref. [5]), the associated variational equations are derived for both, the displacement $\bar{X}_\tau$ and the velocity $\bar{Y}_\tau$ by means of the set-up

$$\bar{X}_\tau = X_{st} + \Delta X_\tau, \quad \text{and} \quad \bar{Y}_\tau = Y_{st} + \Delta Y_\tau. \tag{22}$$

Inserting both perturbations into Eq. (21) gives the variational equations

$$d(\Delta Y_\tau) = -2D[1 + 2\bar{\delta}|Y_{st} - \nu U_{st}|]\Delta Y_\tau d\tau - [1 + \tag{23}$$

$$d(\Delta X_\tau) = \Delta Y_\tau \, d\tau \,, \qquad +3\bar{\gamma}(X_{st} - Z_{st})^2]\Delta X_\tau d\tau. \tag{24}$$

Herein, X_{st}, Y_{st}, Z_{st} and U_{st} are stationary car vibrations and road excitations determined by Eq. (21). Following [6] the polar coordinates A_τ and Φ_τ are introduced into the perturbation equations (23) and (24). Therewith, both are transformed to the phase and ln-amplitude equation

$$d\Phi_\tau = -(1 + T_\tau \cos \Phi_\tau)d\tau, \qquad d(\ln A_\tau) = -T_\tau \sin \Phi_\tau d\tau, \tag{25}$$

$$T_\tau = 2D(1 + 2\bar{\delta}|Y_{st} - \nu U_{st}|) \sin \Phi_\tau + 3\bar{\gamma}(X_{st} - Z_{st})^2 \cos \Phi_\tau. \tag{26}$$

According to the multiplicative ergodic theorem [7], the ln-amplitude equation is integrated in the time domain giving the top Lyapunov exponent

$$\lambda_{top} = \lim_{t\to\infty} \frac{1}{t} \ln \frac{A_t}{A_0} = \lim_{t\to\infty} \frac{1}{t} \int_0^t (-T_\tau \sin \Phi_\tau)d\tau. \tag{27}$$

Fig. (8) shows numerical evaluations of the top Lyapunov exponent λ_{top} for the values $D = 0.05$, $\bar{\delta} = 0$ and $\bar{\delta} = 0.05$ as well as for $\bar{\gamma} = 0.1$ and $\bar{\gamma} = 0.12$. The scan rate selected was $\Delta\tau = 0.001$ applied for 10^7 samples. The obtained numerical results confirm the Kramers effect [8] that for weak but still positive damping stationary system vibrations can be destabilized even for white noise excitations. Obviously, this is also true for road excitations. In the trivial case $\bar{\delta}, \bar{\gamma} = 0$ of linear road-vehicle systems, the top Lyapunov

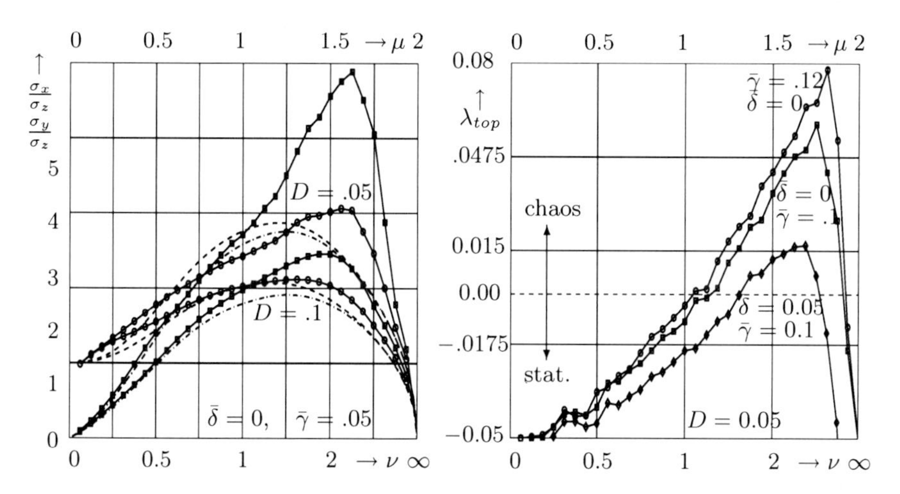

Fig. 7 Overcritical rms-magnification **Fig. 8** Top exponents vs car speed

exponent is given by $\lambda_{top} = -D$. For increasing quadratic damping $\bar{\delta} > 0$ the stationary car vibrations are stabilized. However, they are destabilized by increasing nonlinearity $\bar{\gamma} > 0$ of the cubic-progressive wheel suspension. Fig. (8) shows the destabilizing effects by the two upper curves marked by circles and squares. They cross the zero line in the over-critical speed range and reach their maximal values near $\nu = 4$.

5 Conclusions

The paper investigates road-vehicle systems. The ride on roads generates vertical car vibrations dependent on the car velocity. Digital simulations are discussed applying iterated integrals and their approximations by time and noise increments. The modelling of vehicles is extended to nonlinear wheel suspensions with cubic-progressive spring and quadratic damping. Linear and quadratic damping mechanism reduce and stabilize the stationary car vibrations. Cubic-progressive springs are only efficient for under-critical car speeds. In over-critical speed ranges, the spring nonlinearity leads to amplification effects, caused by instabilities of the nonlinear car vibrations. For sufficiently weak but still positive damping they become unstable and bifurcate into stochastic chaos.

References

1. Wedig, W.: Dynamics of Cars Driving on Stochastic Roads. In: Spanos, P.D., Deodatis, G. (eds.) Proceedings of Computational Stochastic Mechanics, pp. 647–654. Millpress, Rotterdam (2003)
2. Sobczyk, K., MacVean, D.B., Robson, J.D.: Response to Profile-Imposed Excitation With Randomly Varying Transient Velocity. Journal of Sound and Vibration 52(1), 37–49 (1977)
3. Talay, D.: Simulation and Numerical Analysis of Stochastic Differential Systems: A Review. In: Krée, P., Wedig, W. (eds.) Probabilistic Methods in Applied Physics 451. Lecture Notes in Physics, pp. 54–96. Springer, Heidelberg (1995)
4. Wedig, W.: Stochastic Dynamics of Road-Vehicle Systems and Related Bifurcation Problems. In: Macucci, M., Basso, G. (eds.) AIP Conference Proceedings 1129 of the 20th International Conference on Fluctuations and Noise. ICFN 2009, pp. 33–36 (2009)
5. Wedig, W.: Vertical Dynamics of Riding Cars under Stochastic and Harmonic Base Excitations. In: Rega, Vestroni (eds.) IUTAM Symposium on Chaotic Dynamics and Control of Systems and Processes in Mechanics, pp. 371–381. Springer, Dordrecht (2005)
6. Khasminskii, R.Z.: Necessary and sufficient conditions for asymptotic stability of linear stochastic systems. Theory Prob. Appl. 12, 144–147 (1974)
7. Oseledec, V.I.: A multiplicative ergodic theorem, Lyapunov characteristic numbers for dynamical systems. Trans. Moscow Math. Soc. 19, 197–231 (1968)
8. Arnold, L., Imkeller, P.: The Kramers Oscillator Revisited. Lecture Notes in Physics, pp. 280–291 (2000)

Real-Time Vehicle Dynamics Using Equation-Based Reduction Techniques

Lars Mikelsons, Thorsten Brandt, and Dieter Schramm

Abstract. Due to the increased computing power in the last decade, more and more complex vehicle models were developed. Nowadays even complex multibody models can be generated via graphical user interfaces in object-oriented simulation tools like Dymola or SimulationX. On the other hand, the available computing power in electronic control units is still limited, mostly by the cost pressure in the automotive industry. Hence, it is not possible to generate a complex model by drag and drop via a graphical user interface and run it in real-time within a desired time cycle on an ECU inside the vehicle. The same holds for HIL-testbeds and driving simulators, where the model must run in real-time as well. Thus, generally the model is adjusted in an iterative process until the model can be integrated in real-time on the particular ECU. In other words, a model has to be generated that is on the one hand complex enough to reproduce the desired physical effects and on the other hand simple enough to fulfill the real-time requirements. As it is easy to generate a complex model nowadays, an algorithm for the automated reduction of the model is required. Equation-based reduction techniques are a tool for the automated reduction of a given DAE-system for a defined error bound. This approach was already adopted and extended to generate vehicle models with an adjustable accuracy. In this contribution, equation-based reduction techniques are extended to generate models, which are guaranteed to run in real-time on a given real-time target within a given real-time cycle.

1 Introduction

Modern driver assistance systems as well as vehicle dynamics control systems base on the use of vehicle models. This fact holds not only for the operation of such a system, but also for the development process. In this phase it is tried to validate as many desired properties as possible by the use of virtual prototypes in order to

Lars Mikelsons · Thorsten Brandt · Dieter Schramm
Institute for Mechatronics and System Dynamics, University of Duisburg-Essen
e-mail: `{mikelsons,brandt,schramm}@imech.de`

G. Stépán et al. (Eds.): Dynamics Modeling & Interaction Cont., IUTAM BOOK SERIES 30, pp. 91–98.
springerlink.com

minimize costs. Hence, complex models which give results close to reality are demanded. For many object-oriented simulation tools special libraries for vehicle dynamics modeling are available, e.g. the Modelica™ Vehicle Dynamics Library [2]. Thus, complex multibody models can be generated using graphical user interfaces. During the development process these models are employed for offline simulations and hence the computing time is no big issue. In contrast, in a driving simulator [11], in a HIL-testbed or on a ECU, the models have to run in real-time. Although the computing power of an electronic control unit is limited, more and more on-board vehicle electronic systems make use of dynamic models instead of look-up table methods, e.g. electronic stability programs (ESP) or direct yaw rate control (DYC) [1]. Hence, a simplified models have to be generated. Since complex models are easily available (in object-oriented simulation tools), a common way is the simplification of existing models. This simplification (or reduction) usually requires an iterative process, i.e. the original model is reduced manually until it can be simulated on the given real-time hardware in the desired time cycle. During that reduction one has to take care that the physical effects under consideration are still reflected by the model in a proper way. This approach has two drawbacks. Typically the real-time capability is checked only for one scenario (parameters, system inputs and initial values) and hence it is generally not guaranteed that the model can be simulated in the desired time cycle for any scenario. Moreover, the manual reduction process requires a lot of process knowledge. These two drawbacks can be overcome by equation-based reduction techniques. Given the symbolic equations, an error bound and a scenario, equation-based reduction techniques deliver a reduced model which is guaranteed to stay within the error bound at least for the given scenario. These reduction techniques were first used for the steady-state analysis and synthesis of non-linear anlog circuits [4]. They were extended to the reduction of arbitrary systems of differential-algebraic equations (DAE systems) in [18] which paved the way for the reduction of complex mechatronic systems. In [15] equation-based reduction techniques were already used for the generation of vehicle models with an adjustable level of detail. This work is extended in this paper to the automated reduction of vehicle models for real-time simulations. Instead of providing an error bound, now a real-time environment (real-time hardware and time cycle) is the input for the reduction algorithm. The result is a reduced model which can be simulated in real-time in the given real-time environment. However, an error bound can be additionally provided. Moreover, a new ranking procedure basing on the adjoint sensitivity analysis is proposed [6]. The contribution is structured as follows. In Sect. 2, equation-based reduction techniques for real-time simulations are introduced. In Sect. 3, the reduction algorithm is applied to a multibody vehicle model and the results are discussed. The paper closes with a conclusion and an outlook.

2 Equation-Based Reduction Techniques

Many model reduction approaches try to identify slow and fast states in order to project the system (in general a DAE system) onto the manifold of the slow states [10] or to approximate the fast states [5]. Equation-based reduction techniques do not operate on the states of the system, but on single terms. Hence, instead of identifying fast states, those manipulations on the terms of the system, which have only a small influence on the solution of the system, are identified. In this context, such a manipulation of a term is called a reduction. Different reduction techniques are, for example, the cancellation of terms or the substitution of terms by constants. Hence, a reduction is the manipulation of a certain term by a certain reduction technique. A procedure which measures the influence of those reductions on the solution of the system is called ranking. A new ranking procedure is proposed in Sect. 2.2. After the ranking is computed, an approximation of the original system is computed. The approximation algorithm is presented in Sect. 2.3. However, first the terminology used is introduced in Sect. 2.1.

2.1 Fundamentals

In general object-oriented modeling leads to a system of differential-algebraic equations of the form

$$\mathbf{F}(\mathbf{x}, \dot{\mathbf{x}}, t) = 0, \tag{1}$$

where

$$\mathbf{F} : \mathbb{R}^n \times \mathbb{R}^n \times I \mapsto \mathbb{R}^n \tag{2}$$

is differentiable and $I \subset \mathbb{R}$ is an interval. In expanded form $\mathbf{F}$ can be rewritten as

$$\mathbf{F}_i(\mathbf{x}, \dot{\mathbf{x}}, t) = \sum_{k=1}^{l_i^1} t_{k_i}^1(\mathbf{x}, \dot{\mathbf{x}}, t), \quad 1 \leq i \leq m, \tag{3}$$

where l_i^1 denotes the number of terms in F_i and $t_{k_i}^1$ is the k-th term in F_i. Each term $t_{k_i}^1$ may again be a function $f_{k_i}^1$, whose argument is a sum of subterms $t_{k_i}^2$. Hence, $t_{k_i}^1$ is said to be a term of the first and $t_{k_i}^2$ is said to be a term of the second level. Thus, in general $t_{k_i}^j$ is the k-th term in the j-th level of F_i. In other words, the level indicates the hierarchy of arguments nested into each other in each single addend. The manipulation of a term is called reduction in the following. The set of all reductions in the j-th level is denoted by $\mathcal{K}^j$. Furthermore, for a DAE system a scenario $\mathcal{S}$ is a set consisting of the system inputs, the parameters and the initial values. Finally, the solution of the DAE system in Eq.1 computed by a numerical integrator $\mathcal{N}$ for a scenario $\mathcal{S}$ is denoted by $\mathcal{N}(F(\mathbf{x}, \dot{\mathbf{x}}, t), \mathcal{S})$.

2.2 Sensitivity Ranking

A ranking procedure measures the influence of reductions on the solution of the system under consideration. Obviously, the best ranking with respect to accuracy is obtained by simulating all possible reduced systems and comparing them with the solution of the original system. Hence, this procedure is called perfect ranking. In practice, the perfect ranking is far too expensive in terms of computational costs.

Thus, faster ranking procedures are required. Typically, a compromise between accuracy and computation time must be made. There are two procedures known from literature [17]. The residual ranking computes a measure for the influence of a reduction (ranking value) by plugging the solution of the original system into the reduced system. The one-step ranking makes use of the solution of the original system in order to compute an estimated solution of the reduced system. While the one step ranking is a rather accurate ranking, the residual ranking is beneficial in terms of computation time. Here, however, a new ranking procedure based on the adjoint sensitivity analysis is proposed. Some modern numerical integration algorithms allow for a sensitivity analysis during the integration of a DAE system [14, 13]. However, in the case of many parameters, the sensitivity analysis leads to very high computational costs. An adjoint sensitivity analysis is much more efficient in these cases. Here, the sensitivities of the solution with respect to the parameters are not calculated directly. Instead an objective function

$$G(\mathbf{x},\mathbf{p}) = \int_0^T g(\mathbf{x},\mathbf{p},t)dt, \tag{4}$$

depending on a smooth function $g(\mathbf{x})$ is chosen. The sensitivities $\frac{\partial G}{\partial p_i}$ with respect to the n_p parameters $p_i (1 \leq i \leq n_p)$ can be calculated by [7]

$$\frac{\partial G}{\partial \mathbf{p}} = \int_0^T (\mathbf{g}_p - \mathbf{F}_p^T \lambda)dt + (\lambda^T \mathbf{F}_{\dot{\mathbf{x}}} \mathbf{x}_p) \mid_{t=0}, \quad \text{where} \tag{5}$$

$$(\mathbf{F}_{\dot{\mathbf{x}}}^T \lambda)' - \mathbf{F}_{\mathbf{x}}^T \lambda = -\mathbf{g}_{\mathbf{x}}. \tag{6}$$

The DAE system in Eq.6 is called adjoint DAE system and is solved backwards in time. Hence, in order to compute the sensitivities the solution of the DAE system, the solution of the adjoint DAE system and the integral in Eq.5 have to be calculated. The effort to solve the adjoint DAE is approximately the same as the effort to solve the original DAE, while the integral in Eq.5 can be efficiently computed without a significant computational effort [12]. Since the sensitivity of the solution of the DAE system with respect to the terms of the DAE is desired, virtual parameters

$$\mathbf{p}_i^v = 1 \tag{7}$$

are introduced. Then every term and subterm is replaced by the product of the term itself and an one-to-one corresponding virtual parameter p_i^v.The sensitivities with respect to the virtual parameters indicate how much the solution will change if the corresponding terms are manipulated. Hence, the sensitivities can be used as ranking values. Moreover, the sensitivity ranking is generally much faster than the one step ranking, while it exhibits a similar accuracy. The objective function is chosen by the user. Common choices are the sum of the output variables or the sum of the squared output variables. Generally the crucial point about the adjoint sensitivity analysis is the initialization (at $t = t_{end}$) of the adjoint DAE. However, the initialization for index-0 and index-1 problems can be done by solving the least square problem

$$\text{minimize} \left\| \begin{bmatrix} \frac{\partial \mathbf{F}}{\partial \dot{\mathbf{x}}} \big|_{t=t_{end}} \dot{\boldsymbol{\lambda}} + \frac{\partial \mathbf{F}}{\partial \mathbf{x}} \big|_{t=t_{end}} \boldsymbol{\lambda} - \mathbf{g}_{\mathbf{x}}^T \\ \frac{\partial \mathbf{F}}{\partial \dot{\mathbf{x}}} \big|_{t=t_{end}} \boldsymbol{\lambda} \end{bmatrix} \right\|_2 . \tag{8}$$

2.3 Approximation Algorithm

For real-time simulations, it must be guaranteed that every integration step is finished within the real-time cycle. The duration of one integration step depends on the complexity of the model, the real-time hardware and the used integration method. In many cases, the real-time hardware (e.g. a microcontroller) is given. Thus, a model must be generated which is, on the one hand, complex enough to reproduce the desired physical effects and, on the other hand, simple enough to be simulated in real-time on the given hardware. Instead of solving this task manually in an iterative process, an extension of the reduction algorithm described [18, 15] can be used. While previously the reduced models were guaranteed to stay within the error bound, now it must be additionally guaranteed that the reduced models can be integrated in real-time. This is done by comparing the required number of Floating Point Operations (FLOPs) for one integration step with the number of Floating Point Operation per Second (FLOPS) $\mathscr{F}_{RT}$ the real-time target is able to perform. The number of FLOPS a real-time target is able to perform can be measured by appropriate benchmarks, e.g. LINPACK [9]. Thus, the number of FLOPs the target is able to perform within the real-time cycle can be easily computed. The determination of the required FLOPs for one integration step is more complicated and depends on the chosen integration method. Since the system under consideration is a system of ODEs, a fast ODE solver is required. In [3] the explicit Euler method is proposed for real-time simulations, due to its simplicity. Here, however, a linear-implicit one-step method of order one is used. This method was (with some modifications) already used for real-time simulations of vehicle models in [16]. Furthermore, in contrast to the explicit Euler method this method is A-stable. Then for one integration step one evaluation of the right hand side and the solution of a linear system of equations is required. Thus, the number of FLOPs $\mathscr{F}_{req}$ can be written as

$$\mathscr{F}_{req} = \mathscr{F}_f + \mathscr{F}_{LSE}, \tag{9}$$

where $\mathscr{F}_f$ is the number of required FLOPs for the evaluation of the right hand side and $\mathscr{F}_{LSE}$ is the number of required FLOPS for the solution of a linear system of equations. The value for $\mathscr{F}_{LSE}$ depends on the method used. For the Gaussian elimination,

$$\mathscr{F}_{LSE} = \frac{1}{3}n^3 + \frac{3}{2}n^2 - \frac{5}{6}n \tag{10}$$

holds for a system of linear equation of dimension n. The approximation algorithm works as follows. After choosing a set of reductions $\mathscr{K}$, (by choosing a reduction technique and a level), the ranking list $\mathscr{L}$ is computed, i.e. the ranking values for all $\kappa \in \mathscr{K}$, are calculated and stored in $\mathscr{L}$ together with the corresponding reductions. After that, $\mathscr{L}$ is sorted in ascending order with respect to the ranking values. This sorted list is denoted by $\mathscr{R}$. In order to obtain a reduced model which requires a minimum number of FLOPs for one integration step, as many reductions as possible must be performed. Therefore, it is beneficial to start with those reductions which appear to produce only a small error. Hence, the simplest approach is to perform the reductions in the same order as they are stored in $\mathscr{L}$. After each reduction κ, it has to be verified that

$$\|\mathscr{N}(F(\mathbf{x},\dot{\mathbf{x}},t),\mathscr{S}) - \mathbf{x}^\star\| < \varepsilon \quad \text{and} \tag{11}$$

$$\mathscr{F}_{req} > \mathscr{F}_{RT}, \tag{12}$$

where $\mathscr{N}$ is the chosen numerical integrator, $\mathbf{x}^\star$ is the reference solution, $\mathscr{S}$ is a scenario and ε the given error bound. If Eq.11 is not satisfied, κ is retracted. If Eq.12 is not satisfied the algorithm is finished. For this verification a simulation has to be performed. Thus, in the case of complex systems, the computational effort can be very high due to the large number of simulations. Here advanced techniques can reduce the number of simulations [18].

3 Results

The reduction algorithm has been implemented in MATLAB using the Maple Toolbox for MATLAB. The emerging systems are solved using IDAS from the SUNDIALS package [12], since IDAS provides the required interfaces and functions for a adjoint sensitivity analysis. In order to test the efficiency of the algorithm a spatial twin track model is employed. The model exhibits three translational and three rotational degrees of freedom. Furthermore, the four wheels are modeled independently with a rotational and a translational degree of freedom. Consequently, the twin track model has fourteen degrees of freedom. Modeling details are neglected here due to space limitations, but can be found in [15]. The twin track model is reduced for realtime simulations on a Sony Playstation 2 (which is approximately equivalent to an IBM 486 at 33Mhz [8]) in a time cycle of 1ms. In addition to the real-time environment an error bound of 10% for the lateral acceleration and the yaw rate was provided. The chosen scenario is a double lane change at the velocity of 40km/h lasting twenty seconds. In Fig.1 the lateral acceleration of the reduced and the original model are plotted as well as the absolute error. The plot of the yaw rate is neglected here, since it shows almost the same nearly overlaying curves up to a factor. In the reduced model the vertical motion of the tires and the vehicle is completely vanished. Moreover, the roll and pitch motion of the vehicle is neglected. The computation of the tire forces is simplified as well, but can not be interpreted easily. The reduction algorithm lowered the number of required FLOPs for one integration step from $1.67 \cdot 10^6$ to 92215 which means a speed up approximately eighteen.

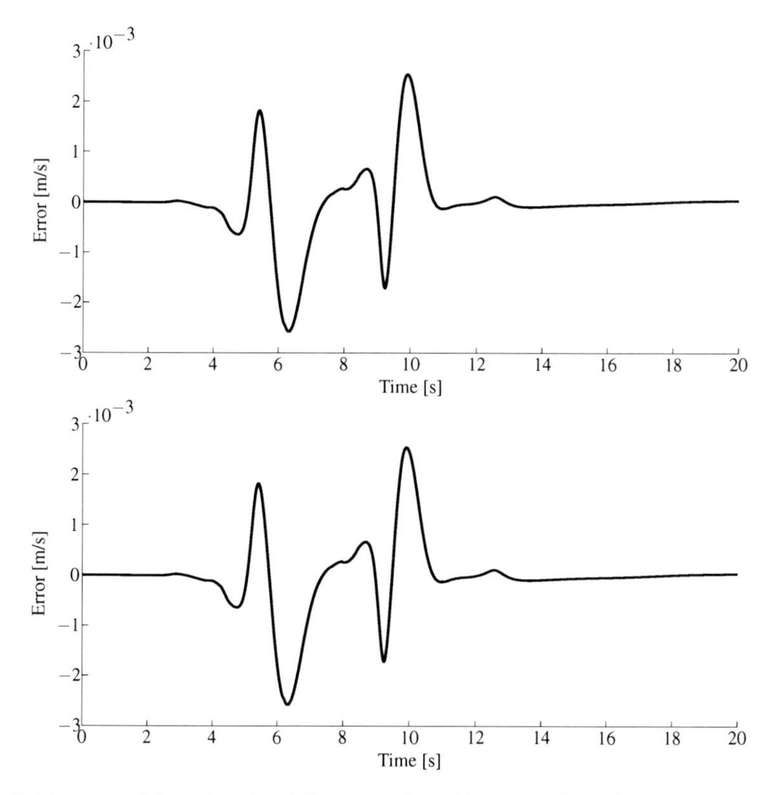

Fig. 1 Yaw rate of the original and the reduced model (top) and the absolute error (bottom)

4 Conclusion and Outlook

In this contribution a concept for the automated generation of real-time capable models for a given real-time environment is presented. The approach extends equation-based reduction techniques. Moreover, a new efficient ranking procedure is proposed. The concept is validated by reduction of a multi-body vehicle model for the real-time simulation on a target with limited computational power. In this example, the reduction algorithm achieves a speed up of an factor of eighteen, while the reduced model exhibits an error below 10%. In the future work the reduction algorithm will be implemented in an OpenModelica compiler.

References

1. Ackermann, J., Sienel, W.: Robust Yaw Rate Control. IEEE Transactions on Control and System Technology 1, 1–15 (1993)
2. Andreasson, J.: Vehicle Dynamics Library. In: 3rd International Modelica Conference (2003)

3. Beutlich, T.: Real-Time Simulation of Modelica-based Models. In: 7th International Modelica Conference (2009)
4. Borchers, C.: Symbolic behavioral model generation of nonlinear analog circuits. IEEE Transactions on Circuits and Systems II: Analog and Digital Signal Processing 45(10), 1362–1371 (1998)
5. Burgermeister, B., Arnold, A., Eichberger, A.: Smooth Velocity Approximation for Constrained Systems in Real-Time Simulation. In: ECCOMAS Thematic Conference Multibody Dynamics (2009)
6. Cao, Y., Li, S., Petzold, L.: Adjoint sensitivity analysis for differential-algebraic equations: Algorithms and software. Journal of computational and applied mathematics 149(1), 171–191 (2002)
7. Cao, Y., Li, S., Petzold, L., Serban, R.: Adjoint sensitivity analysis for differential-algebraic equations: The adjoint DAE system and its numerical solution. SIAM Journal on Scientific Computing 24(3), 1076 (2003)
8. Dongarra, J.: Performance of Various Computers Using Standard Linear Equations Software. Technical Report of the Electrical Engineering and Computer Science Department, University of Tennessee (2009)
9. Dongarra, J., Luszczek, P., Petitet, A.: The LINPACK Benchmark: past, present and future. Concurrency and Computation: Practice and Experience 15(9), 803–820 (2003)
10. Gorban, A., Kazantzis, N., Kevrekidis, I., Ottinger, H., Theodoropoulos, C.: Model Reduction and Coarse-Graining Approaches for Multiscale Phenomena (2006)
11. Hesse, B., Hiesgen, G., Brandt, T., Schramm, D.: Ein Fahrsimulator als Werkzeug zur frühzeitigen Eigenschaftsabsicherung von Mensch-zentrierten mechatronischen Systemen. VDI Mechatronik (2009)
12. Hindmarsh, A., Brown, P., Grant, K., Lee, S., Serban, R., Shumaker, D., Woodward, C.: SUNDIALS: Suite of nonlinear and differential/algebraic equation solvers. ACM Transactions on Mathematical Software (TOMS) 31(3), 363–396 (2005)
13. Li, S., Petzold, L.: Software and algorithms for sensitivity analysis of large-scale differential algebraic systems. Journal of computational and applied mathematics 125(1-2), 131–145 (2000)
14. Maly, T., Petzold, L.: Numerical methods and software for sensitivity analysis of differential-algebraic systems. Applied Numerical Mathematics 20(1), 57–82 (1996)
15. Mikelsons, L., Unterreiner, M., Brandt, T.: Generation of Continuously Adjustable Vehicle Models using Symbolic Reduction Methods. In: ECCOMAS Thematic Conference Multibody Dynamics (2009)
16. Rill, G.: Simulation von Kraftfahrzeugen. Vieweg (1994)
17. Wichmann, T.: Transient Ranking Methods for the Simplification of Nonlinear DAE Systems in Analog Circuit Design. PAMM 2(1), 448–449 (2003)
18. Wichmann, T.: Symbolische Reduktionsverfahren für nichtlineare DAE-Systeme. In: Berichte aus der Mathematik. Shaker Verlag, Germany (2004)

Mathematical Modeling of Oscillatory Systems

Fractional order calculus in dynamical modeling of oscillatory systems, chaotic solutions that may arise in contact tasks, bifurcations of non-smooth systems and discrete element modeling of grains are all covered in these six papers. The paper of Z. Wang presents recent advances in using fractional-order derivative to describe the damping force in oscillatory systems. The solutions proposed by M. Kuroda enable the fractional-order-derivative responses to be measured by a combination of signals obtained by existing sensors. The papers by F. Angulo and G. Olivar address bifurcations arising in nonlinear systems with hysteresis and saturations. The work presented by J. Ing investigates the complex nonlinear response of a piecewise linear oscillator by means of experiments and simulation. Finally, the paper by I. Keppler combines semi-analytical and discrete element models to optimize the vibration parameters of screening.

Solution and Stability of a Linear Fractionally Damped Oscillator

Z.H. Wang

Abstract. This paper presents some recent advances in the generalized Bagley-Torvik equation that uses fractional-order derivative to describe the damping force. It addresses the conventional form of solution that is easy for engineers to understand, as well as the stability analysis of the fractionally damped oscillator.

1 Introduction

Dynamical systems with elements described by fractional-order derivatives, called fractional dynamical systems for simplicity, draw an increasing interest of researchers working in different fields. The field of most extensive applications of fractional calculus is probably the viscoelaticity theory [1]-[3] and the control theory [1][4]-[6]. In viscoelastity theory, the fractional-order derivative models have been shown to be substantially more accurate and appropriate for modeling some real materials for both narrow and broad frequency ranges, they require less parameters and their fittings converge quickly, say, aircraft tire [3], frequency-dependent viscoelastic isolators [7]. In control theory, some fractional-order controllers have been proposed, they result in better control accuracy, optimal stability and better robustness against the gain changes, due to their memory effect, say [1][4]. In [8], it has been proved that fractional-order difference feedback receives an optimal stability for a SDOF vibration system. In addition, a fractional differential equation may admit more complicated nonlinear dynamics than classical systems. It is well known that chaos cannot occur in continuous systems of total order less than three, while in [9], it shows that chaos does exist in a fractional-order simplified Lorenz system with a wide range of fractional orders, less than 3.0 or greater than 3.0, and the lowest order for this system leading to chaos is found to be 2.62.

Z.H. Wang
Institute of Vibration Engineering Research,
Nanjing University of Aeronautics and Astronautics,
210016 Nanjing, China
e-mail: zhwang@nuaa.edu.cn

G. Stépán et al. (Eds.): Dynamics Modeling & Interaction Cont., IUTAM BOOK SERIES 30, pp. 101–108.
springerlink.com

A number of methods are available for solving fractional differential equations analytically or numerically [1][10]-[15]. These methods, however, usually yield solutions in terms of Mittag-Leffler functions or some other kind of power series, which are not easy for engineers in understanding. For instance, as presented in [1], the generalized Bagley-Torvik equation described below regarding as a three-term fractional differential equation has a very complicated form of solution.

This paper focuses on the generalized Bagley-Torvik equation with a fractional-order derivative of order $0 < \alpha < 2$ in dimensionless form

$$\ddot{x}(t) + \mu\, {}_0D^{\alpha}x(t) + x(t) = f(t), \qquad (0 < \alpha < 2,\ \mu > 0) \tag{1}$$

where ${}_0D^{\alpha}x(t)$ is the Caputo's fractional-order derivative defined by

$${}_0D^{\alpha}x(t) = \frac{1}{\Gamma(m+1-\alpha)} \int_0^t \frac{x^{(m+1)}(\tau)}{(t-\tau)^{\alpha-m}} \mathrm{d}\tau \tag{2}$$

where $m = [\alpha] \in \mathbb{N}$ is the greatest integer not larger than α, $x^{(m)}(t)$ is the m-order derivative of $x(t)$, and $\Gamma(z)$ is the Gamma function. It presents some recent advances in the solution and its stability of Eq.(1) without external excitation, focusing on the conventional form of solution that is easy for understanding.

2 Solution of the Free Vibration Equation

It has been proved in [16][17] that if $0 < \alpha < 2$, then the characteristic equation

$$p(s) := s^2 + \mu s^{\alpha} + 1 = 0, \qquad (\mu > 0) \tag{3}$$

has exactly one pair of conjugate complex roots $\beta \pm \mathrm{i}\omega$ with $\beta < 0$, $\mathrm{i}^2 = -1$. Moreover, by means of Laplace transform and residue calculation, it shows that the solution of free vibration equation has the following conventional form

$$x(t) = A\mathrm{e}^{\beta t}\cos(\omega t) + B\mathrm{e}^{\beta t}\sin(\omega t) - x^*(t) \tag{4}$$

where the coefficients A, B use the complex coordinates of $\beta \pm \mathrm{i}\omega = r\mathrm{e}^{\mathrm{i}\theta}$ and depend on the initial values, and $x^*(t)$ is a definite integral, which vanishes if $\alpha = 1$ and it decays to zero as as $t \to +\infty$ if $0 < \alpha < 1$ or $1 < \alpha < 2$. The equilibrium $x = 0$ is asymptotically stable if $\mu > 0$, due to $\beta < 0$.

2.1 The Dominated Approximation of the Solution

One observation is that for $0 < \alpha < 2$, the solution of the free vibration equation is dominated by the "eigenfunction expansion" only, namely,

$$x(t) \approx A\mathrm{e}^{\beta t}\cos(\omega t) + B\mathrm{e}^{\beta t}\sin(\omega t), \qquad (t \gg 1) \tag{5}$$

a form similar to the solution of ordinary differential equations with constant coefficients. In fact, for $1 < \alpha < 2$ as an example, one has

$$|x^*(t)| \leq \tilde{x}(t) := \frac{\mu}{\pi}\int_0^{+\infty} \frac{(\xi^{\alpha-1}|x_0| + \xi^{\alpha-2}|x_1|)|\sin(\alpha\pi)|}{(\xi^2+1)^2 + 2\mu\xi^\alpha(\xi^2+1)\cos(\alpha\pi) + \mu^2\xi^{2\alpha}} e^{-\xi t}d\xi$$

$$\begin{aligned}
\frac{\pi|\tilde{x}'(t)|}{\mu e^{\beta t}} &= \int_0^{+\infty} \frac{\xi(\xi^{\alpha-1}|x_0| + \xi^{\alpha-2}|x_1|)|\sin(\alpha\pi)|}{(\xi^2+1)^2 + 2\mu\xi^\alpha(\xi^2+1)\cos(\alpha\pi) + \mu^2\xi^{2\alpha}} e^{-(\xi+\beta)t}d\xi \\
&> \int_0^{-\beta} \frac{\xi(\xi^{\alpha-1}|x_0| + \xi^{\alpha-2}|x_1|)|\sin(\alpha\pi)|}{(\xi^2+1)^2 + 2\mu\xi^\alpha(\xi^2+1)\cos(\alpha\pi) + \mu^2\xi^{2\alpha}} e^{-(\xi+\beta)t}d\xi \\
&= e^{-(\kappa+\beta)t}\int_0^{-\beta} \frac{\xi(\xi^{\alpha-1}|x_0| + \xi^{\alpha-2}|x_1|)|\sin(\alpha\pi)|}{(\xi^2+1)^2 + 2\mu\xi^\alpha(\xi^2+1)\cos(\alpha\pi) + \mu^2\xi^{2\alpha}} d\xi \\
&\gg |\beta|\sqrt{A^2+B^2}, \quad (1 < \alpha < 2, \ t \gg 1)
\end{aligned}$$

where $-(\kappa+\beta) > 0$ and $\kappa \in (0, -\beta)$is a fixed number due to the mean-value theorem for integral. It means that $\tilde{x}(t)$ decay to zero much faster than $\sqrt{A^2+B^2}e^{\beta t}$. Thus, $x^*(t)$ can be neglected in the stationary solution. This is true for more general systems under mild assumptions. With such an approximate solution, dynamics analysis of fractional dynamical systems usually can be simplified.

2.2 *The Function of the Fractional Damping*

Note that if $s = re^{i\theta}$, $(r > 0)$, is a root of $p(s) = 0$, then separating the real and imaginary parts of $p(s) = 0$ gives

$$\begin{cases} r^2\cos(2\theta) + \mu r^\alpha \cos(\alpha\theta) + 1 = 0 \\ r^2\sin(2\theta) + \mu r^\alpha \sin(\alpha\theta) = 0 \end{cases} \tag{6}$$

Obviously, $p(s)$ has no positive roots. In addition, the second equation in Eq.(6) does not hold in the cases of $\theta = \pi$, $\theta = \pm\pi/2$, $0 < \theta < \pi/2$ and $-\pi/2 < \theta < 0$. Thus, the characteristic roots are possible only in $\pi/2 < \theta < \pi$ and $-\pi < \theta < -\pi/2$. When $\pi/2 < \theta < \pi$, one has $\sin(2\theta) < 0$, $\sin(\alpha\theta) > 0$, thus, one solves r uniquely from the second equation in Eq.(6) as following

$$r = \left(\mu\frac{\sin(\alpha\theta)}{-\sin(2\theta)}\right)^{1/(2-\alpha)} > 0 \tag{7}$$

In the case of $0 < \mu \ll 1$, one has $\theta \approx \pi/2$ and $r \approx 1$. Thus, Eq.(7) implies that

$$\mu\sin(\alpha\pi/2) \approx \sin(2(\theta - \pi/2)) \approx 2(\theta - \pi/2)$$

It follows that $\beta = r\cos\theta = O(\mu)$. Consequently,

$$x(t) \approx \rho\sin(\omega t + \phi), \qquad (t \gg 1) \tag{8}$$

where ρ and ϕ vary in slow time μt. It means that when $0 < \mu \ll 1$, the solution of the free vibration equation can be assumed to have the form (8) with constant ρ and ϕ, if $t \gg 1$. Then, using ${}_0D^\alpha x(t) \approx {}_{-\infty}D^\alpha x(t)$ (Short Memory Principle[1]), one has

$$\begin{aligned} {}_0D^\alpha x(t) \approx & {}_{-\infty}D^\alpha(\rho \sin(\omega t + \phi)) = \rho\omega^\alpha \sin\left(\omega t + \phi + \alpha\frac{\pi}{2}\right) \\ = & \rho\omega^\alpha \cos(\alpha\frac{\pi}{2})\sin(\omega t + \phi) + \rho\omega^\alpha \sin(\alpha\frac{\pi}{2})\cos(\omega t + \phi) \end{aligned} \tag{9}$$

It means that $-\mu\,{}_0D^\alpha x(t)$ consists of two parts, the part with $\cos(\omega t + \phi)$ plays the same role as the classical damping force $-\mu\dot{x}(t)$ if $0 < \alpha < 2$, and the term with $\sin(\omega t + \phi)$ plays the same role as the elastic restoring force or the inertial force. If $0 < \alpha < 1$, the fractional-order damping increases the elastic restoring force, it can also be regarded as decreasing the inertial force. If $1 < \alpha < 2$, the fractional-order damping increases the inertial force, it can also be regarded as decreasing the elastic restoring force. That is to say, the fractional-order damping not only stabilizes the undamped oscillator $\ddot{x}(t) + x(t) = 0$, but also changes its intrinsic frequency.

Moreover, $\beta = O(\mu)$ if $0 < \mu \ll 1$, then the dominated approximate solution given in Eq.(5) is in two time scales: $T_0 = t = \mu^0 t$ and $T_1 = \mu^1 t$. Thus, the solution of the fractionally damped oscillator can be assumed a form in two time scales, a form that is fundamental when the method of multiple scales is used for finding approximate solution of fractional differential equations, say, as shown in [18].

2.3 *The Energy Analysis*

Moreover, the approximate solution (8) enables one to analyze the energy of the free oscillator. In fact, the energy function is

$$E = \frac{1}{2}\dot{x}^2 + \frac{1}{2}x^2 \tag{10}$$

If $0 < \mu \ll 1$, then by using Eq.(8), one has

$$\dot{E} = -\mu\dot{x}(t)\,{}_0D^\alpha x(t) \approx -\mu\rho\cos(\omega t + \phi)\rho\omega^\alpha \sin(\omega t + \phi + \alpha\frac{\pi}{2})$$

Because $\dot{E} \approx 0$ and the right hand is periodic in t, it can be simplified by averaging

$$\begin{aligned} \dot{E} \approx & -\frac{1}{2\pi/\omega}\int_0^{2\pi/\omega}\left(\mu\rho^2\omega^\alpha \cos(\omega t + \phi)\sin(\omega t + \phi + \alpha\frac{\pi}{2})\right)\mathrm{d}t \\ = & -\mu\rho^2\omega^\alpha \sin(\alpha\frac{\pi}{2}) \end{aligned}$$

Hence, $\dot{E} < 0$ if $0 < \alpha < 2$. The dissipation of the energy means that the equilibrium $x = 0$ of the free vibration equation is asymptotically stable.

2.4 Solution Based on Fractional-Order Exponential Function

From Section 2.1, one can see that once the unique pair of conjugate roots $\beta \pm \mathrm{i}\,\omega$ are in hand, the solution of the free vibration equation can be expressed approximately as the linear combination of two linearly independent functions $\mathrm{e}^{\beta t}\cos(\omega t)$ and $\mathrm{e}^{\beta t}\sin(\omega t)$. However, it is worthy to point out that actually $\mathrm{e}^{(\beta \pm \mathrm{i}\,\omega)t}$ are *not* solutions of the free vibration equation, because [10]

$$ {}_0D^{\alpha}\mathrm{e}^{st} = s^{\alpha}\mathrm{e}^{st} + \frac{\sin(\alpha\pi)}{\pi}\int_0^{+\infty}\frac{\tau^{\alpha}\mathrm{e}^{-\tau t}}{s+\tau}\mathrm{d}\tau \tag{11} $$

In order that e^{st} is a solution of the free vibration equation, it is required to have ${}_0D^{\alpha}\mathrm{e}^{st} = s^{\alpha}\mathrm{e}^{st}$, which holds only if $\alpha = 1$ for our concern. Instead, one has

$$ {}_{-\infty}D^{\alpha}\mathrm{e}^{st} = s^{\alpha}\mathrm{e}^{st} \tag{12} $$

in the sense of Riemann-Liouville's fractional-order derivative or Caputo's fractional-order derivative [1][19]. When $t \gg 1$, the integral in Eq.(11) decays rapidly with time, and it can be neglected as compared with the first term in Eq. (11) [10], namely ${}_0D^{\alpha}\mathrm{e}^{st} \approx {}_{-\infty}D^{\alpha}\mathrm{e}^{st}$. It concludes once that $\mathrm{e}^{(\beta \pm \mathrm{i}\,\omega)t}$, or in the form of $\mathrm{e}^{\beta t}\cos(\omega t)$ and $\mathrm{e}^{\beta t}\sin(\omega t)$, are *approximate* solutions of the free vibration equation, and they offer a satisfactory accuracy if one seeks for a stationary solution.

Alternatively, one can also seek for a solution in terms of fractional-order exponential functions [13]-[15]. For Caputo's fractional-order derivative, the α-exponential function is defined by [13]

$$ E_{\alpha}^{\lambda(t-a)} = \sum_{k=0}^{\infty}\frac{\lambda^k (t-a)^{\alpha k}}{\Gamma(k\alpha+1)}, \quad (t \geq a) \tag{13} $$

It satisfies the following fractional-order differential equation [13]

$$ {}_aD^{\alpha}x(t) = \lambda\, x(t), \quad (t > a) \tag{14} $$

Take $\alpha = 1/2$ as an example, a case that has been discussed by many authors. The free vibration equation (1) can be transformed into a sequential fractional-order differential equation as following

$$ {}_0D^{4\alpha}x(t) + \mu\, {}_0D^{\alpha}x(t) + x(t) = 0 \tag{15} $$

where ${}_0D^{k\alpha} = {}_0D^{\alpha}({}_0D^{(k-1)\alpha})$, $(k = 2,3,4)$. Thus, one has

$$ {}_0D^{4\alpha}E_{\alpha}^{\lambda t} + \mu\, {}_0D^{\alpha}E_{\alpha}^{\lambda t} + E_{\alpha}^{\lambda t} = q(\lambda)E_{\alpha}^{\lambda t} $$

where

$$ q(\lambda) = \lambda^4 + \mu\,\lambda + 1 \tag{16} $$

is called the characteristic polynomial of Eq. (15). Therefore, Eq. (15) has a solution $cE_\alpha^{\lambda t}$ if and only if λ is a root of $q(\lambda) = 0$. Based on this fact, as done for ordinary differential equations with constant coefficients, a general solution in terms of fractional-order exponential functions can be easily obtained for Eq.(15) [13]. Come back to the free vibration equation, using two natural prerequisites $|\dot{x}(0)| < \infty$ and $|\ddot{x}(0)| < \infty$, one can reduce the four free constants in the general solution of Eq.(15) to two free constants, so that the general solution of the free vibration equation is determined fully by the initial conditions $x(0) = x_0$, $\dot{x}(0) = x_1$ [14][15].

3 Stability Analysis via Stability Switches

The stability of a fractional differential equation can be determined by the root location of its characteristic roots [6][20][21]. In fact, as shown in Section 2.1, the equilibrium $x = 0$ of the free vibration equation (1) is asymptotically stable, due to $\beta < 0$ for the case of $\mu > 0$ and $0 < \alpha < 2$. In general, for fractional dynamical systems, that the characteristic roots with negative real parts only corresponds to BIBO (bounded-input bounded-output) stability. The graphical method given in [20] works effectively for testing the BIBO stability of fractional delay systems with fixed parameters. The testing algorithm proposed in [6] can be used to check the BIBO stability of fractional delay systems with a uncertain parameter in a very simple way, but it may fail for some fractional delay systems. In what follows, the stability analysis of the generalized Bagley-Torvik equation is taken as an example to show the key idea of the method of stability switches reported in [21], see also the paragraph after Eq. (6) in Section 2.2.

Without loss of generality, the fractional-order α in Eq.(1) is assumed a rational number: $\alpha = k/n = k\beta$, with $1 \leq k < 2n$ and $k \neq n$, to meet the conditions $0 < \alpha < 2$ and $\alpha \neq 1$. Then, referring to [1][13], the unique equilibrium $x = 0$ of the free vibration equation is stable if and only all the roots of the characteristic function (which can be obtained by using the fractional-order exponential function defined in Section 2.4)

$$\Delta(\lambda) := \lambda^{2n} + \mu\lambda^k + 1 = 0 \tag{17}$$

stay within the sector in the complex plane defined by

$$|\arg(\lambda_j)| > \frac{\beta\pi}{2} = \frac{\pi}{2n}, \qquad (j = 1, 2, \cdots, 2n) \tag{18}$$

Here, if $\lambda = s^{1/n}$, then $\Delta(\lambda) = p(s)$. The key point of stability switch is the continuous dependence of each branch of the characteristic root λ on the parameter μ. As μ varies, the stability of $x = 0$ changes only if a branch of the root λ of $\Delta(\lambda) = 0$ comes to the boundary of the stability domain defined by Eq.(18), satisfying

$$|\arg(\lambda)| = \frac{\pi}{2n} \qquad \Leftrightarrow \qquad \lambda = r\mathrm{e}^{\pm \mathrm{i}\pi/(2n)}, \ (r > 0) \tag{19}$$

Separating the real and imaginary parts of $\Delta(r\mathrm{e}^{\pm \mathrm{i}\pi/(2n)}) = 0$ yields

$$-r^{2n} + \mu r^k \cos\frac{k\pi}{2n} + 1 = 0, \qquad \pm\mu r^k \sin\frac{k\pi}{2n} = 0 \tag{20}$$

which holds only if $\mu = 0$, due to $1 \leq k < 2n$ and $k \neq n$. It means that the stability of the equilibrium $x = 0$ keeps unchanged for all $\mu > 0$. At $\mu = 0$, let

$$\lambda_0 = \cos\frac{\pi}{2n} + \sin\frac{\pi}{2n}\mathrm{i}$$

then, except for $\lambda_0, \overline{\lambda}_0$, which stay on the boundary of the stability domain, all the roots of $\lambda^{2n} = -1 = \mathrm{e}^{\mathrm{i}(\pi+2j\pi)}$, $(j = 0, \pm1, \pm2, \cdots, \pm2(n-1))$, stay in the stability domain. When μ increases from 0 to a very small number $0 < \varepsilon \ll 1$, the branch of characteristic root λ passing through λ_0 becomes

$$\lambda = \lambda_0 + \varepsilon(a + b\mathrm{i}) + o(\varepsilon)$$

where the constants a, b are found to be

$$a = \frac{1}{2n}\cos\frac{(k+1)\pi}{2n}, \qquad b = \frac{1}{2n}\sin\frac{(k+1)\pi}{2n}$$

from $\Delta(\lambda) = 0$. Therefore, when $0 < \varepsilon \ll 1$, the stability condition holds

$$|\arg(\lambda)| \approx \arctan\frac{2n\sin\frac{\pi}{2n} + \varepsilon\sin\frac{(k+1)\pi}{2n}}{2n\cos\frac{\pi}{2n} + \varepsilon\cos\frac{(k+1)\pi}{2n}} > \arctan\frac{\sin\frac{\pi}{2n}}{\cos\frac{\pi}{2n}} = \frac{\pi}{2n} \tag{21}$$

As a result, the equilibrium $x = 0$ is stable for all $\mu > 0$.

4 Conclusions

This paper studies the solution and stability of the generalized Bagley-Torvik equation describing an oscillator with a fractional damping. Firstly, it shows that the fractional damping, which is essentially different from the classical linear damping, changes both the stability of the undamped oscillator and its intrinsic frequency. Then, the solution of the free vibration is decomposed to two parts, one is an eigenfunction expansion, and the other is a definite integral decaying to zero. It shows that the eigenfunction expansion is accurate enough to approximate the stationary solutions and thus it can be used for energy analysis as well as for nonlinear dynamics. This fact is true for general fractional dynamical systems under mild assumptions. Finally, the method of stability switch is shown effective in the stability analysis of the fractional oscillator involving a parameter.

Acknowledgements. This work was supported by NSF of China under Grant 10825207.

References

1. Podlubny, I.: Fractional Differential Equations. Academic Press, San Diego (1999)
2. Bagley, R.L., Torvik, P.J.: On the appearance of the fractional derivative in the behavior of real materials. ASME Journal of Applied Mechanics 51, 294–298 (1984)
3. Eldred, L.B., Baker, W.P., Palazotto, A.N.: Kelvin-Voigt vs fractional derivative model as constitutive relations for viscoelastic materials. AIAA Journal 33(3), 547–550 (1995)
4. Podlubny, I.: Fractional-order systems and $PI^{\lambda}D^{\mu}$-controllers. IEEE Transactions on Automatic Control 44(1), 208–214 (1999)
5. Luo, Y., Chen, Y.-Q.: Fractional order [proportional derivative] controller for a class of fractional order systems. Automatica 45(10), 2446–2450 (2009)
6. Farshad, M.B., Masoud, K.G.: An efficient numerical algorithm for stability testing of fractional-delay systems. ISA Transactions 48, 32–37 (2009)
7. Coronado, A., Trindade, M.A., Sampaio, R.: Frequency-dependent viscoelastic models for passive vibration isolation systems. Shock and Vibration 9, 253–264 (2002)
8. Wang, Z.H., Zheng, Y.G.: The optimal form of the fractional-order difference feedbacks in enhancing the stability of a sdof vibration system. Journal of Sound and Vibration 326(3-5), 476–488 (2009)
9. Sun, K.H., Wang, X., Sprott, J.C.: Bifircation and chaos in fractional-order simplified Loreenz system. International Journal of Bifurcation and Chaos 20(4), 1209–1219 (2010)
10. Narahari Achar, B.N., Hanneken, J.W., Clarke, T.: Response characteristics of a fractional oscillator. Physica A 309, 275–288 (2002)
11. Suarez, L.E., Shokooh, A.: An eigenvector expansion method for the solution of motion containing fractional derivatives. ASME Journal of Applied Mechanics 64, 629–635 (1997)
12. Li, Y.L.: Solving a nonlinear fractional differential equation using Chebyshev wavelets. Communications in Nonlinear Science and Numerical Simulation (2010), doi:10.1016/j.cnsns.2009.09.020
13. Bonilla, B., Rivero, M., Trujillo, J.J.: Linear differential equations of fractional orders. In: Sabatier, J., Agrawal, O.P., Tenreiro Machado, J.A. (eds.) Advances in Fractional Calculus, pp. 77–91. Springer, Dordrecht (2007)
14. Wang, Z.H., Wang, X.: General solution of the Bagley-Torvik equation with fractional-order derivative. Communications in Nonlinear Science and Numerical Simulation 15, 1279–1285 (2010)
15. Wang, Z.H., Wang, X.: General solution of a vibration system with damping force of fractional-order derivative. In: Luo, A.C. (ed.) Dynamics and Vibrations with Discontinuity, Stochasticity and Time-Delay Systems, pp. 1–8. Springer, Heidelberg (2010)
16. Naber, M.: Linear fractionally damped oscillator. International Journal of Differential Equations, 2010, 12, Article ID 197020 (2010), doi:10.1155/2010/197020
17. Wang, Z.H., Du, M.L.: Asymptotical behavior of the solution of a SDOF linear fractionally damped vibration system. Shock and Vibration (2010), doi:10.3233/SAV-2010-0566.
18. Rossikhin, Y.A., Shitikova, M.V.: New approach for the analysis of damped vibrations of fractional oscillators. Shock and Vibration 16, 365–387 (2009)
19. Das, S.: Functional Fractional Calculus for System Identification and Controls. Springer, Berlin (2008)
20. Buslowicz, M.: Stability of linear continuous-time fractional order systems with delays of the retarded type. Bulletin of the Polish Academy of Sciences: Technical Sciences 56, 319–324 (2008)
21. Wang, Z.H., Hu, H.Y.: Stability of a linear oscillator with damping force of fractional-order derivative. SCIENCE CHINA: Physics, Mechanics & Astronomy 53(2), 345–352 (2010)

The Fractional Derivative as a Complex Eigenvalue Problem

Masaharu Kuroda

Abstract. For the dynamics described by an equation of motion including fractional-order-derivative terms, the fractional-order-derivative responses cannot be measured directly through experiments. In the present study, three solutions are proposed that enable the fractional-order-derivative responses to be measured by a combination of signals obtained by existing sensors. Specialized sensors or complicated signal processing are not necessary. Fractional-order-derivative responses at a certain point on a structure can be expressed through linear combinations of the displacement signal and the velocity signal at each point on the structure. Although their calculation processes are different, all three methods eventually reach the same result.

1 Introduction

Fractional calculus is a calculus in which the order of the derivative or the integral is not limited within the integer numbers and is extended to non-integer numbers, such as rational numbers or irrational numbers. This concept can be traced back to Leibniz (1695) [1] and dates back almost to the establishment of classical calculus. Liouville (1832) started the fundamental research into fractional calculus, and Riemann (1847) provided the definition of fractional calculus. Post (1919) demonstrated that Liouville's definition was identical to Riemann's definition [1].

As mentioned above, fractional calculus can provide a more fertile "expressive power" to differential equations and may describe much more precisely physical phenomena that have been difficult to express by conventional differential equations. A few techniques have been proposed to obtain fractional-order-derivative responses of the dynamics of mechanical structures [2, 3]. However, these methods require specialized electrical circuits or digital filters.

This report presents three new methodologies in which fractional-order-derivative responses at a certain point can be obtained using measurement signals at multiple sensing points on the structure [4-8]. Based on these methods, it is sufficient for existing accelerometers or displacement sensors to be prepared as the sensors.

Masaharu Kuroda
National Institute of Advanced Industrial Science and Technology (AIST), Tsukuba, Japan

G. Stépán et al. (Eds.): Dynamics Modeling & Interaction Cont., IUTAM BOOK SERIES 30, pp. 109–117.
springerlink.com

2 Fractional Calculus

Although there are a few definitions of the fractional derivative/integral, the Riemann-Liouville fractional derivative/integral is adopted for use in the present study:

$$ {}_aD_t^{\,p} f(t) = \left(\frac{d}{dt} \right)^{m+1} \int_a^t (t-\tau)^{m-p} f(\tau) d\tau , \tag{1} $$

where D is the differential operator, a is the initial value of the integral, t is the time, and m is the integer such that $0 \le m < p < m+1$.

In addition to the normal integer-order derivative, the fractional-order derivative satisfies the linearity condition [9-13]:

$$ {}_aD_t^{\,\alpha} [ax(t)+by(t)] = a\,{}_aD_t^{\,\alpha} [x(t)] + b\,{}_aD_t^{\,\alpha} [y(t)] \cdot \tag{2} $$

Furthermore, the fractional derivative satisfies the composition rule for the condition whereby all initial values are set to zero until reaching the necessary order for the fractional derivative [6, 8]:

$$ {}_aD_t^{\,p} \left({}_aD_t^{\,q} f(t) \right) = {}_aD_t^{\,q} \left({}_aD_t^{\,p} f(t) \right) = {}_aD_t^{\,(p+q)} f(t), \tag{3} $$

$$ f^{(k)}(a) = 0 , \left(k = 0,1,2,\cdots,r-1 \right), \tag{4} $$

where $0 \le m < p < m+1$, $0 \le n < q < n+1$, and $r = \max(m,n)$. This relationship for the composition rule is essential to the present study. Therefore, all mathematical analyses are performed under the above described zero initial conditions. Subscripts a and t assigned to the operator ${}_aD_t^p$ are omitted hereinafter.

3 Method of the Expanded System-Matrix

The first method to obtain fractional-order-derivative responses is based on the extension of system matrix $\boldsymbol{A}$ in a state space representation of an equation of motion [14, 15].

Treating a spring-mass-dashpot system with three degrees of freedom as an example, we explain the specific procedure to obtain the fractional-derivative responses of a structure.

As shown in Fig. 1, an external force is applied to mass point 1:

$$ f = (f_1, \;\; 0, \;\; 0)^T . \tag{5} $$

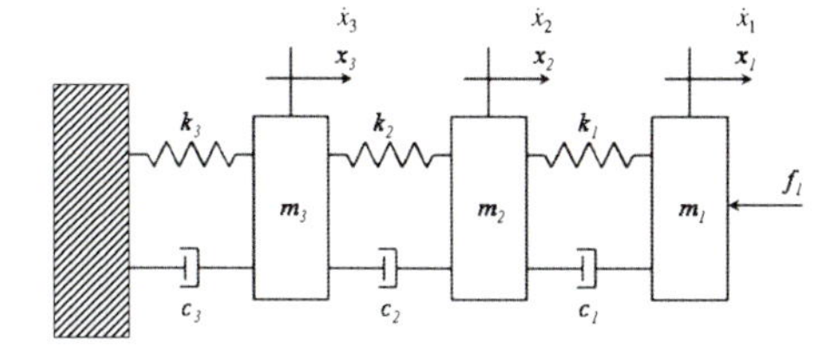

Fig. 1 Schematic diagram of the dynamic system with three degrees of freedom

Sensors are assumed to be located at masses 1, 2, and 3. At each sensor location, the displacement and velocity are assumed to be detected. The system response can be described using the following displacement vector:

$$\boldsymbol{x} = \left(x_1, \quad x_2, \quad x_3\right)^T. \tag{6}$$

The equation of motion for this vibratory system with the external excitation can then be obtained as follows:

$$\boldsymbol{M}\ddot{\boldsymbol{x}} + \boldsymbol{C}\dot{\boldsymbol{x}} + \boldsymbol{K}\boldsymbol{x} = \boldsymbol{f}, \tag{7}$$

where the matrix $\boldsymbol{M}$ is the mass matrix, the matrix $\boldsymbol{C}$ is the damping matrix, and the matrix $\boldsymbol{K}$ is the stiffness matrix. In this example, the components of each matrix are given as follows:

$$\boldsymbol{M} = \begin{pmatrix} m_1 & 0 & 0 \\ 0 & m_2 & 0 \\ 0 & 0 & m_3 \end{pmatrix}, \quad \boldsymbol{C} = \begin{pmatrix} c_1 & -c_1 & 0 \\ -c_1 & c_1 + c_2 & -c_2 \\ 0 & -c_2 & c_2 + c_3 \end{pmatrix}, \quad \boldsymbol{K} = \begin{pmatrix} k_1 & -k_1 & 0 \\ -k_1 & k_1 + k_2 & -k_2 \\ 0 & -k_2 & k_2 + k_3 \end{pmatrix}. \tag{8}$$

In this example, the matrix components are given as follows: m_1 = 1.0 kg, m_2 = 0.5 kg, m_3 = 0.25 kg, c_1 = 0.1 Ns/m, c_2 = 0.01 Ns/m, c_3 = 0.001 Ns/m, and $k_1 = k_2 = k_3$ = 1,500 N/m. Figure 2 depicts the Bode diagram for each mass point to the external force f. This vibratory system is understood to have three sharp resonant peaks: at 3.15, 11.1, and 19.0 Hz.

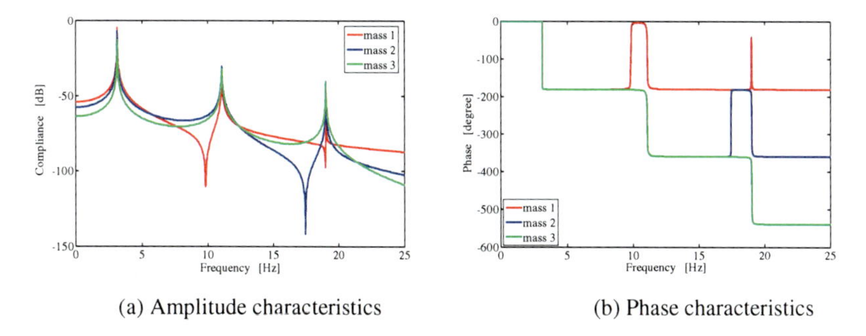

(a) Amplitude characteristics (b) Phase characteristics

Fig. 2 Bode diagrams for the vibratory system shown in Fig. 1

Next, the problem in which $\alpha = 1/2$ is considered. Here, the $D^{1/2}x$ response and the $D^{3/2}x$ response are not directly measurable in experiments. Therefore, it is quite natural to consider how to obtain $D^{1/2}x$ and $D^{3/2}x$ from $x(=D^{0/2}x)$, $\dot{x}(=D^{2/2}x)$, and $\ddot{x}(=D^{4/2}x)$.

We can use the composition rule of the fractional derivative repeatedly to formally expand the equation of motion as follows:

$$D^{\frac{1}{2}}\boldsymbol{M}\left(D^{\frac{3}{2}}+D^{\frac{2}{2}}+D^{\frac{1}{2}}+D^{\frac{0}{2}}\right)\boldsymbol{x}-\boldsymbol{M}\left(D^{\frac{3}{2}}+D^{\frac{2}{2}}+D^{\frac{1}{2}}\right)\boldsymbol{x}$$

$$+D^{\frac{1}{2}}\boldsymbol{C}\left(D^{\frac{1}{2}}+D^{\frac{0}{2}}\right)\boldsymbol{x}-\boldsymbol{C}D^{\frac{1}{2}}\boldsymbol{x}+\boldsymbol{K}D^{\frac{0}{2}}\boldsymbol{x}=\boldsymbol{0}. \tag{9}$$

Finally, the expanded equation of motion, which overcomes the aforementioned difficulty of not being able to obtain the fractional-order-derivatives directly through experiments, can be obtained as follows:

$$D^{1/2}\widetilde{\boldsymbol{M}}\widetilde{\boldsymbol{x}}+\widetilde{\boldsymbol{K}}\widetilde{\boldsymbol{x}}=\widetilde{\boldsymbol{B}}\widetilde{\boldsymbol{u}}, \tag{10}$$

$$\widetilde{\boldsymbol{x}}=\left(D^{\frac{3}{2}}\boldsymbol{x}^T \quad D^{\frac{2}{2}}\boldsymbol{x}^T \quad D^{\frac{1}{2}}\boldsymbol{x}^T \quad D^{\frac{0}{2}}\boldsymbol{x}^T\right)^T,\ \widetilde{\boldsymbol{u}}=\boldsymbol{f},\ \widetilde{\boldsymbol{B}}=\left(\boldsymbol{0} \quad \boldsymbol{0} \quad \boldsymbol{0} \quad \boldsymbol{I}\right)^T, \tag{11}$$

where the pseudo-mass matrix and the pseudo-stiffness matrix, respectively, are written as follows:

$$\widetilde{\boldsymbol{M}}=\begin{bmatrix} \boldsymbol{0} & \boldsymbol{0} & \boldsymbol{0} & \boldsymbol{M} \\ \boldsymbol{0} & \boldsymbol{0} & \boldsymbol{M} & \boldsymbol{0} \\ \boldsymbol{0} & \boldsymbol{M} & \boldsymbol{0} & \boldsymbol{C} \\ \boldsymbol{M} & \boldsymbol{0} & \boldsymbol{C} & \boldsymbol{0} \end{bmatrix},\ \widetilde{\boldsymbol{K}}=\begin{bmatrix} \boldsymbol{0} & \boldsymbol{0} & -\boldsymbol{M} & \boldsymbol{0} \\ \boldsymbol{0} & -\boldsymbol{M} & \boldsymbol{0} & \boldsymbol{0} \\ -\boldsymbol{M} & \boldsymbol{0} & -\boldsymbol{C} & \boldsymbol{0} \\ \boldsymbol{0} & \boldsymbol{0} & \boldsymbol{0} & \boldsymbol{K} \end{bmatrix}. \tag{12}$$

The formulation given by Eq. (10) has an advantage whereby the solution can be determined through the use of the conventional solution to the eigenvalue problem. The eigenvalues are complex conjugate pairs, and this system has 3×2 pairs of eigenvalues. The eigenvalues are mapped onto the two sheets of the Riemann surface for the function $\lambda = s^{1/2}$. Three pairs of complex conjugate eigenvalues appear on each sheet.

The modal development of the expanded system response can be written in the following form:

$$\widetilde{\boldsymbol{x}}(t)=\boldsymbol{\Phi}\boldsymbol{\xi},\ \boldsymbol{\xi}=\left[\xi_1 \quad \bar{\xi}_1 \quad \xi_2 \quad \bar{\xi}_2 \quad \xi_3 \quad \bar{\xi}_3\right]^T, \tag{13}$$

where the vector $\boldsymbol{\xi}$ is the column vector indicating the modal coordinate of the system. The mode matrix $\boldsymbol{\Phi}$ is constructed from the conjugate pairs of the eigenvectors that are relevant to the conjugate eigenvalues on the principal sheet of the Riemann surface, as shown by the following equation:

$$\boldsymbol{\Phi} = \begin{bmatrix} \varphi_1 & \overline{\varphi}_1 & \varphi_2 & \overline{\varphi}_2 & \varphi_3 & \overline{\varphi}_3 \end{bmatrix}. \tag{14}$$

Equation (13) extracts the information necessary to construct the conventional (integer-order) state vectors and the fractional-order state vectors in the system.

The row vectors corresponding only to the integer-order state vectors can be extracted from the modal matrix $\boldsymbol{\Phi}$. Consequently, a new smaller matrix $\boldsymbol{\phi}_Z$ can be formed:

$$z = \boldsymbol{\phi}_z \xi, \quad z = \begin{bmatrix} x_1 & x_2 & x_3 & \dot{x}_1 & \dot{x}_2 & \dot{x}_3 \end{bmatrix}^T. \tag{15}$$

On the other hand, the row vectors corresponding only to the fractional-order state vector at mass point 1 can also be extracted from the modal matrix $\boldsymbol{\Phi}$. As a result, another smaller matrix $\boldsymbol{\phi}_w$ can be formed:

$$\boldsymbol{w} = \boldsymbol{\phi}_w \xi, \quad \boldsymbol{w} = \begin{bmatrix} D^{\frac{0}{2}} x_1 & D^{\frac{1}{2}} x_1 & D^{\frac{2}{2}} x_1 & D^{\frac{3}{2}} x_1 \end{bmatrix}^T. \tag{16}$$

Consequently, the relationship between the integer-order state vectors and the fractional-order state vectors can be obtained as follows:

$$\boldsymbol{w} = \boldsymbol{\phi}_W \xi = \boldsymbol{\phi}_W \boldsymbol{\phi}_Z^{-1} z = \boldsymbol{\Psi} z, \tag{17}$$

where the matrix $\boldsymbol{\Psi}$ is the state conversion matrix between the integer-order states and the fractional-order states. Based on this equation, we can implement a sensor system for fractional derivatives from the measurement information concerning the displacement and velocity at each sensing location.

Eventually, the relationship between the integer-order state vectors and the fractional-order state vectors can be determined as follows:

$$\begin{bmatrix} D^{\frac{0}{2}} x_1 \\ D^{\frac{1}{2}} x_1 \\ D^{\frac{2}{2}} x_1 \\ D^{\frac{3}{2}} x_1 \end{bmatrix} = \begin{bmatrix} 1 & 0 & 0 & 0 & 0 & 0 \\ 3.81 & -0.83 & -0.14 & 0.14 & 0.02 & 0.0047 \\ 0 & 0 & 0 & 1 & 0 & 0 \\ -147.75 & 111.74 & 7.70 & 3.80 & -0.82 & -0.14 \end{bmatrix} \begin{bmatrix} x_1 \\ x_2 \\ x_3 \\ \dot{x}_1 \\ \dot{x}_2 \\ \dot{x}_3 \end{bmatrix}. \tag{18}$$

Using this equation, we can obtain the fractional-order temporal-differentiation response terms of the observed data from a linear combination of the displacement signal and the velocity signal at each sensor point. For example, Fig. 3 shows the Bode diagram of the fractional-order-derivative responses at mass point 1.

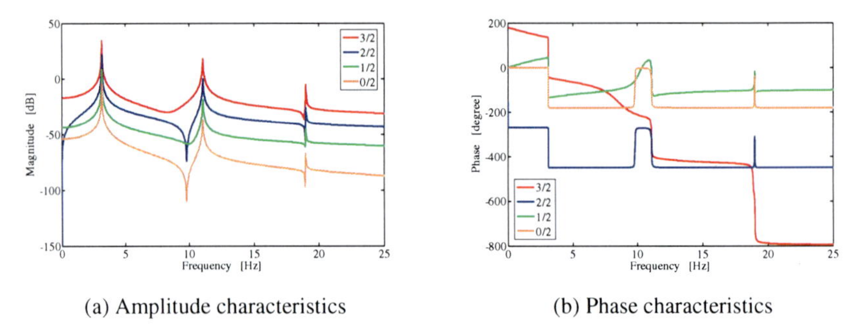

(a) Amplitude characteristics (b) Phase characteristics

Fig. 3 Bode diagrams for $D^{3/2}x_1$, $D^{2/2}x_1$, $D^{1/2}x_1$, and $D^{0/2}x_1$ for the vibratory system shown in Fig. 1

4 Method of the *n*-th Root of the System Matrix

The second method is to directly obtain the n-th root of the system matrix in the state space representation of an equation of motion:

$$D^1\begin{pmatrix} \boldsymbol{x} \\ \dot{\boldsymbol{x}} \end{pmatrix} = \begin{pmatrix} \dot{\boldsymbol{x}} \\ \ddot{\boldsymbol{x}} \end{pmatrix} = \begin{pmatrix} \boldsymbol{O} & \boldsymbol{I} \\ -\boldsymbol{M}^{-1}\boldsymbol{K} & -\boldsymbol{M}^{-1}\boldsymbol{C} \end{pmatrix}\begin{pmatrix} \boldsymbol{x} \\ \dot{\boldsymbol{x}} \end{pmatrix} = \boldsymbol{A}\begin{pmatrix} \boldsymbol{x} \\ \dot{\boldsymbol{x}} \end{pmatrix}. \tag{19}$$

The first-order derivative operator D^1 can be assumed to be equivalent to the one time multiplication of the system matrix $\boldsymbol{A}$. In this case, the $1/n$-th fractional-order-derivative operator $D^{1/n}$ is equal to the $1/n$ time multiplication of system matrix $\boldsymbol{A}$, namely, one time multiplication of the matrix $\boldsymbol{A}^{1/n}$.

Then, for the case in which matrix $\boldsymbol{A}$ is given, how can we derive the n-th root of the system matrix $\boldsymbol{A}$, i.e., the matrix $\boldsymbol{A}^{1/n}$? The matrix $\boldsymbol{U}$ is the diagonalizing matrix comprised of the eigenvectors of matrix $\boldsymbol{A}$, and the matrix $\boldsymbol{D}$ is the diagonal matrix composed of the corresponding eigenvalues:

$$\boldsymbol{U}^{-1}\boldsymbol{A}\boldsymbol{U} = \boldsymbol{D}. \tag{20}$$

Then, we can raise both sides of the above equation to the $1/n$-th power to obtain the following equation:

$$\boldsymbol{U}^{-1}\boldsymbol{A}^{\frac{1}{n}}\boldsymbol{U} = \boldsymbol{D}^{\frac{1}{n}}. \tag{21}$$

In practice, matrix $\boldsymbol{D}$ is a diagonal matrix. Therefore, the $1/n$-th power of matrix $\boldsymbol{D}$ can be calculated very easily, as follows:

$$\boldsymbol{A}^{\frac{1}{n}} = \boldsymbol{U}\boldsymbol{D}^{\frac{1}{n}}\boldsymbol{U}^{-1}. \tag{22}$$

The important result here is that all of the components of matrix $\boldsymbol{A}^{1/n}$ become real numbers. More importantly, the state conversion matrix $\boldsymbol{\Psi}$ used in the method of the expanded system-matrix agrees with the coefficient matrix derived by the method of the n-th root of the system matrix.

5 Method of the Complex Eigenvalue Problem

The third method is a technique based on the solution of a complex eigenvalue problem. Generally, the case in which the damping matrix in an equation of motion is not expressed by a proportional damping matrix is referred to as a complex eigenvalue problem, and both the eigenvalues and the corresponding eigenvectors are expressed using complex numbers.

The equation of motion for the non-proportional damping case can be described as follows:

$$\boldsymbol{M}\ddot{\boldsymbol{x}} + \boldsymbol{C}\dot{\boldsymbol{x}} + \boldsymbol{K}\boldsymbol{x} = \boldsymbol{0}. \tag{23}$$

The eigenvalue equation for a vibratory system with non-proportional damping is written as follows:

$$\left(\lambda^2 \boldsymbol{M} + \lambda \boldsymbol{C} + \boldsymbol{K}\right)\boldsymbol{x} = \boldsymbol{0}. \tag{24}$$

The solutions for the eigenvalue equation, namely, λ and $\boldsymbol{x}$, are both complex values [16].

By adding a trivial equation $\boldsymbol{y} = \lambda \boldsymbol{x}$ to the equation of motion, we obtain the following equation:

$$\begin{pmatrix} \boldsymbol{M} & \boldsymbol{0} \\ \boldsymbol{0} & -\boldsymbol{K} \end{pmatrix}\begin{pmatrix} \boldsymbol{y} \\ \boldsymbol{x} \end{pmatrix} - \lambda \begin{pmatrix} \boldsymbol{0} & \boldsymbol{M} \\ \boldsymbol{M} & \boldsymbol{C} \end{pmatrix}\begin{pmatrix} \boldsymbol{y} \\ \boldsymbol{x} \end{pmatrix} = \begin{pmatrix} \boldsymbol{0} \\ \boldsymbol{0} \end{pmatrix}. \tag{25}$$

Therefore, we have the following equations:

$$\boldsymbol{P}\boldsymbol{u} - \lambda \boldsymbol{Q}\boldsymbol{u} = \boldsymbol{0}, \quad \boldsymbol{u} = \left(\boldsymbol{y}, \quad \boldsymbol{x}\right)^T, \tag{26}$$

where the matrices $\boldsymbol{P}$ and $\boldsymbol{Q}$ are matrices of size $(2n \times 2n)$, and the vector $\boldsymbol{u}$ is a vector of dimension $(2n)$.

Since the matrix $\boldsymbol{Q}$ is a non-singular matrix, $\boldsymbol{Q}^{-1}$ can be calculated. Accordingly, Eq. (26) can be transformed into the normal eigenvalue problem expressed as follows:

$$\hat{\boldsymbol{A}}\boldsymbol{u} = \lambda \boldsymbol{u}, \quad \hat{\boldsymbol{A}} = \boldsymbol{Q}^{-1}\boldsymbol{P}. \tag{27}$$

The method used to obtain eigenvalues of this type is useful here. Eventually, the eigenvalues and eigenvectors of Eq. (27), and therefore those of Eq. (26), become n pairs of complex conjugates.

In addition, complex eigenvectors produce eigenmodes. In the physical interpretation, the fact that the eigenvectors are expressed by complex numbers means that the entire system does not oscillate in phase or in reversed-phase at the eigenfrequency. Each point of the system vibrates with a phase lead or lag in relation to other points.

6 Conclusions

The fractional calculus has a hidden potentiality to provide a wider variety of dynamics and their control laws than before. The present paper revealed that the fractional derivative problem was a complex eigenvalue problem. In addition, three methods were proposed in order to obtain the fractional derivative responses in the vibratory system with multiple degrees of freedom. Whenever any of the three methods is applied, the fractional-order states are represented by complex vibration modes and, accordingly, each point of the system oscillates with a phase difference (lead or lag), i.e., neither in-phase nor in reversed-phase.

Since future studies will demonstrate the effectiveness of these methods through experiments, the representation accuracy of the fractional derivative responses depending on a change in the number of sensors (or the number of mass points) and the proposed design of the fractional-derivative sensor depending on the discreet system or the continuous system are enumerated.

References

[1] Oldham, K.B., Spanier, J.: The Fractional Calculus, pp. 1–15. Dover, New York (2002)
[2] Motoishi, K., Koga, T.: Simulation of a Noise Source with 1/f Spectrum by Means of an RC Circuit. IEICE Trans. J65-A(3), 237–244 (1982) (in Japanese)
[3] Chen, Y., Vinagre, B.M., Podlubny, I.: A New Discretization Method for Fractional Order Differentiators via Continued Fraction Expansion. In: Proc. ASME IDETC/CIE 2003, DETC2003/VIB 48391, pp. 761–769 (2003)
[4] Kuroda, M., Kikushima, Y., Tanaka, N.: Active Wave Control of a Flexible Structure Formulated Using Fractional Calculus. In: Proc. 74th Annual Meeting of JSME, vol. (I), pp. 331–332 (1996) (in Japanese)
[5] Kuroda, M.: Active Vibration Control of a Flexible Structure Formulated Using Fractional Calculus. In: Proc. ENOC 2005, pp. 1409–1414 (2005)
[6] Kuroda, M.: Formulation of a State Equation Including Fractional-Order State-Vectors. In: Proc. ASME IDETC/CIE 2007, DETC2007-35273, pp. 1–10 (2007)
[7] Kuroda, M.: Active Wave Control for Flexible Structures Using Fractional Calculus. In: Sabatier, J., Agrawal, O.P., Tenreiro Machado, J.A. (eds.) Advances in Fractional Calculus: Theoretical Developments and Applications in Physics and Engineering, pp. 435–448. Springer, Dordrecht (2007)
[8] Kuroda, M.: Formulation of a State Equation Including Fractional-Order State Vectors. J. Computational and Nonlinear Dynamics 3, 021202-1–021202-8 (2008)
[9] Podlubny, I.: Fractional Differential Equations. Academic Press Inc., San Diego (1999)
[10] Hilfer, R. (ed.): Applications of Fractional Calculus in Physics. World Scientific, Singapore (2000)
[11] West, B.J., Bologna, M., Grigolini, P.: Physics of Fractal Operators. Springer, New York (2003)
[12] Kilbas, A.A., Trujillo, J.J.: Differential Equations of Fractional Order: Methods, Results and Problems. II. Applicable Analysis 81(2), 435–493 (2002)

[13] Kilbas, A.A., Srivastava, H.M., Trujillo, J.J.: Theory and Applications of Fractional Differential Equations. Elsevier, Amsterdam (2006)
[14] Yang, D.L.: Fractional State Feedback Control of Undamped and Viscoelastically damped Structures. Master's Thesis, AD-A-220-477, Air Force Institute of Technology, pp. 1–98 (1990)
[15] Sorrentino, S., Fasana, A.: Finite element analysis of vibrating linear systems with fractional derivative viscoelastic models. J. Sound and Vibration 299, 839–853 (2007)
[16] Nagamatsu, A., et al. (eds.): Dynamics Handbook, pp. 111–112. Asakura, Tokyo (1993) (in Japanese)

Discontinuity-Induced Bifurcations Due to Saturations

Gustavo A. Osorio, Fabiola Angulo, and Gerard Olivar

Abstract. This paper regards the study of systems of non-smooth differential equations with parameters and saturation discontinuities in the vector field. Once the problem is stated, we define an appropriate solution concept. This regards mainly to the framework of Filippov systems and sliding solutions in the control literature. Second, we consider linear systems with one saturation linear manifold in a two-dimensional state space and deduce several properties that will do analytical bifurcation analysis easier. This allows us to characterize the invariant sets of these systems and thus perform efficient bifurcation analysis. Non-smooth bifurcations due to interaction of an invariant set with a switching-impact-sliding manifold are also known as discontinuity-induced bifurcations. When we consider two-dimensional linear systems depending on parameters trace and determinant of the matrix, and two additional significant parameters, we can compute equilibrium points, pseudo-equilibriums and (smooth and non-smooth) periodic orbits analytically, depending on the parameters. We completely classify the dynamical systems according to the invariant sets that we found. We extend our analysis to two saturation linear manifolds, and generalize our method to several saturation non-linear manifolds and non-linear differential equations through Lie-derivatives framework. This is applied to a simple oscillator circuit which is usually used in control strategies of mechanical systems.

1 Introduction

Systems with saturations are very common in applications. Mechanical, electronic, chemical systems have saturations either in the state space or in the control variables or actuators in order to restrict orbits configuration or control efforts to an admissible region. The mathematical framework for such systems is the general non-smooth

Gustavo A. Osorio · Fabiola Angulo · Gerard Olivar
Department of Electrical and Electronics Engineering & Computer Sciences,
Campus La Nubia, Universidad Nacional de Colombia, sede Manizales
e-mail: {gaosoriol, fangulog, golivart}@unal.edu.co

G. Stépán et al. (Eds.): Dynamics Modeling & Interaction Cont., IUTAM BOOK SERIES 30, pp. 119–126.
springerlink.com

dynamics. As some parameters are varied, non-smooth bifurcations appear which lead to a vast and still not-completely explored non-smooth phenomena.

Several classifications of non-smooth systems have been stated in the literature, mainly regarding to the degree of the non-smoothness. Thus, a well-known classification distinguishes among systems with impacts, sliding systems, and piecewise-smooth continuous systems, which corresponds basically to a classification based on the strongness of the non-smooth mechanism. All these systems assume that there exists one or more manifolds in the state space which divides it into different behavior regions. The dynamics on each region is described by a set of smooth differential equations. Depending on the degree of non-smoothness between two consecutive regions, we find impact, sliding or piecewise-smooth continuous systems.

Impact systems assume that on one side of the manifold the dynamics are totally forbidden and thus no orbit can appear on such side. Thus the dynamics can only exist on one side of the manifold. A reset function is then defined on the manifold which describes where the orbit re-starts once it hits the manifold. Since very usually this makes that the orbits hitting the manifold have discontinuities in the trajectories, this sort of non-smoothness is considered as the strongest one [1].

Sliding systems have a not-so-strong non-smoothness. For sliding systems, which are part of the so-called Filippov systems, a manifold Σ divides the dynamics. On one side of the manifold, say region M_1, we have a smooth vector field F_1 and on the other side, say region M_2, we have a different vector field F_2 such that

$$F_1(X) \neq F_2(X)$$

on the manifold Σ. This condition defines Filippov systems. Moreover, for sliding systems one additional condition must be imposed. This is that F_1 points towards region M_2 along Σ and F_2 points towards region M_1 along Σ. Of course, these intuitive picture can be stated with derivatives formalism which involve the equations for Σ and the vector fields F_1 and F_2 [3].

Finally, piecewise-smooth continuous systems are defined similarly as Filippov systems, but without the condition that

$$F_1(X) \neq F_2(X)$$

on the manifold Σ. Thus they verify that

$$F_1(X) = F_2(X)$$

on Σ and then no sliding orbits can exist. In such an environment, the manifold is a switching one which allows that the trajectories pass through it smoothly.

Anyway, there exist other non-smooth systems which do not fit with impact, sliding or piecewise-smooth continuous systems, such hysteresis or saturated systems, which exhibit similar but different behavior. In this paper we deal with saturated systems. We will focus on linear saturated systems but as we will show in the final part of the paper, many interesting properties can be extended to non-linear saturated systems as well.

Thus next section deals with systems with saturations. In Section 3 we state the main results for linear planar systems with one or with two saturations, as well as the generalization for n-dimensional systems and non-linear systems. Section 4 deals with an application to an electronic circuit which the authors generally use for control of mechanical systems, and finally on Section 5 we state the final conclusions.

2 Systems with Saturations

Assume first that we have a planar linear homogeneous system

$$\dot{X} = F(X) \equiv AX,$$

where $X^T = (x,y)$ and a linear saturation manifold Σ in the state space such that the dynamics are forbidden on one side of the manifold and, in the other (admissible) side, the dynamics are described by the linear system.

Specifically, we have

$$A = \begin{pmatrix} a & b \\ c & d \end{pmatrix} \tag{1}$$

and we can assume without losing generalization that the saturation is defined only in the second component of the state vector by the line,

$$y = h,$$

where h is a parameter value and the admissible region is $y \leq h$. In so doing, we are defining implicitly the boundary Σ for the admissible region as the zero level set of an scalar function the we will call $\sigma(x,y) = h - y$, as follows:

$$\Sigma := \{(x,y) \in \mathscr{R}^2 / \sigma(x,y) = 0\}.$$

There may be a constrained (sliding) dynamical mode whenever the vector field F pushes against the saturation. We define that the system will be constrained to the saturation while $y = h$ and

$$c \cdot x + d \cdot y \geq 0$$

The constrained (sliding) dynamics will be given by

$$\dot{X} = A_s X \tag{2}$$

being

$$A_s = \begin{pmatrix} a & b \\ 0 & 0 \end{pmatrix}. \tag{3}$$

Some numerical simulations with one saturation corresponding to a saddle equilibrium of the linear system are shown in Fig.1.

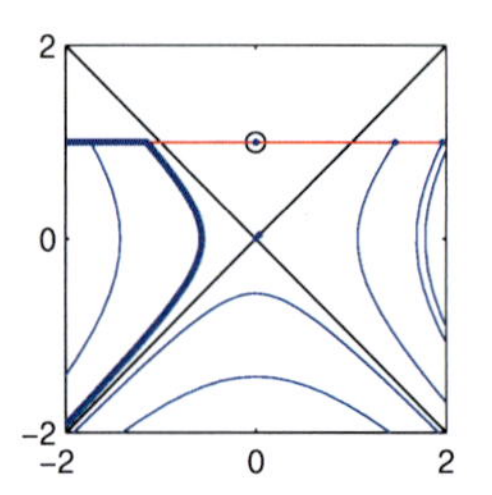

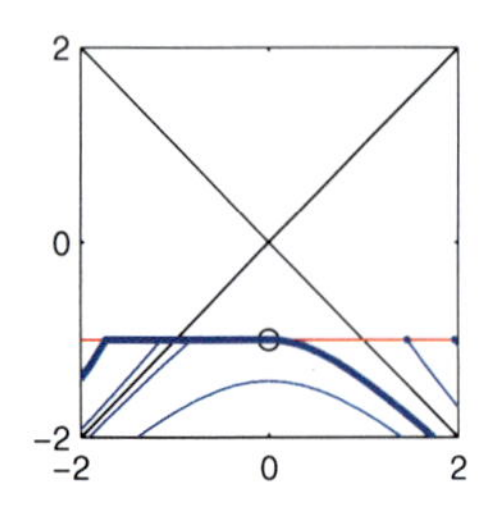

Fig. 1 Left: Saddle equilibrium linear system with one saturation manifold. Right: Saddle equilibrium linear system with the equilibrium passing by the saturation manifolds.

We are interested on the analysis of the qualitative dynamics and bifurcating behavior of limit invariant sets under variation of the parameters h, a, the determinant $\delta := Det(A)$ and the trace $\tau := Tr(A)$.

3 Main Results

In this section we state our main results regarding existence of equilibriums, pseudo-equilibriums, sliding dynamics, non-smooth periodic orbits, stability properties and bifurcations.

3.1 *Existence of an Equilibrium and Bifurcations*

We split the analysis into three cases: $h > 0$, $h = 0$ and $h < 0$ as follows:

When $h > 0$ the system defined by Eq.1 and 2 will show an equilibrium (that we will note as X^*) at the origin, with a dynamical behavior qualitatively described by δ and τ. In general, one gets a saddle, a stable or unstable focus, a stable or unstable node, a center, or a continuous of equilibria. As we make $h = 0$ there is a border-collision of equilibria, and if we further move h to negative values, the only invariant set that persists after varying parameter h from positive to negative values(*i.e.* $dh < 0$), is the **continuous of equilibria**. Any other type of equilibria will vanish under these conditions.

3.2 *Sliding and Pseudo-equilibria (Existence and Bifurcations)*

It is well known from the theory of non-smooth dynamical systems that several dynamics and invariant sets may appear as a result of the interaction between the vector field and the discontinuity manifold [2]. For linear systems with saturation, as defined in Eq.1 and Eq.2, it is possible to find pseudo-equilibria (that we will write as $X_P^* = (x_P^*, y_P^*)$) associated to the constrained (sliding) dynamics if the following conditions are satisfied:

- **CondPs-1.** $X_P^* = (x_P^*, y_P^*) \in \Sigma$, or $\sigma(x_P^*, y_P^*) = 0$.
- **CondPs-2.** $\dot{X}(X_P^*) = A_s X_P^* = 0$.
- **CondPs-3.** $\mathscr{L}_{(h-y=0)}(AX)(X_P^*) < 0$.

The first condition (CondPs-1.) guarantees that the pseudo-equilibrium X_P^* lies on the saturation. For the linear case above, this can be written as:

CondPs-1 (Linear). $y_P^* = h$

The second condition (CondPs-2.) defines points where the vector field is exactly perpendicular to the saturation boundary, and hence for our example this is equivalent to:

CondPs-2 (Linear). $a \cdot x_P^* + b \cdot y_P^* = 0.$

The third condition (CondPs-3.) defines the sliding set for those points at which the first *Lie* derivative is negative. This is equivalent to say that, at those points, the vector field pushes against the saturation boundary (manifold) *i.e.*

CondPs-3 (Linear). $c \cdot x_P^* + d \cdot y_P^* > 0.$

We can summarize the first two conditions doing the substitution:

$$y_P^* = h, \qquad and \qquad \dot{x} = 0.$$

where, it can be verified that there is only one point that satisfies both conditions, and that can be written as:

$$X_P^* = (x_P^*, y_P^*) = (-\frac{b}{a}h, h) \qquad for \qquad a \neq 0.$$

It is easy to test that the third condition is equivalent to

$$\frac{\delta}{a}h > 0. \tag{4}$$

Condition in the Eq.4 guarantees the existence of a pseudo-equilibrium for our example.

All possible qualitative dynamics are characterized by eight different scenarios according to the sign of δ, a and h. In general for any combination of δ, h and a, if Eq.4 is satisfied, then we have an isolated pseudo-equilibrium on the saturation manifold. For example if we assume that $\delta < 0$, $a < 0$ and $h > 0$, then we have that condition Eq.4 is always fulfilled.

Stability of pseudo-equilibria

Since the pseudo-equilibria is associated to the saturated dynamics defined by Eq.2, and we have that eigenvalues of A_s are $\lambda_{S1} = 0$ and $\lambda_{S2} = a$, the stability of any pseudo-equilibrium will be given by the sign of a. If we have that $a < 0$, then the pseudo-equilibrium is asymptotically stable.

3.3 Non-smooth Periodic Orbits

Given an initial value problem by a linear diferencial equation like the one in Eq.1 and an initial condition $x(0) = x_0$, it can be shown that for this particular case the analytical solution can be written as:

$$x(t) = e^{At}x_0 \tag{5}$$

where the exponential matrix for a second order system is:

$$e^{At} = \mathscr{L}^{-1}\{(sI - A)^{-1}\}. \tag{6}$$

If the spectrum of A is formed by complex eigenvalues, we can write the exponential matrix as follows:

$$e^{At} = e^{\alpha t}\left(I\cos(\beta t) + \beta(\alpha I - A)^{-1}\sin(\beta t)\right) \tag{7}$$

where I is the identity matrix, $\alpha := \tau_A/2$, $\beta := \sqrt{|\tau_A^2 - 4\delta_A|}/2$, $\tau_A = a + d$ and $\delta_A = a \cdot d - b \cdot c$.

Then we can provide a test function which allows us to determine the existence of an isolated periodic orbit as follows:

1. Verify that the spectrum of A has complex eigenvalues with positive real part.
2. Use the solution given by Eq.5 and Eq.7 in order to find the image of the unique tangent orbit leaving from the saturating boundary. In so doing we will have the image $\tilde{X}_T$ as follows:

$$\tilde{X}_T = e^{\alpha t}\left(I\cos(\theta) + \beta(\alpha I - A)^{-1}\sin(\theta)\right)X_T$$

   where $\theta = \arctan(\frac{\beta}{d-\alpha}) + \pi$, and we define the tangent orbit to the saturating boundary through the point $X_T = (-\frac{d}{c}h\ ,\ h\ )^t$.
3. If the image $\tilde{X}_T$ belongs to the admissible region before reaching the saturating boundary, then we can guarantee that there will be an isolated periodic non-smooth orbit.

3.4 Non-smooth Bifurcations

We will now show the structural stability of the system under variation of a vector parameter $\mu = (h, a)^T$. We will emphasize on discontinuity induced bifurcations (DIBs), avoiding the presentation of standard phenomena.

Bifurcations of equilibria

The equilibria X^* ceases to exist at $h^* = 0$, under variation from positive to negative values of the parameter (*i.e.* $dh < 0$). This type of bifurcation is well-known as a border collision of equilibria in the theory of DIBs.

Bifurcations of pseudo-equilibria

It is posible to distinguish between two different bifurcation scenarios, as follows:

1. The pseudo-equilibria X_p^* ceases to exist at $a^* = 0$, under variation from negative to positive values of the parameter ($i.e.$ $da > 0$). In this case it is possible to verify that $\lim_{a \to 0^-} x^* = -\infty$.
2. The pseudo equilibria X_p^* ceases to exist at $a^* = \frac{b*c}{d}$. It is worth to mention that for this value of the parameter a, we have $\Delta A = 0$.

4 Application to an Oscillator

Our application regards to an oscillatory circuit (Wien bridge) which we usually use in mechanical systems such as robotic arms or pneumatic position systems (see Fig.2). Such a circuit is shown in Fig.3.

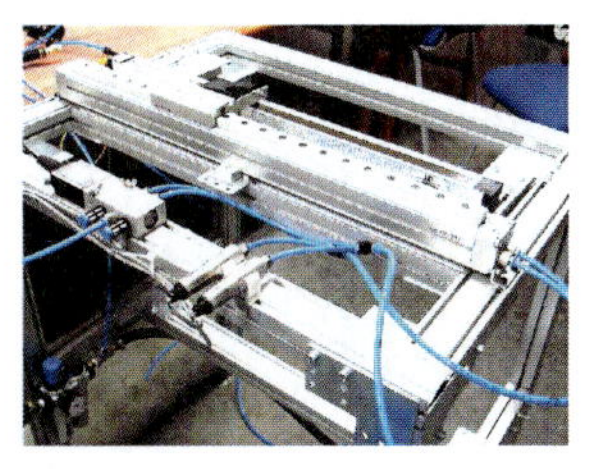

Fig. 2 Left: Robotic arm Scorbotter plus. Right: Pneumatic position system Festo

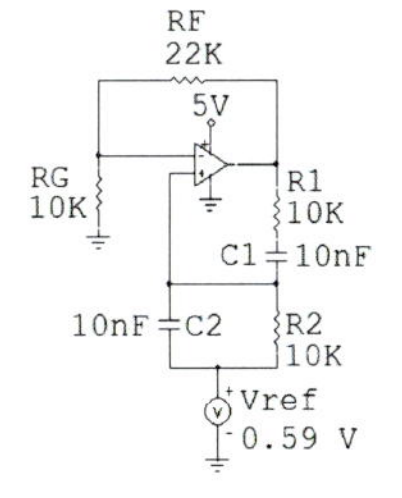

Fig. 3 Left: Scheme of the used Wien bridge oscillatory circuit.Right: Wien bridge oscillatory circuit

As parameter V_{ref} is varied, this oscillator undergoes several non-smooth bifurcation due to the interaction with the saturation manifold. Several non-smooth saturated periodic orbits are obtained (some saturated at a 5V and some saturated at 0V). Some orbits have two saturated pieces (one at 5V and one at 0V) showing a non-smooth codimension-two bifurcation point. Experimental simulations are shown in Fig.4.

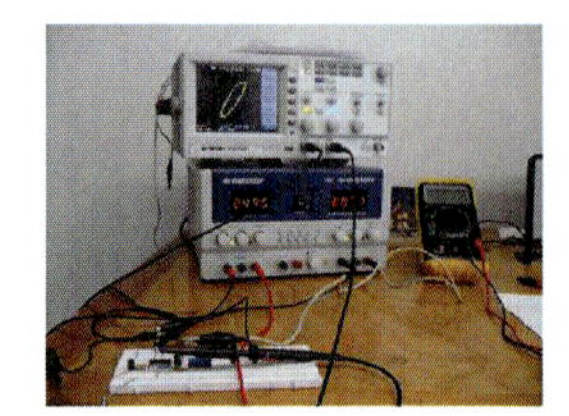

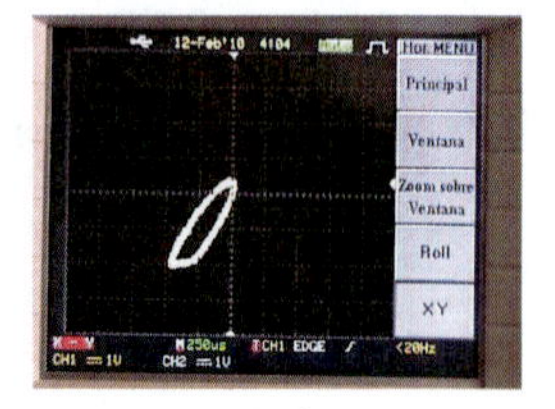

Fig. 4 Left: Periodic orbit with saturation at 5V. Right: Periodic orbit with saturation at 0V

5 Conclusions

We elaborated some theory for non-smooth dynamical systems with saturations. When the system and the saturation are linear, precise analytical results can be obtained regarding existence and stability of pseudo-equilibria and periodic orbits, as well as non-smooth bifurcations. This can be applied to several control mechanisms for mechanical systems. Concretely, we performed experiments and numerics for a Wien bridge electronic circuit used in robotic arms and position systems.

References

1. Di Bernardo, M., Budd, C.J., Champneys, A.R., Kowalczyk, P.: Piecewise-smooth Dynamical Systems: Theory and Applications. Springer, Heidelberg (2007)
2. Di Bernardo, M., Budd, C.J., Champneys, A.R., Kowalczyk, P., Nordmark, A.B., Olivar, G., Piiroinen, P.T.: Bifurcations in Nonsmooth Dynamical Systems. SIAM Review 50, 629–701 (2008)
3. Kuznetsov Yu, A., Rinaldi, S., Gragnani, A.: One-parameter Bifurcations in Planar Filippov Systems. International Journal of Bifurcation and Chaos 13, 2157–2188 (2003)

Bifurcations in Hysteresis Systems Due to Vibrations and Impacts

Fabio A. Leyton, Jorge E. Hurtado, and Gerard Olivar

Abstract. Operators of industrial machinery, in their daily work, are affected by vibrations produced by equipment. These vibrations are transmitted to the body, with a considerable health risk amount for the long-time period, and leading to a heavy environmental workplace. Vibrations can also cause damage to the machinery, misadjusting pieces and rising up maintenance costs. In order to correctly modeling these nonlinear systems we use the well-known Bouc-Wen model, which shows a good agreement between numerical simulations and experiments. Other models like Masing, Biot, and Spencer, which is a generalization of Bouc-Wen, can also be considered in order to display the same hysteretic phenomena. The Bouc-Wen model, due to its simplicity and versatility has been extensively studied in the scientific community, and successfully applied to several hysteresis problems. In the work we will show in our paper, we study the bifurcation diagrams of an application of the Bouc-Wen model to industrial machinery. We found transitions from periodicity to quasiperiodic and chaotic motion. Moreover, recently described big-bang codimension-two bifurcation has also been found. Close to the codimension-two point, an infinite number of periodic orbits exists.

1 Introduction

In their daily routine, any operator of industrial machinery is subject to withstand the vibrations produced by the team with which he works. These vibrations are

Fabio A. Leyton · Jorge E. Hurtado
Department of Civil Engineering,
Campus La Nubia, Universidad nacional de Colombia, sede Manizales
e-mail: {faleytonm,jehurtadog}@unal.edu.co

Gerard Olivar
Department of Electrical and Electronics Engineering & Computer Sciences, Universidad nacional de Colombia, sede Manizales
e-mail: golivart@unal.edu.co

G. Stépán et al. (Eds.): Dynamics Modeling & Interaction Cont., IUTAM BOOK SERIES 30, pp. 127–134.
springerlink.com

transmitted to his body, affecting long-term health, and giving a little suitable work environment to perform their working activities comfortably. These vibrations can also cause damage to the machinery itself, leading to mismatch of some parts and are therefore raising maintenance costs. This is why the study of these vibrations is crucial for the improvement of conditions in work areas in industries. They pose a threat both to the operator health and to spending on repairs and maintenance.

Mathematically, vibrations correspond to nonlinear phenomena such as periodic orbits, quasiperiodicity or chaos. Chaos control in such systems, leading to a reduction of the orbit amplitude would allow preventing unwanted vibrations in machinery. If the phenomenon is well-modeled and characterized, control systems can be designed to solve this problem optimally, and the implementation of passive control devices for this purpose can be explored.

In the new era of implementation of smart materials magneto-rheological fluids (MR) must be considered, as they have the advantage of reacting to a magnetic field and do not need a great source of power. MR fluids were developed in the 1940s mainly by Winslow and Rainbow. Over the 90's, MR dampers have been recognized by a number of attractive features for use in vibration control applications.

Several references that address analysis of vibrations in machinery [2] are based on the design of the structure. Taking into account that all types of machinery produce different effects, mathematical models are provided for the representation of these systems. Approximations through hysteretic models are considered and phenomenon analysis is performed.

For the modeling of the systems considered, the Bouc-Wen model is one of the best which fits these phenomena, since it allows for nonlinear and non-smooth behavior. Other models can also be considered, which also show hysteretic behavior, such as that of Masing, Biot, or Spencer, which is simply an extension of the Bouc-Wen model. The Bouc-Wen model has been examined in many studies worldwide, such as in [1], [4]-[13], where its use lead to satisfactory results.

At present, there are very advanced computational tools for the analysis of nonlinear behavior, and it is also common to find research on new methods for determining characteristics of chaotic systems. Bifurcation analysis considers the system behavior when parameters are varied.

2 Bouc-Wen Model

In industry, there are many situations where it is possible to reduce vibrations. The use of vibration isolators elements reduce the transmission of excitation forces or vibrations themselves among the different parts of the system, and it has recently attracted much research. When the excitation frequency matches a natural frequency of the system, a phenomenon called resonance appears. One of the most important feature of resonance is that it gives rise to large displacements and amplifies the system vibrations. In most mechanical systems, the presence of large displacements is an undesirable phenomenon, since it causes the appearance of large stresses and

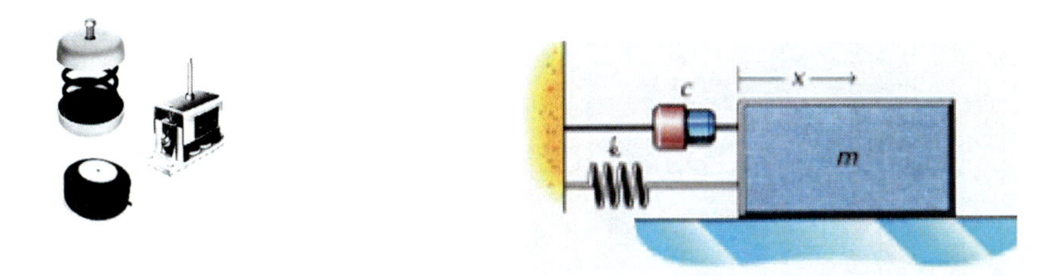

Fig. 1 Left: Insulators. Right: degree-1 model

strains that can lead some parts of the system to collapse. Consequently, the resonance conditions must be avoided in the design and construction of mechanical systems. The rigidity of the system is one of the parameters that can be modified in order to change the natural frequencies of a mechanical system.

When a rotating machine is attached directly on a rigid foundation, it will be under the action of an harmonic force due to imbalance of the rotating machine, which will be superimposed on the static charge associated with the weight. Therefore, an elastic element (insulator) is placed between the machine and the foundation that seeks to reduce the forces transmitted to the latter. The system can be idealized as a one degree of freedom one. The insulator element incorporates both elastic stiffness (k) as a damping (c) (see Fig. 1).

Magneto-rheological dampers (MR) consist of a suspension of iron particles in a viscous medium such as oil. The application of a magnetic field in the liquid causes the iron particles to align, increasing the resistance of the fluid passing a liquid state to a semisolid state. MR dampers are relatively inexpensive to manufacture because the fluid properties are not sensitive to pollutants. Other attractive features include its small power requirements, reliability and stability. They need only 20-50 watts of power and can operate with a battery, eliminating the need for a great source of energy. Since the device strength can be adjusted by varying the magnetic field strength, mechanical valves are not necessary, which makes a very reliable device. The fluid responds in milliseconds.

Hysteresis models are based on empirical parametric approaches using linear and nonlinear elements. They allow simulation of system memory and energy dissipation. Energy dissipation must be included in the model, through, for example, friction elements, springs and other mechanical elements. The classical model of Masing hysteresis combines Hooke elements (based on an elastic body model) and a number of St. Venant elements (based on models of plastic body), in parallel. The Biot hysteresis model includes Newton elements (Newton flow model) instead of St. Venant ones. Often, mechanical seals are also used, like in Spencer's model, which is an extension of the Bouc-Wen model.

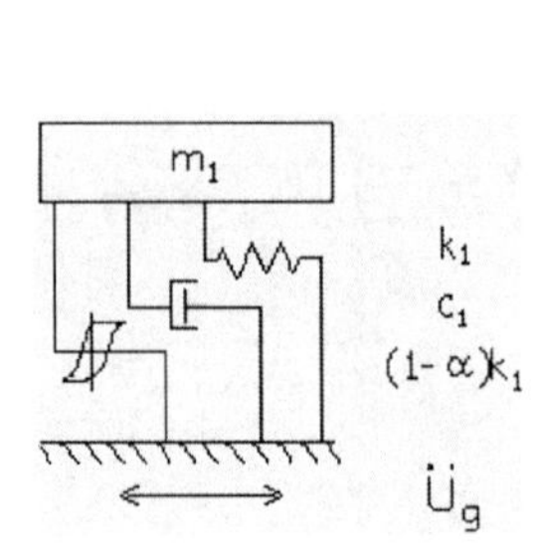

Fig. 2 Bouc-Wen model

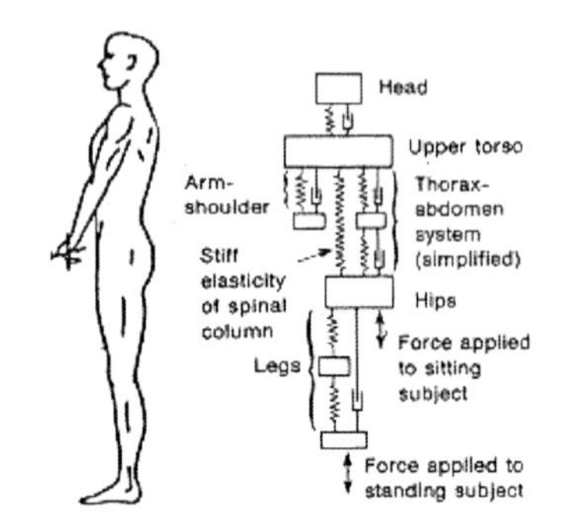

Fig. 3 Rheological model for human body

2.1 Model Equations

The dynamics of Bouc-Wen model is governed by the following set of differential equations (see Figs. 2 and 3);

$$\begin{aligned} m\ddot{x} + c\dot{x} + k(\alpha x + (1-\alpha)z) &= F\cos\omega t \\ \dot{z} &= A\dot{x} - \beta|\dot{x}||z|^{n-1}z - \gamma\dot{x}|z|^n \end{aligned} \tag{1}$$

with $0 < \alpha < 1$, $\beta + \gamma > 0$, $\beta - \gamma \geq 0$ and $n \geq 1$.

Transferability is defined as the resultant of the components due to the spring and damper (see Fig. 2). The isolation is achieved when the transmissibility is less than 1. It may be noted that this requires that the excitation frequency is, at least $\sqrt{2}$ times the system's natural frequency. Nearly unity, the system acts not as an insulator, but as an amplifier, transmitting the displacements efforts far beyond the originals. This may reduce the value of transmissibility, reducing the natural frequency of the system. As regards damping, transmissibility can also be reduced by decreasing the damping ratio Anyway some cushioning is needed to avoid infinitely large vibration amplitudes when the system passes through resonance.

We are interested now on applying the hysteresis Bouc-Wen model to a machine supported on an inertia-block, with vibration isolated from the foundation (see Fig 4).

Writing the equations for the machine and the hysteresis model we get

$$\begin{aligned} m_1\ddot{x}_1 + k_1(\alpha(x_1 - x_2) + (1-\alpha)z) &= F \\ \dot{z} &= A\dot{x} - \beta|\dot{x}||z|^{n-1}z - \gamma\dot{x}|z|^n \\ m_2\ddot{x}_2 + c_2\dot{x}_2 + k_2x_2 - k_1(\alpha(x_1 - x_2) + (1-\alpha)z) &= 0 \end{aligned} \tag{2}$$

where subindex 1 is related to the machinery and subindex 2 to the foundation.

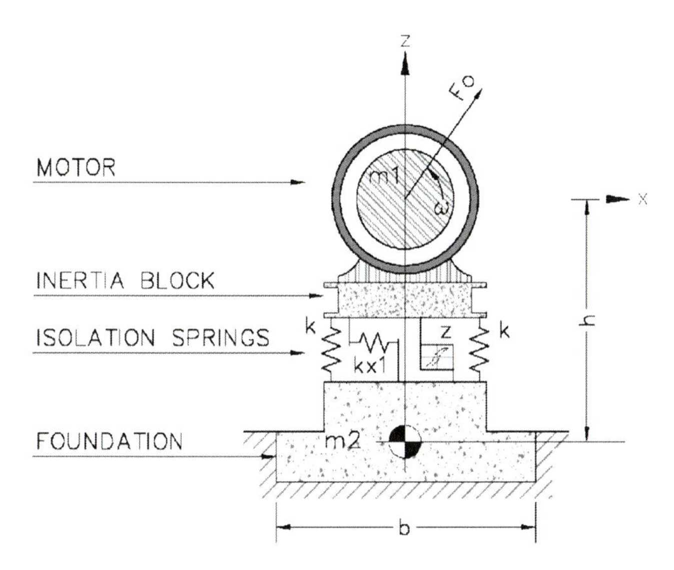

Fig. 4 Scheme of the machine-isolation blocks

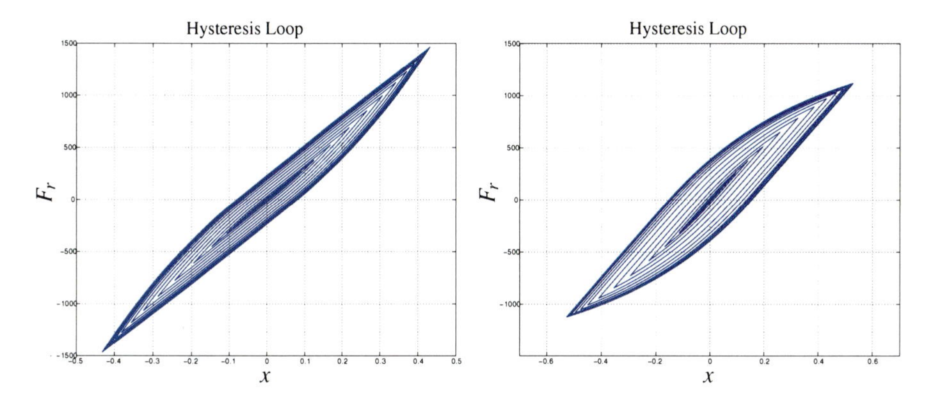

Fig. 5 Left: hard hysteresis. Right: soft hysteresis

2.2 *Numerical Simulations and Bifurcations*

We are now interested on the numerical simulation of the Bouc-Wen model given by Eq.2. Parameters β and γ vary in the ranges corresponding to Fig. 6. The hysteresis effect can be soft or hard (see Fig. 5). Since the second equation involves an absolute value, care must be taken in the numerical simulations. Being a non-smooth system, an event-driven scheme was used.

keeping constant all parameters of the model, but β and γ (related to the hysteresis) we found periodic and quasiperiodic orbits (see Fig. 6).

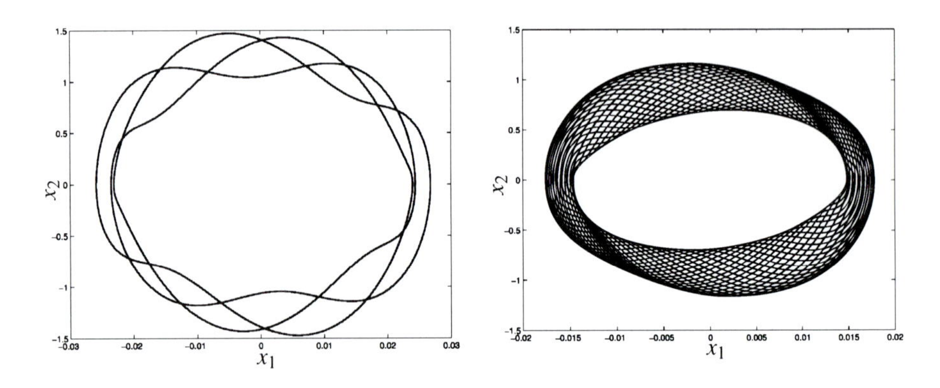

Fig. 6 Periodic and quasiperiodic orbits

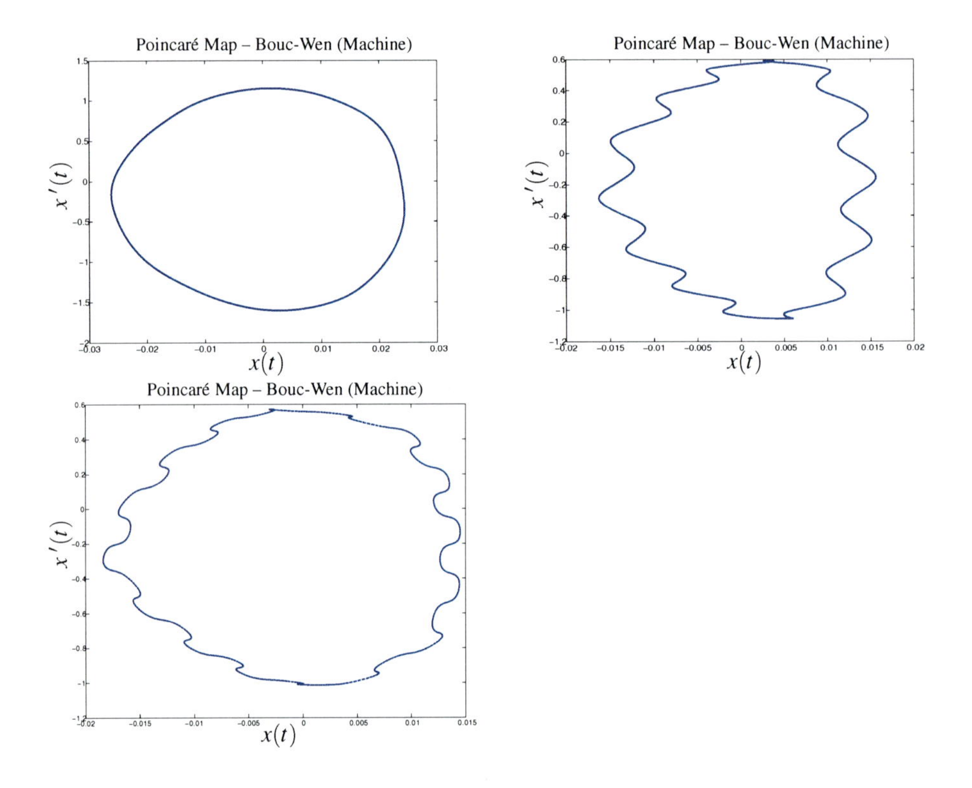

Fig. 7 Quasiperiodic evolution to torus destruction $\beta = 45$, $\gamma = 44.00, 44.50, 44.67$

Numerical simulations of this model show that there is a transition from periodicity to quasiperiodicity, and quasiperiodic motion evolves to torus destruction in a well-known route to chaotic motion (see Fig. 7).

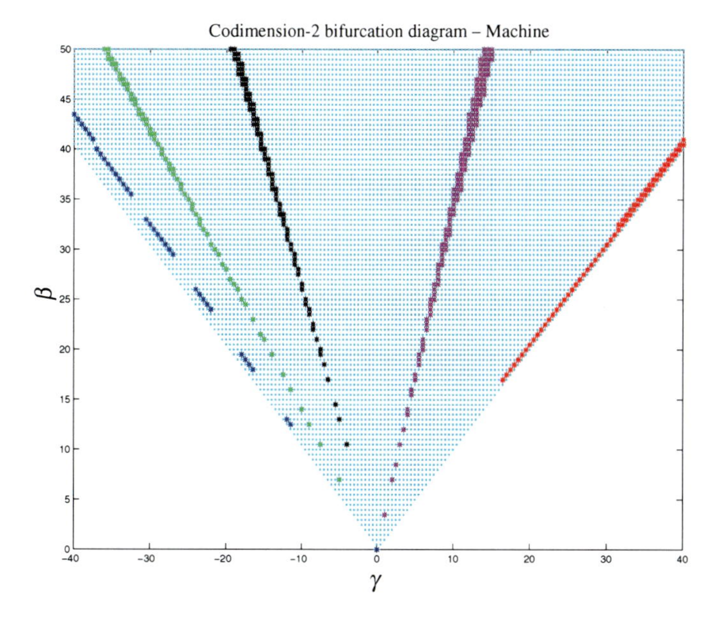

Fig. 8 Two-dimensional bifurcation diagram showing a codimension-two big-bang bifurcation point

A two-dimensional bifurcation diagram shows that there exists a codimension-two point in the parameter space where rays of periodicity converge. This codimension-two point has been named a big-bang bifurcation [3]. Close to this point periodic orbits of all periodicities exist, and thus it is a source of quasiperiodic and chaotic phenomena (see Fig. 8).

3 Conclusion

A Bouc-Wen model, which corresponds to mechanical systems with hysteresis, such as in seismic effects in buildings, has been numerically simulated. Simulations show that a codimension-two big-bang bifurcation point exists in the parameter space, where periodic orbits coexist. This is a source of aperiodic motion. Future work is directed towards chaos and quasiperiodicity control through isolators. Also, an extended model which includes coupling of vibrating machinery and the human body will be studied.

Acknowledgements. G. Olivar acknowledges financial support from DIMA, CeiBA Complexity and Vicerectoría de Investigación from Universidad Nacional de Colombia.

References

1. Al-Bender, F., Symens, W.: Dynamic characterization of hysteresis elements in mechanical systems I. American Institute of Physics 15, 1–11 (2005)

2. Arya, S., O'Neill, M., Pincus, G.: Design of Structures and Foundations for Vibrating Machines. Gulf Publishing Company (1979)
3. Avrutin, V., Schanz, M., Banerjee, S.: Codimension-three bifurcations: Explanation of the complex one-, two-, and three-dimensional bifurcation structures in nonsmooth maps. Physics Review E 75, 066205 (2007)
4. Christopoulos, C.: Frequency Response of Flag-Shaped Single Degree-of-Freedom Hysteretic Systems. Journal of Engineering Mechanics (ASCE) 8, 894–903 (2004)
5. Chung, S.-T., Loh, C.-H.: Identification and Verification Different Hysteretic Models. Journal of Earthquake Engineering 6, 331–355 (2002)
6. Clarke, R.P.: Non-Bouc Degrading Hysteresis Model for Nonlinear Dynamic Procedure Seismic Design. Journal of Structural Engineering (ASCE) 2, 287–291 (2005)
7. Hornig, K.H.: Parameter Characterization of the BoucWen Mechanical Hysteresis Model for Sandwich Composite Materials by Using Real Coded Genetic Algorithms. Auburn University (2003)
8. Ikhouane, F., Rodellar, J.: On the Hysteretic Bouc-Wen Model, Part I: Forced Limit Cycle Characterization. Nonlinear Dynamics 42, 63–78 (2005)
9. Ikhouane, F., Rodellar, J.: On the Hysteretic Bouc-Wen Model, Part II: Robust Parametric Identification. Nonlinear Dynamics 42, 79–95 (2005)
10. Ikhouane, F., Rodellar, J., Hurtado, J.E.: Analytical Characterization of Hysteresis Loops Described by the Bouc-Wen Model. Mechanics of Advanced Materials and Structures 13, 463–472 (2006)
11. Ikhouane, F., Rodellar, J., Hurtado, J.E.: Variation of the hysteresis loop with the Bouc-Wen model parameters. Nonlinear Dynamics 48, 361–380 (2007)
12. Marano, G.C., Greco, R.: Damage and Ductility Demand Spectra Assessment of Hysteretic Degrading Systems Subject to Stochastic Seismic Loads. Journal of Earthquake Engineering 10, 615–640 (2006)
13. Song, J., Der Kiureghian, A.: Generalized Bouc-Wen Model for Highly Asymmetric Hysteresis. Journal of Engineering Mechanics (ASCE) 6, 610–618 (2006)

Complex Nonlinear Response of a Piecewise Linear Oscillator: Experiment and Simulation

James Ing, Ekaterina Pavlovskaia, and Marian Wiercigroch

Abstract. In this work an experimental piecewise linear oscillator is presented. This consists of a linear mass-spring-damper undergoing intermittent contact with a slender beam, which can be modelled as providing stiffness support only. Experimental bifurcation diagrams are presented in which a complex response is observed. Smooth bifurcations are recorded, but more typically rapid transitions between attractors were observed. All of the following are shown to occur: coexisting attractors, basin erosion associated with grazing trajectories, pairs of unstable periodic orbits, chaotic response, loss of stability through grazing and saddle node bifurcations. All of these come into play to generate the observed experimental bifurcation scenarios. It is shown that a global analysis is important in understanding the system response, especially close to grazing conditions. Although grazing is known to cause a local change in stability, it was found that more often grazing of one orbit would cause a change in the other orbits which resulted in boundary crisis and annihilation of the attractor.

1 Introduction

In mechanical systems impacts or intermittent contacts are very common. This results in overall nonlinearity of the system response, as well as more exotic bifurcations associated with the nonsmoothness of the describing equations of motion. Much work has been done on the nature of the normal forms for both hard (instantaneous) [1, 2] and soft (yielding) impacts [3, 4]. Classification of the bifurcations has shown that both smooth bifurcations are possible, along with more exotic ones that are unique to nonsmooth systems. This is associated with the singularity in the trace of the Jacobian matrix when grazing is approached from one side. In previous work

---

James Ing · Ekaterina Pavlovskaia · Marian Wiercigroch
Centre for Applied Dynamics Research, School of Engineering,
King's College, University of Aberdeen, Aberdeen, AB24 3UE, U.K.
e-mail: `j.ing@abdn.ac.uk`

G. Stépán et al. (Eds.): Dynamics Modeling & Interaction Cont., IUTAM BOOK SERIES 30, pp. 135–143.
springerlink.com

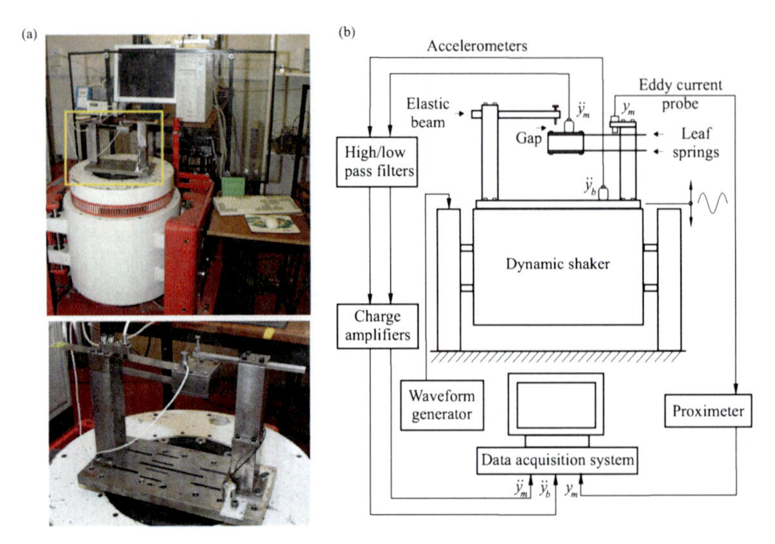

Fig. 1 Photographs (a) and schematic (b) of the experimental rig. Adopted from [5]

it has been shown that the singularity is restricted to the systems with a discontinous force [6], therefore abrupt bifurcations typical for nonsmooth systems should be be caused by smooth, albeit possibly nonlocal bifurcation. This paper shows the mechanism for the abrupt bifurcations observed experimentally, and shows how the stability is eroded through grazing events. The paper is organised as follows: Section 2 introduces the experimental setup, section 3 the mathematical model, section 4 shows the experimental bifurcation structure and that from the corresponding model, finally conclusions are given in section 5.

2 Experimental Setup

The experimental rig was designed and built at the University of Aberdeen [7, 8]. It consists of a block of mild steel supported by parallel leaf springs. These provide the primary stiffness while preventing the mass from rotation. The secondary support consists of a beam, mounted on a separate column, which prevents large displacements of the mass. The gap is controlled by an adjustable bolt mounted on the beam. Harmonic excitation of the system via the base is then generated by an electomagnetic shaker. For further design details see [7, 8]. Measurement is conducted to monitor the displacement of the mass, and the base and mass accelerations. The time history is then plotted, and the Savitsky-Golay algorithm used for polynomial smoothing of the data. As a by product of this the first derivative is available, which enables direct plotting of the phase portrait. The base excitation is monitored precisely in order that stroboscopic Poincaré sections can be constructed and, in turn, bifurcation diagrams.

3 Mathematical Model

The secondary support can be considered as massless, and damping can be neglected due to the short time of contact (both of these can be measured and justified experimentally). The system then becomes piecewise linear, see figure 2(a). The subspaces are divided by the impact surfaces which occur at $x = e$. The nondimensionalised equation of motion is then given by,

$$\begin{aligned} x' &= v \\ v' &= \omega^2 a \sin(\omega\tau) - 2\xi v - x - \beta \begin{cases} 0 & x \le e \\ (x-e) & x \ge e \end{cases} \end{aligned} \tag{1}$$

where $x = y/y_0$ is the nondimensionalised vertical displacement of the mass, $v = x'$ is the nondimensionalised velocity, $\tau = \omega_n t$ is the nondimensional time, $\omega_n = \sqrt{k_1/m}$, $\beta = k_2/k_1$ is the stiffness ratio, $e = g/y_0$ is the nondimensional gap, $a = A/y_0$ and $\omega = \Omega/\omega_n$ and the nondimensional forcing amplitude and frequency, $\xi = c/2m\omega_n$ is the damping ratio, $y_0 = 1$ mm, $'$ denotes differentiation with respect to τ. For the system under consideration there are two discontinuity surfaces which, along with the zero velocity surface, separate the phase space into two subspaces. These are detailed in figure 2(b). Additionally this shows a possible arrangement of local maps, which are formed from the solutions of each linear equation in the subspace in which it is valid, and the concept of a global map is introduced, which is an appropriate composition of the local maps to construct the flow of the underlying differential equation. Local maps can be between the discontinuity surfaces, or onto the stroboscopic surface into order to obtain the Poincaré map, and the number and order of the maps is not known *a priori*, unless a particular solution is being sought. Periodic solutions of a particular period are found

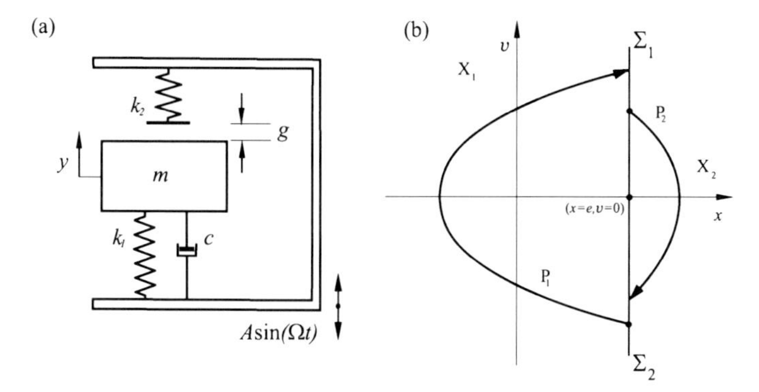

Fig. 2 A piecewise linear model of the stiffness (a) allows local solutions to be constructed, which are then mapped between the subspaces to form a global solution (b)

via the Newton method. The stability can be determined from the Jacobian which is composed of the Jacobians of each of the local maps, for further details see [8].

4 Results

In order to get a flavour of the response, a parameter plot was constructed to show the first nonlinear resonance, the grazing conditions of the linear system, and the types of behaviour near grazing for the 1-sided system. This is shown in figure 3, where the solid line shows the parameters for a grazing trajectory. Each colour represents a different period of response. The initial conditions for figure 3 were chosen from the response of the linear oscillator corresponding to a grazing trajectory roughly in the centre of the parameter plot. Due to the co-existence of solutions close to grazing it was desirable to choose initial conditions not too far from a grazing trajectory in order to see a response more typical of that for a system driven slowly towards grazing. Although some typical structures like period doubling can be seen, figure 3 shows most clearly that there are many discontinuous jumps in the system response,

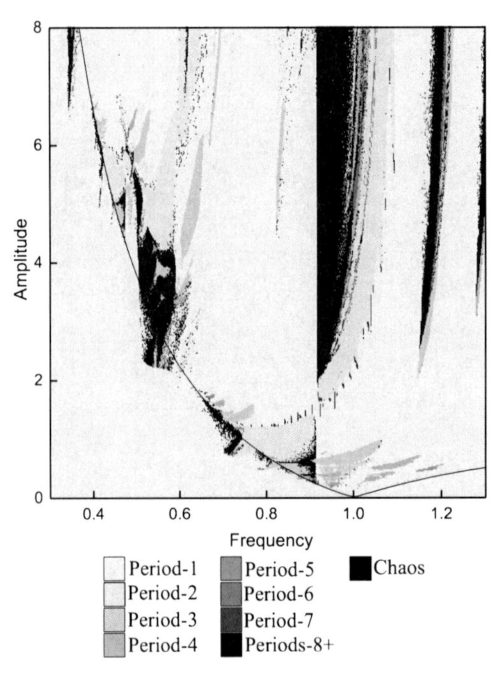

Fig. 3 Parameter plot constructed for the initial conditions, (-0.041,-0.47) showing the periodicity of the resulting response according to shade. The solid line indicates parameters for which the resulting linear response is grazing. The complexity of the response and the lack of clear boundaries between the different motions very close to grazing indicate the complexity of the phase plane for these parameters.

especially near grazing trajectories. Some of these will be probed experimentally and numerically in more detail in the following discussion.

Figure 4 shows an experimental and a numerical bifurcation diagram which is a cross-section through figure 3 at the following parameters; $\beta = 29$, $\xi = 0.01$, $a = 0.47$, $e = 1.26$ and $\omega_n = 9.38$ Hz. The response shows abrupt transitions from nonimpacting period-1 to period-4 then period-3 followed by a jump to period-2 and a reverse period doubling. Since the system has continuous eigenvalues, these transitions must be understood using conventional bifurcation theory. A closer look near the grazing of the period-1 orbit, and plotting all the periodic orbits in this region reveals that the attractors converge near this point. The local stretching which

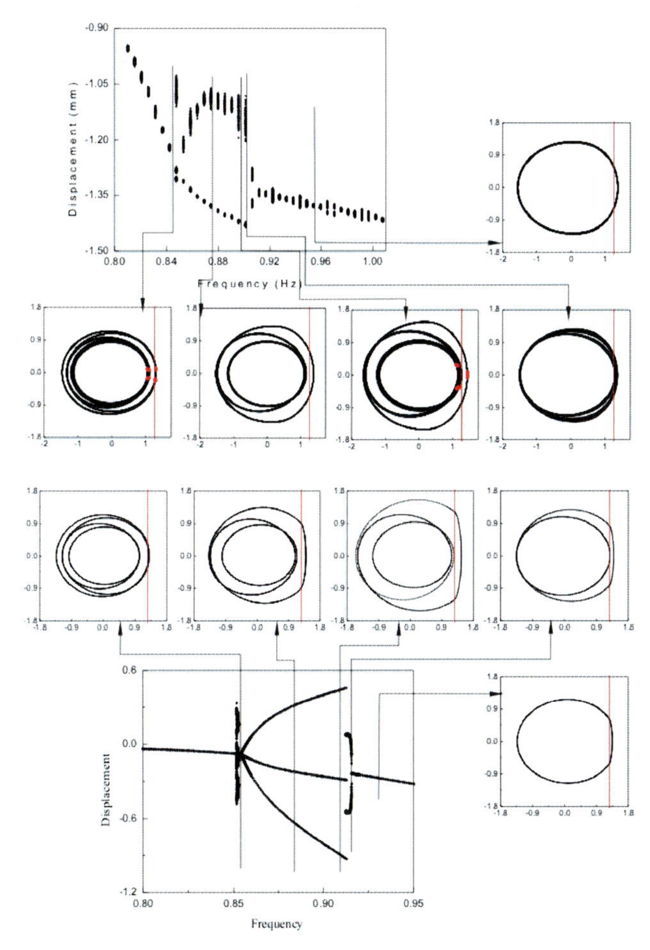

Fig. 4 Experimental and numerical bifurcation diagram showing the response of the system as it passes through grazing, $a = 0.47$. Both cases show transitions from nonimpacting period-1 to period-4 then period-3 followed by a jump to period-2 and a reverse period doubling. Adopted from [10].

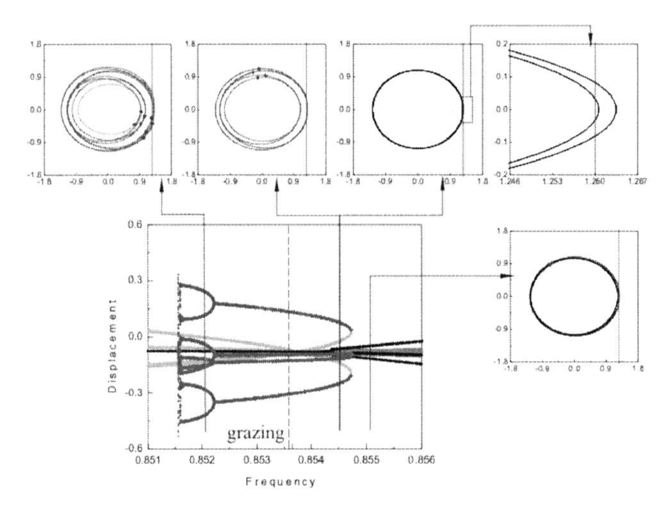

Fig. 5 Numerical bifurcation diagram showing the response of the system as it passes through grazing. The unstable period-3 and period-4 orbits, shown in different shades of light grey, converge onto the period-1 orbit at grazing which results in the transitions observed experimentally. Stable period-1 and period-3 orbits are shown in black and the stable period-4 orbit in dark grey.

occurs due to the impact results in a loss of the structural stability of the orbit, and as the period-1 orbit begins to impact, it undergoes a smooth period doubling bifurcation which results in a boundary crisis and a jump to the coexisting period-4 attractor. This is caused by the convergence of the unstable periodic orbits into the neighbourhood of the period-1 orbit. This scenario is repeated for the period-3 orbit and the trajectory must end up on the period-2 orbit, which then undergoes a smooth period doubling bifurcation.

A different cross-section through figure 3 is shown in figure 6. In this case there is an short interval of chaos followed by a period-2 attractor which loses stability and again a short interval of chaos is observed before another period-2 attractor is seen, which undergoes a reverse period doubling bifurcation. The transitions near grazing can be understood from examination of the period-3 orbits shown in figure 7. Although born in a saddle-node bifurcation far from grazing, it still influences the dynamics since the unstable orbit converges on the grazing period-1 orbit and results in the loss of its structural stability. Since the period-3 orbit itself is unstable the response is trapped on the unstable manifold of the saddle, which explains the transition to chaos observed experimentally. This is the general case for all experimental scenarios encountered, and therefore it is clear that a global analysis is required to understand the dynamics of such systems, especially with a view to engineering applications since boundary crisis can be a dangerous bifurcation.

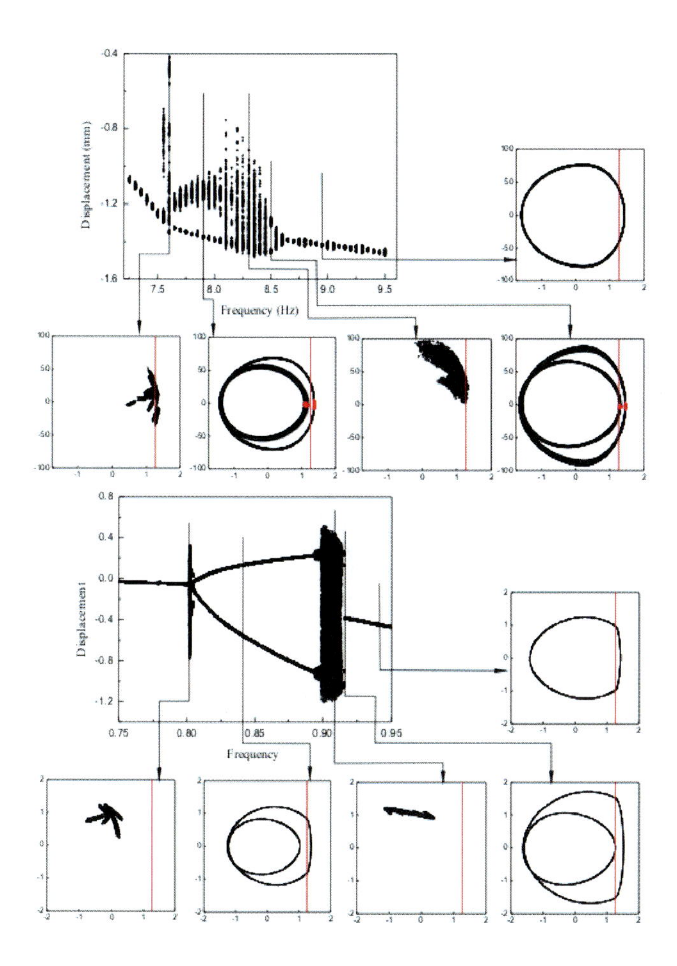

Fig. 6 Experimental and numerical bifurcation diagram showing the response of the system as it passes through grazing, $a = 0.7$. Both cases show a narrow band of chaos close to the grazing period-1 orbit, followed by a period-2 orbit, another interval of chaos, and a period-2 which reverse period doubles. Adopted from [10].

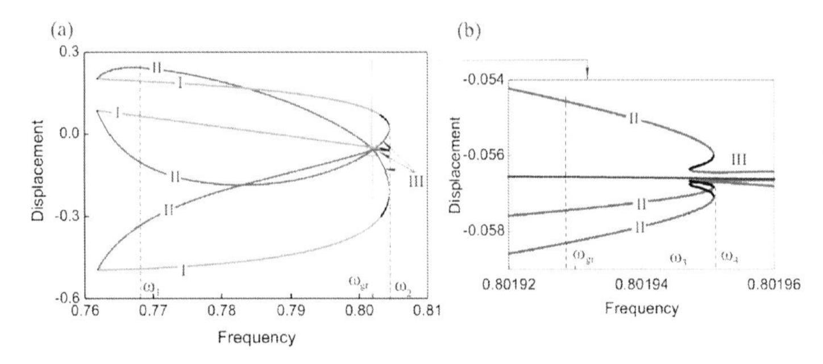

Fig. 7 Evolution of periodic orbits. Stable period-3 orbits are shown in black, unstable in light and dark grey. The period-1 orbit is shown in very dark grey. Adopted from [9].

5 Conclusions

A piecewise linear oscillator was presented. Experimental and numerical studies have shown a loss in the stability near grazing, which was caused not by a local instability, but by the global dynamics. Specifically coexisting periodic orbits are shown to be common for all parameters studied. They can be stable, resulting in a transition to a different periodic orbit, or unstable, resulting in a chaotic response as the orbit is trapped on the unstable manifold of the saddle. These bifurcations are dangerous, since they cannot be reversed by a small parameter change once the orbit is established. Therefore a global analysis is systems undergoing impact is recommended since local analysis is not sufficient.

References

1. Nordmark, A.B.: Non-periodic motion caused by grazing incidence in an impact oscillator. Journal of Sound and Vibration 145, 279–297 (1991)
2. Chin, W., Ott, E., Nusse, H.E., Grebogi, C.: Grazing bifurcations in impact oscillators. Phys. Rev. E 50, 4427–4444 (1994)
3. Ma, Y., Agarwal, M., Banerjee, S.: Border collision bifurcations in a soft impact system. Physics Letters A 354(4), 281–287 (2006)
4. Banerjee, S., Grebogi, C.: Border collision bifurcations in two-dimensional piecewise smooth maps. Physical Review E 59, 4052–4061 (1999)
5. Pavlovskaia, E., Ing, J., Wiercigroch, M., Banerjee, S.: Complex dynamics of bilinear oscillator close to grazing. International Journal of Bifurcation and Chaos (to appear)
6. Ma, Y., Ing, J., Banerjee, S., Pavlovskaia, E., Wiercigroch, M.: The nature of the normal form map for soft impacting systems. International Journal of Non-Linear Mechanics 43(6), 504–513 (2008)
7. Wiercigroch, M., Sin, V.W.T.: Experimental study of a symetrical piecewise base-excited oscillator. Journal of Applied Mechanics 65, 657–663 (1998)
8. Ing, J., Pavlovskaia, E., Wiercigroch, M., Banerjee, S.: Experimental study of impact oscillator with one-sided elastic constraint. Phil. Trans. R. Soc. 366, 679–704 (2008)

9. Banerjee, S., Ing, J., Pavlovskaia, E.E., Wiercigroch, M., Reddy, R.K.: Invisible grazings and dangerous bifurcations in impacting systems: the problem of narrow-band chaos. Physical Review E 79, 037201 (2009)
10. Ing, J., Pavlovskaia, E., Wiercigroch, M., Banerjee, S.: Bifurcation analysis of an impact oscillator with a one-sided elastic constraint near grazing. Physica D 239(6), 312–321 (2010)

Optimization of a Vibrating Screen's Mechanical Parameters

Béla Csizmadia, Attila Hegedűs, and István Keppler

Abstract. The efficiency of sizing, the energy consumption, noise and vibrational pollution is highly affected by the vibrational parameters of screens. To allow the smooth and energy efficient functioning of the screen, frequency optimization is inevitable. In this paper, the optimal vibrational parameters were determined partly by using analytical methods. The analytical model deals only with the movement of one spherical grain in a non-inertial reference frame. By using the analytical model, an optimal frequency range for screening was determined, but with this method it is still not possible to take into account the interaction between the particles. In order to deal with this problem, discrete element method was used to model the vibrational screen, and to take into account the collision between particles. By using the discrete element model, the domain of optimal vibrational parameters for efficient screening procedure were more precisely determined.

Keywords: vibrating screen, optimal vibrating frequency, granular material, sizing, discrete element method.

1 Introduction

Sizing of granular materials is an important task in many fields of engineering. Not only in agricultural engineering, but also in mining-, building- and chemical industry. This paper will focus only on vibrating screens [1]. The most important part of such a machine is a plane screen, which is in alternating motion, and the granules falling on the screen are being moved as well.

There are two types of screening: one is the so called *picking* and the other is *cleaning*. We will focus our attention to cleaning.

Béla Csizmadia · Attila Hegedűs · István Keppler
Szent István University, Faculty of Engineering,
H-2103 Gödöllő, Páter Károly u. 1, Hungary
e-mail: keppler.istvan@gek.szie.hu

G. Stépán et al. (Eds.): Dynamics Modeling & Interaction Cont., IUTAM BOOK SERIES 30, pp. 145–152.
springerlink.com

Until now, the determination of the optimal vibrating frequency is done by experiment. This renders the screen design to be more difficult and expensive. In this paper, the optimal vibrating frequency will be determined by using analytical and numerical methods. The purpose of vibrating frequency optimization is to allow the smooth and energy efficient functioning of the screen.

First, by using analytical methods, the optimal vibrating parameter region for one grain will be determined and from this region the most efficient point will be chosen by using numerical methods. For this we have to take into account the collision between the grains, and the screen's walls also.

2 The Efficient Screening

By analyzing the motion of one spherical grain in two dimensions it is possible to determine *analytically* the conditions for efficient screening. Two conditions must be taken into account: the grain must *slide* downwards on the screen, not to bounce on it, and most of the contaminants must *fall through* the slots of the screen.

2.1 Sliding without Bouncing

The motion of a grain on the screen can be sliding forward (in the direction of the forwarding), sliding for- and backwards without leaving the screen plate, or by bouncing on the screen. To make the screening efficient, the motion must be sliding without bouncing. We proved previously in [2] by analyzing the relative motion of a spherical grain on a screen alternating by the law $A\sin\omega t$ and in an incline α to the horizontal that types of the motion can be grouped by using the equations

$$\frac{A\omega^2}{g} < \cot\beta, \tag{1}$$

$$\frac{A\omega^2}{g} > -\frac{\sin\beta - \mu_0\cos\beta}{\cos\beta + \mu_0\sin\beta}, \tag{2}$$

$$\frac{A\omega^2}{g} < \frac{\mu_0\cos\beta + \sin\beta}{\cos\beta - \mu_0\sin\beta}. \tag{3}$$

Here A is the amplitude of the screen motion, ω is its frequency, $\alpha = -\beta$ is the angle of the screen to the horizontal, μ_0 is the coefficient of static friction and β is the direction of the oscillation.

Using equations (1), (2) and (3) the domains for the different types of motion can be determined (*fig.* 1). At zone I., there is no motion at all, the grain is in relative rest on the moving screen. Zone II. means upward sliding, zone III. means sliding up and downwards, in zone IV. the particle is bouncing. Zone III. is the domain representing efficient screening.

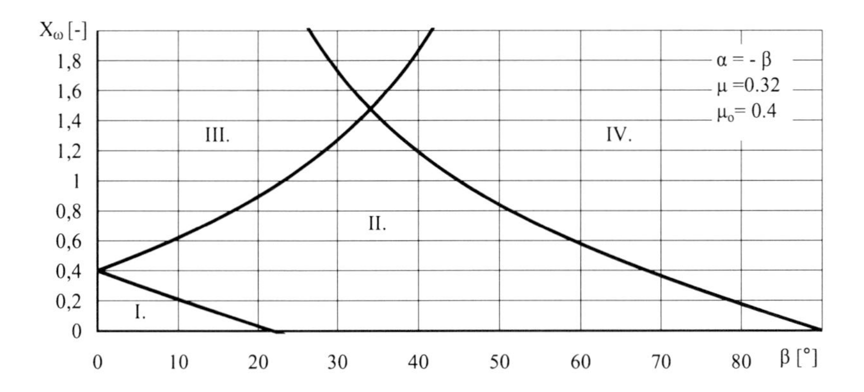

Fig. 1 Domains for different types of motion: at zone I., there is no motion at all. Zone II. means upward sliding, zone III. means sliding up and downwards, in zone IV. the particle is bouncing.

2.2 *Falling through the Screen*

The alternating sliding motion on the screen is not enough for efficient screening, the contaminants must fall out the screen. This will give an other barrier for the maximal relative velocity of the grain [4], [5].

Suppose, that at time $t = 0$ the grain is in relative rest on the plane. If condition

$$\frac{A\omega^2}{g} > \frac{\sin\alpha + \mu_0 \cos\alpha}{\cos\beta + \mu_0 \sin\beta} \tag{4}$$

becomes valid at a given time t_1 the grain starts to slide downwards on the screen. t_1 can be determined by using equation

$$\sin\omega t_1 = \frac{g}{A\omega^2} \frac{\sin\alpha + \mu_0 \cos\alpha}{\cos\beta + \mu_0 \sin\beta}. \tag{5}$$

From this time the velocity of the relative motion is (using $\rho = \arctan\mu_0$):

$$v_r = \dot{x} = -\frac{\sin(\alpha+\rho)}{\cos\rho} g(t - t_1) + \frac{\cos(\beta-\rho)}{\cos\rho} A\omega\left(\cos\omega t_1 - \cos\omega t\right). \tag{6}$$

The maximal relative velocity can be determined by deriving equation (6):

$$\dot{v}_r = -\frac{\sin(\alpha+\rho)}{\cos\rho} g + \frac{\cos(\beta-\rho)}{\cos\rho} A\omega^2 \sin\omega t = 0. \tag{7}$$

From here, the time t_2 becomes

$$\sin \omega t_2 = \frac{\sin(\alpha + \rho)}{\cos(\beta - \rho)} \frac{g}{A\omega^2}, \tag{8}$$

and using this time, the maximal velocity is

$$v_{r\max} = -\frac{\sin(\alpha + \rho)}{\cos \rho} g(t_2 - t_1) + \frac{\cos(\beta - \rho)}{\cos \rho} A\omega(\cos \omega t_1 - \cos \omega t_2). \tag{9}$$

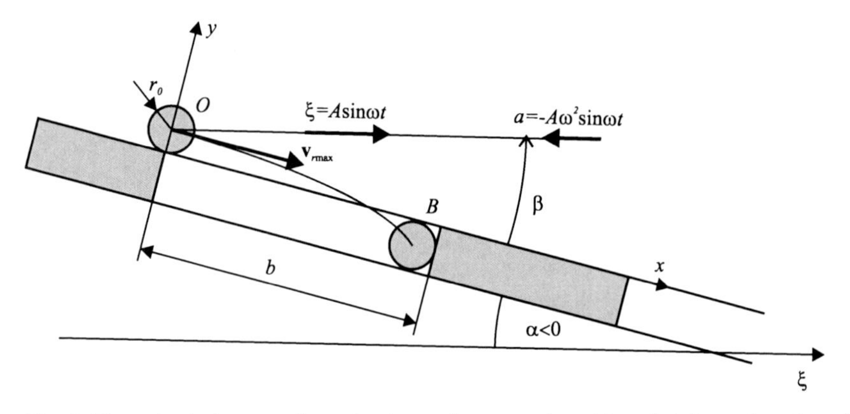

Fig. 2 The spherical contaminant having radius r_0 and positioned right at the edge of the screen hole must fall through the screen

We suppose, that the spherical contaminant having radius r_0 is positioned right at the edge of the screen hole (*fig.* 2). The maximal velocity can be determined by the condition that the contaminant must fall through the screen from this starting position. By using the equations of relative motion and Newton's second law, we get a system of five equations.

We use the following dimensionless parameters before writing the system of five equations to be solved as

$$X_\omega = \frac{A\omega^2}{g} \tag{10}$$

as the most important parameter to be determined containing the frequency of screening and parameters

$$X_1 = \omega t_1, \qquad X_2 = \omega t_2, \qquad X_b = \omega t_b, \qquad X_v = \frac{v_{r\max}}{A\omega}, \tag{11}$$

where t_b is the time when the particle is reaching the screen again. The following parameters will be used:

- parameters depending on the material screened:
  - ρ_g density of the grain,
  - μ_0 coefficient of static friction between the grain and the screen,
  - $\chi = \frac{b}{r_0}$ ratio of screen orifice b and radius of contaminant r_0.

- parameters depending on the machine construction:
  - α angle of screen's inclination to the horizontal,
  - $\kappa = \frac{b}{A}$ the ratio of screen orifice diameter and amplitude of sinusoidal motion.

By using these parameters, the system of five equations determining the maximal relative velocity of the screen are:

$$X_\omega \sin X_1 = \frac{\sin\alpha + \mu_0 \cos\alpha}{\cos\alpha - \mu_0 \sin\alpha}, \tag{12}$$

$$X_\omega \sin X_2 = \tan(\alpha + \rho), \tag{13}$$

$$X_v - \frac{\cos(\alpha+\rho)}{\cos\rho}(\cos X_1 - \cos X_2) = \frac{\sin(\alpha+\rho)}{\cos\rho}\frac{X_1 - X_2}{X_\omega}, \tag{14}$$

and

$$\begin{aligned}&[(\chi-1)\cos\alpha - \sin\alpha](X_b - X_2)^2 - \\ -2X_\omega\{&[(\chi-1)\sin\alpha + \cos\alpha]\cos X_2 + X_v\}(X_b - X_2) + \\ &2X_\omega[(\chi-1)\sin\alpha + \cos\alpha](\sin X_b - \sin X_2) = 0,\end{aligned} \tag{15}$$

$$\frac{2\kappa}{\chi}X_\omega = (X_b - X_2)^2\cos\alpha + 2X_\omega(\sin\alpha\cos X_2)(X_b - X_2) - 2\sin\alpha(\sin X_b - \sin X_2). \tag{16}$$

This system of equation can not be solved manually. The numerical solutions can be drawn in the form of nomograms (*fig.* 3). In a nomogram like this, the value of $X_\omega = \frac{A\omega^2}{g}$ can be found. The nomogram shown on *fig.* 3 is placed into zone III. of *fig.* 1. In this way the minimal and maximal value of ω screening frequency can be determined, where the screening is possible.

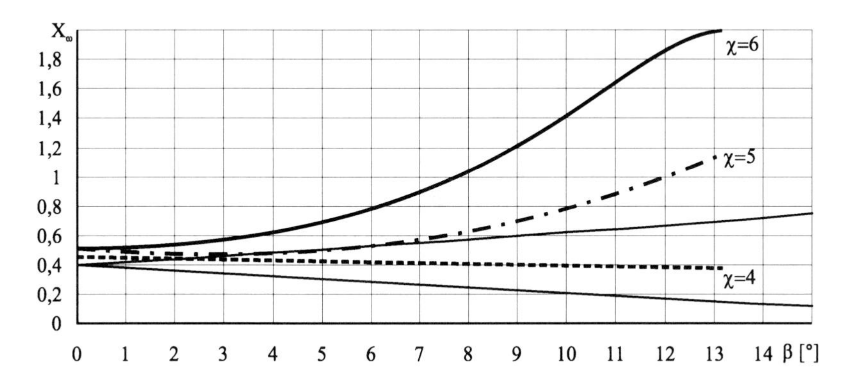

Fig. 3 Nomogram for the minimal and maximal screening frequencies for the case $\kappa = 0.1$, and some points as the parameters used for testing the screen (see at the end of the paper)

The efficiency of the screening procedure can not be analyzed further analytically. For further analysis we have to use a numerical method capable to take into account the three dimensional motion of the particles and the interactions, collisions between them.

3 Discrete Element Model of Screening

The discrete element method is a numerical method for computing the motion of large number of particles based on their equation of motion. Cundall applied it first in rock mechanics [3]. Nowadays two main programs are available to use, one of them is the program called PFC3D (this is useful for theoretical calculations), and EDEM, which seems to be easier to use in the engineering practice (because of easy import of complex geometries). EDEM 2.1 was used for our calculations. By using this method, the followings can be taken into account:

- The problem is not two dimensional.
- The grain particles do collide with each other during the screening procedure.
- The length of the slit on the screen has longitudinal extent also.

The following micromechanical parameters of the granular material were used for the discrete element screening model:

- $\nu = 0.4$ Poisson's ratio of the grain.
- $G = 1.1 \cdot 10^7$ Nm^{-2} shear modulus of the grain.
- $\rho_g = 1500$ kgm^{-3} density of the grain.
- $C_r = 0.6$ coefficient of restitution.
- $\mu_0 = 0.4$ coefficient of static friction.
- $\mu_r = 0.01$ coefficient of rolling friction.

The 3D model of the screen was created. The model screen's geometrical properties can be seen in *fig.* 4

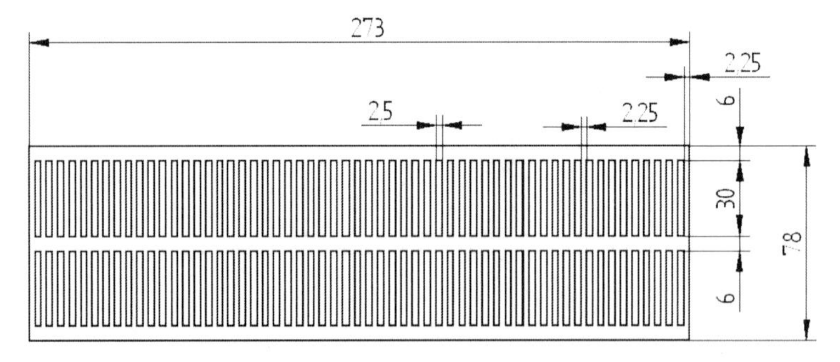

Fig. 4 The geometry of the model screen used for DEM simulations

The contact model was Hertz-Mindlin No-slip contact model with linear cohesion. In the Hertz-Mindlin model, the normal force acting between two particles in contact is calculated as

$$F_N = \frac{4}{3} E \sqrt{R} \delta_n^{\frac{3}{2}}.$$

Here E is the particle's Young modulus, R is its radius, δ_n is the normal overlap. Additionally there is a damping force calculated as

$$F_n^D = -2\sqrt{\frac{5}{6}} \beta_1 \sqrt{s_n m} v_{\mathrm{rel}},$$

where m is the particle mass, v_{rel} is the normal component of the relative velocity and

$$\beta_1 = \frac{\ln c_r}{\sqrt{\ln^2 c_r + \pi^2}},$$

and s_n is the normal stiffness. The simulation time step has a crucial effect of the stability of the model i.e. the use of too high time step cause the "explosion" of the whole model because of large value of δ_n overlap. 25% of the Rayleigh time step

$$T_R = (0.1631\nu + 0.8766)^{-1} \pi R \sqrt{\frac{\rho_g}{G}}.$$

was used as simulation time step.

In our first simulations we tested the validity of the analytical results. We forced the screen to translate sinusoidally, poured on it 100 particles. The parameters used for testing the screen's working can be seen on *fig.* 3.The test simulations were carried out by using the data in *table* 1. The simulations proved the results of analytical calculations. The screens is able to transfer and clean the material poured on it using the frequencies designated within the zone evaluated by analytical method. Some points outside the domain of screening were also tested, these points resulted motions in good correlation with those types of motion listed on *fig.* 1.

We measured the efficiency of screening by defining the screen efficiency factor $e_s = \frac{N_f}{N}$, where N_f is the number of grain fallen out through the orifices on the screen, and $N = 100$ was the total number of particles.

Table 1 Simulation parameters and efficiency results

	β[°]	χ [-]	r_0 [mm]	X [-]	ω $\frac{\mathrm{Rad}}{s}$	f [Hz]	e_s [-]	Var. [-]
1	7	6	0.16 b	0.85	20	3.18	0.79	0.03
2	7	5	0.2 b	0.6	17	2.70	0.71	0.02
3	7	4	0.25 b	0.4	14	2.23	0.76	0.03
4	10	6	0.16 b	1.5	27	4.30	0.78	0.02
5	4	6	0.16 b	0.7	18.5	2.94	0.81	0.03

It can be seen on the table, that angle α of the screen to the horizontal has an important impact on the effectiveness of screening. The reason is simple: β has an influence on the residence time i.e. the time, how long the particle is within the screen box. And as high the residence time, so high the chance of falling through the orifices. The residence time of 100 contaminants varied from $T_r^{\alpha=4^\circ} = 2.5$s to $T_r^{\alpha=10^\circ} = 0.9$s. The high residence time has an other effect on the screening: the performance of the cleaning machine is decreasing with the increasing residence time. By modeling the screening of a grain assembly containing particles having radius $r_0 < r < b$, is was possible to find the optimal angle of screening simply by monitoring the videos made from simulations related to different screen angle values. In this way, we were able to determine the optimal screening angle for the given size of contaminant as $\alpha = 7^\circ$.

4 Results

We determined the optimal working parameters of vibrating screens by the combination of analytical and numerical calculations. By writing the equations of motion of a spherical grain moving on the plate we defined the domain of screening as function of screen angle and the frequency of screening. Within the region of screening we determined highest possible frequency enabled for the screen operation. With these data it was possible to define all the borders of the domain of screening parameters. By using discrete element method, we determined the efficiency of screening procedure for points of screening parameters within this region, and this made it possible to pick the probably most efficient combination of screening parameters.

Acknowledgements. Financial support from the Bolyai Scholarship of the Hungarian Academy of Sciences is gratefully acknowledged by István Keppler Ph.D.

References

1. Letosnyev, M.N.: Mezőgazdasági gépek elmélete. Akadémiai Kiadó, Budapest (1951)
2. Hegedűs, A., Biró, I., Brndeu, L., Deák, L., Orgovici, I.: Some Queestions of the Vibrational Transport and Sizing. In: Bul. IX-a Conf. de Vibr., Timioara (1999)
3. Cundall, P.A., Strack, O.D.L.: A discrete numerical model for granular assemblies. Geotechnique 29, 47–65 (1979)
4. Csizmadia, B., Hegeds, A.: Optimalization the flexible elements of two-mass vibratory screen. Hung. Agr. Eng. 15(2002)
5. Chiriac, A., Bereteu, L., Biro, I., Nagy, R., Boltosi, A.: On the Dynamics of a Plane Sieve for Seed Sorting. On the Dynamics of a Plane Sieve for Seed Sorting, Driven by an Electric Motor, XXXII. MTA -AMB, pp. 261–265. Gödöllő (2008)

Biomechanics and Rehabilitation

These papers cover biomechanical models and applications ranging from the molecular level to whole body human models. The discussed applications include surgery, human chest modeling in automotive simulation systems, and rehabilitation robotics. A. Bibo *et al.* propose an overdamped visco-elastic rigid body model, linked with springs for the motor protein responsible for the contraction of muscle tissue. The paper by W. Lacarbonara presents a model of the cochlea (the part of the inner ear dedicated to hearing) and investigates its nonlinear vibration behavior in detail. The questions of kinematic modeling and parameter estimation of the human knee joint are addressed in the work by I. Bíró. The paper of P. Olejnik showed that by controlling the supporting force acting on the thorax, the relative compression of the chest due to an elastic impact can successfully be minimized. Biomechanical aspects of the sit-to-stand transfer are investigated by V. Pasqui. In her paper a rehabilitation robotic device is presented providing the minimization of the verticalisation effort of elderly.

Internal Lever Arm Model for Myosin II

András Bibó, Mihály Kovács, and György Károlyi

Abstract. Motor proteins are special enzymes capable of transforming chemical energy into mechanical work. One of the best known motor proteins belongs to the family of myosins found in eukaryotic tissues. The most studied — and first discovered — type of myosin is the skeletal muscle myosin (myosin II) which is the motor protein responsible for the contraction of muscle tissue. In the present study we propose an overdamped visco-elastic rigid body model, linked with springs, for myosin II, where an internal lever arm is added to the previously investigated lever arm model. Our model is based on the significant difference of two springs' stiffness, which is responsible for the sequence of steps of the motion (binding to actin vs. lever arm swinging). The model provides exponential time–displacement curves which are consistent with experimental measurements.

1 Introduction

Enzymes play a crucial role in living organisms. These macromolecules take part in chemical reactions where their spatial structure (conformation) might temporarily

András Bibó
Department of Structural Mechanics,
Budapest University of Technology and Economics,
Műegyetem rkp. 3., 1111 Budapest, Hungary
e-mail: biboan@gmail.com

Mihály Kovács
Department of Biochemistry, Eötvös Loránd University,
Pázmány Péter sétány 1/C, 1117 Budapest, Hungary

György Károlyi
Department of Structural Mechanics,
Budapest University of Technology and Economics,
Műegyetem rkp. 3., 1111 Budapest, Hungary
e-mail: karolyi@tas.me.bme.hu

G. Stépán et al. (Eds.): Dynamics Modeling & Interaction Cont., IUTAM BOOK SERIES 30, pp. 155–163.
springerlink.com

change but is retained by the end of the reaction: they can be considered catalysts speeding up certain reactions. Motor proteins, like myosin II, are a special types of enzymes. Besides catalyzing reactions, they transfer energy from chemical to mechanical form with great efficiency. Chemical energy is stored in adenosine triphosphate (ATP), and myosin II helps ATP to release a phosphate ion to become adenosine diphosphate (ADP) while energy is liberated. Myosin II uses this energy to provide muscle contraction.

Human and other skeletal muscles can have different shapes and sizes, but their main structure is the same. Muscle tissue is built up of repeating functional segments called *sarcomeres* [1]. Sarcomeres consist of actin and myosin filaments, that can slide with respect to each other, providing the contraction of muscle, see Fig. 1.

Under a nervous impulse, release of calcium ions frees the binding sites of the actin filaments so that the myosin filaments attach to the them by forming cross-bridges [4]. The cross-bridges themselves are myosin II molecules which, after hydrolysing ATP, can pull the actin filament inward thus shortening the muscle fiber. When the phosphate ion and ADP are released, a fresh ATP binds to the protein, then myosin II releases the actin filament. The cycle starts over when the new ATP is converted to ADP again [4].

The main structure of myosin II molecules can be revealed from experiments. The sliding displacement between the myosin and the actin filament is performed by the rotation of the *lever arm*. To localize the pivotal point of the lever arm, early studies suggested that the actin binding site would be the center of rotation [2]. By now, it has been established that myosin II consists of a *motor domain* that can attach to the actin filament and can bind ATP, and of a *lever arm* that is connected to the myosin filament, and can rotate around 50° with respect to the motor domain [3]. Despite the increasing amount of X-ray crystallographic and other observations, the mechanical force and displacement transduction between the different parts of the motor domain is yet to be clarified [5].

There are many approaches to understand the complicated behavior of myosin II. The enzyme kinetic approach is based on observations concerning the time dependence of the concentration of components in solutions containing isolated components of sarcomeres. Enzyme kinetic descriptions are based on the assumption of distinct chemical states with various transition rates that give the probability of

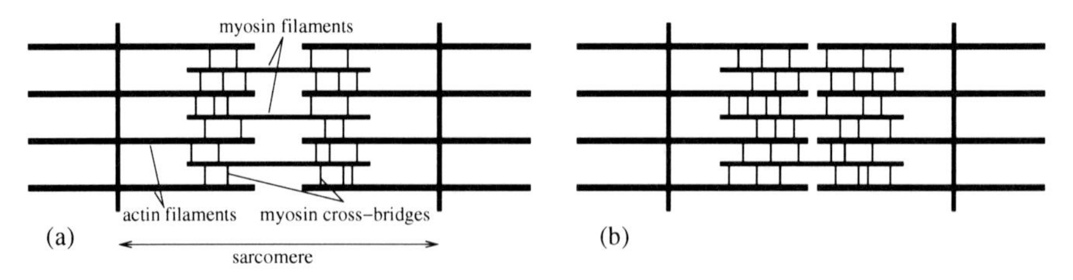

Fig. 1 Sketch of a sarcomere in (a) resting and in (b) contracted state. Myosin II cross-bridges drag the myosin filaments along the actin filaments, which results in a shortening of the sarcomere and hence in the contraction of the muscle tissue.

chemical changes per unit time. Even though this type of models can reveal some mechanical aspects of muscle contraction, they require an arbitrary selection of hypothetical states, which makes their validity quite ambiguous.

Another type of models are mechano-chemical models. Introducing mechanical elements, they try to establish a simple, low degree of freedom mechanical system that describes the behavior of myosin II. The chemical cycles are coordinated by mechanical quantities and vice versa. An example for this type of models is the *elastic lever arm model* [10], which has been set up for myosin V, another member of the family of myosin proteins, responsible for vesicle transport within a cell. By unifying the mechanical and chemical approaches, this model can provide a deeper insight into the complicated events of the enzymatic cycles, meanwhile it relies on the same arbitrary selection of hypothetical states as the enzyme kinetic approach.

Recently, molecular dynamical simulations, taking the dynamics and interactions of almost all the thousands of atoms into account, attempt to give a more detailed explanation of the enzymatic processes of myosin II. Besides the enormous computational demand of these calculations, they are not yet capable of simulating all the myosin molecules acting together in a sarcomere.

In the following Sec. 2, we present a low degree of freedom model that can be used to build up a model to simulate few hundreds of myosin molecules interacting with each other within the sarcomere. Then in Sec. 3 we present some numerical results on the behavior of the model. Finally, in Sec. 4 we draw our conclusions.

2 Internal Lever Arm Model

We propose a new, purely mechanical model for myosin II. This model is based on the lever arm hypothesis, but introduces an *internal lever arm* besides the conventional (external) lever arm. This internal lever arm is inside the myosin motor domain, and it has a major role in splitting the time scales for the motion of the lever arm and for the mechanism that binds the motor domain to the actin filament. The sketch of the model is shown in Fig. 2a. The springs model the interactions within motor domain.

Essentially, the model has three distinct phases according to the relation of the motor domain to the actin filament. The motor domain can either be separated from the actin, it can be binded to a binding site on the actin filament, or it can 'slide' along the filament while trying to reach the binding site. According to its ATP content, the model can either be in an excited state after 'burning' ATP, or it can be in a relaxed state before converting ATP into ADP. The state of a myosin molecule can be described by the following variables: y is the displacement of the motor domain towards the actin filament, φ is the rotation of the lever arm, and x is the displacement of the attachment point of the lever arm to the myosin filament, see Fig. 2a. When the model is in the excited state, the terminal end of spring k_H is lifted by a distance y_0 relative to the motor domain, hence the location of this terminal end is either $\Delta = 0$ (relaxed state) or $\Delta = y_0$ (excited state). A further variable which, as we shall see, is not independent from the previous ones, is y_H that gives the position of the internal lever arm.

In biology, the behavior of myosin II consists of four main steps, called *strokes*, that lead the system through its various stages. Although these strokes can be identified in our model, we will see that they are not completely distinct steps of the motion. The first stroke begins when energy is liberated, that is, when ATP is converted to ADP. To facilitate the storage of energy, the terminal end of spring k_H is shifted upwards by y_0 ($\Delta = y_0$), hence energy is stored now as elastic energy in the stretched spring k_H. During the first stroke, the myosin motor domain is separated from the actin filament, that is, y is smaller than the initial gap between the motor domain and the actin filament. Because spring k_H has become stretched, the system is no longer in equilibrium, and the internal lever arm starts to rotate (y_H grows).

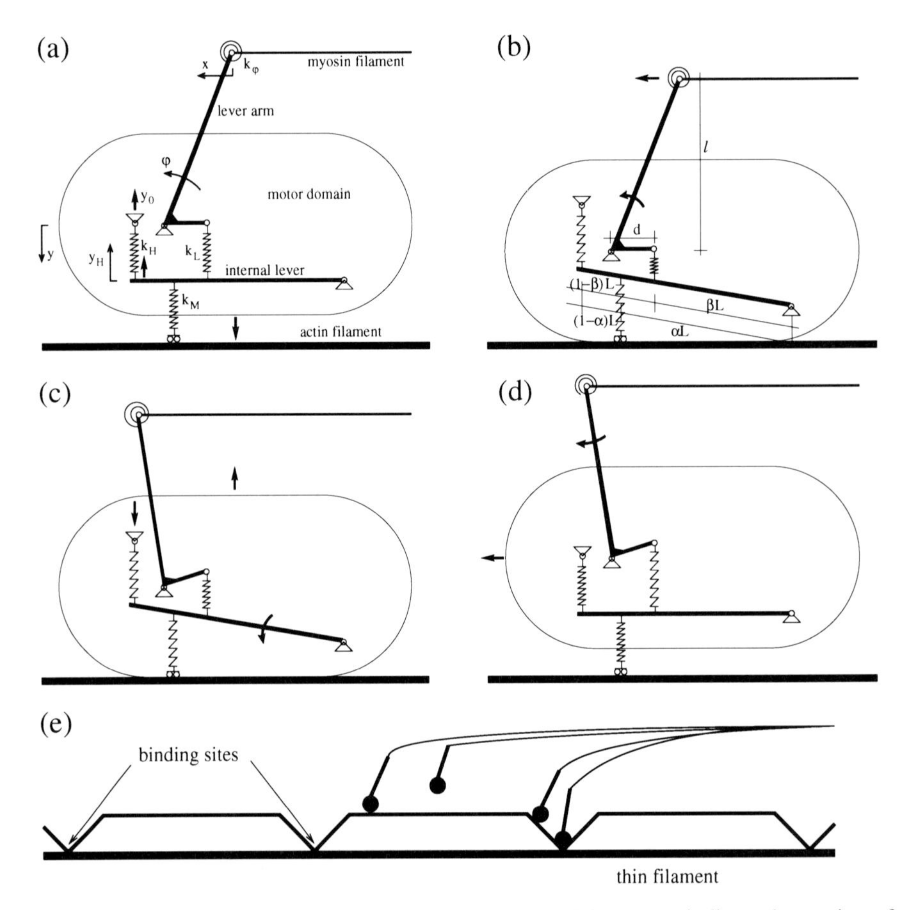

Fig. 2 Sketch of the strokes of the internal lever arm model. Arrows indicate the motion of the system during the following phase. The variables are indicated in (a), where the spring stiffnesses are also shown. The distances are marked in (b). A simplified constraint used to mimic the non-binding part of the actin filament is shown in (e).

This rotation exerts some force on the main lever arm through spring k_L and also pulls the motor domain closer to the actin via spring k_M. The stiffness of the springs can be set such that the motion towards the actin filament is much faster than the motion of the (external) lever arm.

This first stroke of the enzymatic cycle lasts while the motor domain approaches the actin filament (from Fig. 2a to Fig. 2b). However, the motor domain may not reach the actin filament exactly at a binding site, so we assume that there is a constraint that holds the motor domain at a certain distance from the actin filament during sliding (cf. Fig. 2e). This models the lack of attractive force between the non-binding length of the actin filament and the motor domain: between two binding sites the force exerted by spring k_M is cancelled out by the reaction force of the constraint. During this stage the motor domain is assumed to be transported along the actin filament by the other binded myosin II molecules performing the power stroke, see below. If the motor domain is close enough to a binding site, it arrives on an inclined segment of the constraint, the separation gradually decreases, and the motor domain becomes bound to the binding site. This initiates the second stroke of motion.

During the second stroke the motor domain is bound to the actin filament, see Fig. 2c. However, the springs within the motor domain are still triggered, so they are able now to rotate further the main lever arm. This is the power stroke: the main lever arm pulls the myosin filament along the actin filament.

As the springs approach their equilibrium configuration, this conformational change facilitates the molecule to release the phosphate ion and ADP and to get a new ATP. When ATP is bound to myosin, the motor domain can no longer attach to the actin filament. In our model the end of spring k_H is restored to its original position to mimic this conformational change. With a convenient choice of the parameters this step does not change the energy of the system: the compression of the spring in the original state is the same as the tension in the final state, hence

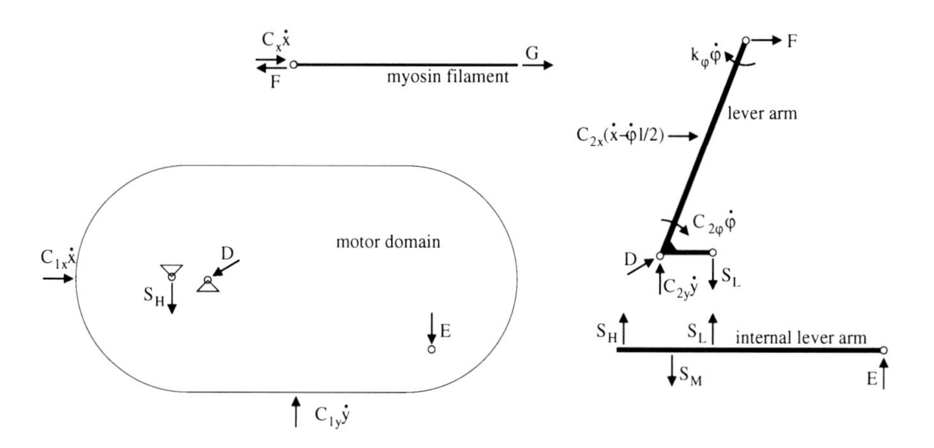

Fig. 3 Sketch of the parts of the model indicating the forces acting on them

the potential energy remains the same. The reset of spring k_H resets the equilibrium configuration of the system, so it starts to move again: it is the third stroke when myosin detaches from actin (from Fig. 2c to Fig. 2d).

The fourth stroke is the recovery stroke, when the original equilibrium configuration is reached again (from Fig. 2d to Fig. 2a). This new conformation, however, allows for the conversion of ATP into ADP, so the whole cycle starts over.

Figure 3 shows the forces that act on each part of the model. Myosin II motion is exceedingly overdamped, the inertial forces are neglected from the equations of motion [6]. In case of the internal lever arm, we assume that both inertial forces and damping are neglected: this item only serves as a force (and time scale) separating device, and, being inside the motor domain, it is assumed to have negligible mass and not to be sensitive to damping. Hence we find from the balance of moments that $S_L \beta L + S_H L - S_M \alpha L = 0$. Here, linearising in displacements, the forces in the springs can be expressed as $S_M = k_M(\alpha y_H - y)$, $S_H = k_H(\Delta - y_H)$ and $S_L = k_L(\varphi d - \beta y_H)$, where Δ is either 0 or y_0, depending on whether the myosin is in excited or in relaxed state. From these formulae we find that the position of the internal lever arm can be expressed by y and φ as

$$y_H = \frac{\beta d k_L \varphi + k_H \Delta + \alpha k_M y}{\alpha^2 k_M + \beta^2 k_L + k_H}. \tag{1}$$

The horizontal velocity of the myosin motor domain is $\dot{x} - \ell\dot{\varphi}$, the vertical velocity is $\dot{y}$, and the corresponding damping forces are $C_{1x}(\dot{x} - \ell\dot{\varphi})$ and $C_{1y}\dot{y}$, dot indicates derivation with respect to time. For the lever arm, we assume a horizontal damping force $C_{2x}(\dot{x} - \dot{\varphi}\ell/2)$, a vertical one $C_{2y}\dot{y}$, and one against rotation $C_{2\varphi}\dot{\varphi}$. The myosin filament is damped by $C_x\dot{x}$. The horizontal and vertical equations of motion for the motor domain–lever arm complex can be written as:

$$\begin{aligned} R_x - F - (C_{1x} + C_{2x})\dot{x} + \left(C_{1x} + \frac{1}{2}C_{2x}\right)\ell\dot{\varphi} &= 0, \\ S_M - R_y - (C_{1y} + C_{2y})\dot{y} &= 0, \end{aligned} \tag{2}$$

where R_x and R_y are the components of the reaction force between the constraint and the motor domain, and F is the force between the myosin filament and the lever arm. The equations for the rotation of the lever arm around its pivotal point and the dragged myosin filament are

$$-S_L d - k_\varphi \varphi - F\ell - C_{2x}\frac{\ell}{2}\dot{x} + \left(C_{2x}\frac{\ell^2}{4} - C_{2\varphi}\right)\dot{\varphi} = 0, \quad F - G - C_x\dot{x} = 0, \tag{3}$$

where G is the load carried by the myosin filament. The unknowns in this set of equations are the variables x, y and φ and the forces F, R_x and R_y. Hence Eqs. (2–3) must be complemented by the following conditions:

- If the motor domain is not on the constraint (far from actin), the reaction forces are zero: $R_x = R_y = 0$.

- If the motor domain is on the constraint, the reaction force is perpendicular to the surface, and the motor domain must remain on the surface: $R_x = DR_y$ and $\dot{y} = D(\dot{x} - \ell\dot{\varphi})$ where D is the slope of the surface at the motor domain.
- If the myosin motor domain is attached to the binding site on the actin filament, we have the kinematic constraints $\dot{y} = 0$ and $\dot{x} = \ell\dot{\varphi}$.

These conditions make Eqs. (2–3) solvable for each case. In our model, when y reaches the constraint, the stage of sliding starts. If during sliding the binding site is reached, binding occurs. If the reaction force R becomes attractive, the motor domain lifts off the constraint. The binding state ends when R becomes attractive, then either sliding along the constraint or lift off from the filament occurs, depending on the direction of R.

3 Simulation Results

Equations (2–3) with the constraints, give a piecewise linear set of first order differential equations. We solve the equations using the fourth order Runge-Kutta method with a small fixed step-size. As an example, Fig. 4a shows how the vertical displacement y of the motor domain and the rotation φ of the (external) lever arm change with time in case of an unloaded myosin filament, $G = 0$. Due to the negligible inertial forces and the large damping the velocities are not continuous, they contain sudden jumps between the different stages. These discontinuities in the velocity cause the breakpoints in Fig. 4a. Between breaks, there are exponential curves.

As seen in Fig. 4a, that only about 30 % of the lever arm rotation φ is performed while the motor domain reaches the actin filament: the solid plateaus at 4.1 nm indicate the completion of the attachment to actin, while the dotted line shows the much slower approach of the equilibrium configuration by the (external) lever arm.

During the first cycle a transient motion can be observed (Fig. 4b). After the motor domain reaches the constraint at around 0.01 ms it moves towards the

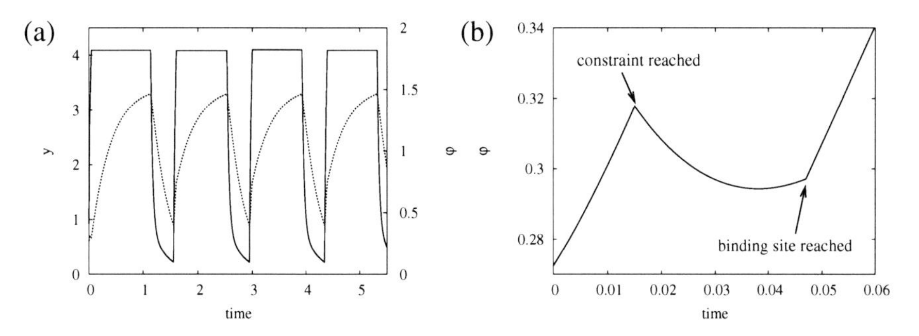

Fig. 4 (a) Time-dependence of the vertical displacement y of the motor domain (solid line, left scale) and the lever arm rotation φ (dotted line, right scale). Time is measured in milliseconds, displacement in nanometers, rotation in radians. (b) Initial behaviour of the lever arm rotation φ.

binding site and hence the lever arm lags behind. The lever arm speeds up again only when the binding site is reached and the power stroke begins. This behavior is also observed in experiments [8].

In macroscopic experimental studies there are two major types of setups. In the first, there is a constant load on the muscle fiber, while the displacement of the end of the fiber is measured (isotonic contraction). This implies a constant force G applied on the myosin filament. In the second, the length of the muscle fiber is kept constant, while force generated by the activated myosin cross-bridges is examined at the end of the fiber (isometric contraction). Hence it is worth investigating both cases with $G = 0$ and $G \neq 0$.

When the myosin filament is loaded, $G \neq 0$, two things happen: the equilibrium rotation angle changes and the force is trying to pull the motor domain away from the actual actin binding site, both leading to a slower enzymatic cycle. This coincides with the observed behavior of myosin molecules [6].

4 Conclusions

The internal lever arm model is a minimal mechanical model that is capable to describe the dynamics of myosin II. It performs the four strokes of the enzymatic cycle described in Ref. [9]. It provides piecewise exponential time-displacement curves consistent with experimental results. As the motion is continuous, there is no need for arbitrary external influences resulting in sudden changes between the different strokes, energy is supplied only once in each cycle when ATP is hydrolyzed. The frequency of cycles decreases as the load increases due to the slow-down of the enzymatic cycle [7]. Our model can reproduce the stall force [6], the maximum value of the load when the cyclic behavior stops. In lack of ATP, myosin II stays strongly bound to actin, and, both in our model and *in vivo*, only a huge force can pull the molecule away from a binding site. Based on experiments, it has been suggested [5] that during power-stroke the contact force between the actin and the motor domain increases, this is also observed in our model. Kinetic experiments have demonstrated a drastic decrease of ATP consumption in the lack of actin [2], showing the energetically efficient behavior of myosin: no 'fuel' is consumed when there is no chance for contraction. Our model recovers this: the enzymatic cycle cannot be completed away from a binding site, inhibiting the molecule to make futile cycles.

Myosin II molecules work like members of a tug-of-war team, communicating through the bundle of myosin filaments during contraction. There are always some motor domains in the power-stroke making the sarcomere capable of continuous sliding. We plan to connect about 300 myosin II molecules acting together in a sarcomere to model the dynamics of a complete sarcomere.

Based on the comparison with experimental data the validity of our model can be justified, which is expected to give a good estimation of the rigidity of the subdomains of the myosin II molecule. As the model is based on the significant difference of two spring's rigidity, which is responsible for the sequence of substeps of the motion (binding to actin vs. lever arm swinging), further investigations can be made to adapt our model to other members of the myosin family.

Acknowledgements. We are grateful for useful discussions with Zs. Gáspár. Financial support from OTKA under grant no. K 68415 is acknowledged.

References

1. Bagshaw, C.R.: Muscle Contraction. Kluwer, Dordrecht (1993)
2. Eisenberg, E., Hill, T.L.: Prog. Biophys. Molec. Biol. 33, 55–82 (1978)
3. Fischer, S., Windshügel, B., Horak, D., Holmes, K.C., Smith, J.C.: Proc. Nat. Acad. Sci. (USA) 102, 6873–6878 (2005)
4. Geeves, M.A., Holmes, K.C.: Advances in Protein Chemistry, vol. 17, pp. 161–193 (2005)
5. Houdusse, A., Sweeney, H.L.: Curr. Opin. Struct. Biol. 11, 182–194 (2001)
6. Howard, J.: Mechanics of Motor Proteins and the Cytoskeleton. Sinauer, Sunderland (2001)
7. Kovács, M., Thirumurugan, K., Knight, P.J., Sellers, J.R.: Proc. Nat. Acad. Sci. (USA) 104, 9994–9999 (2007)
8. Linari, M., Dobbie, I., Reconditi, M., Koubassova, N., Irving, M., Piazzesi, G., Lombardi, V.: Biophysical Journal 74, 2459–2473 (1998)
9. Lymn, R.W., Taylor, E.W.: Biochemistry 10, 4617–4624 (1971)
10. Vilfan, A.: Biophysical Journal 88, 3792–3805 (2005)

Nonlinear Wave Propagation in the Cochlea with Feed-Forward and Feed-Backward

Walter Lacarbonara and Charles R. Steele

A nonlinear active feed-forward/feed-backward mechanism exciting the basilar membrane (through the activity of the outer hair cells) has been highlighted in previous works as responsible for remarkable amplification in the cochlea. A saturation phenomenon in the transduction process has further been pointed out as a significant nonlinear feature. In this work, a candidate nonlinearity is accounted for through the stretching effect in a one-dimensional string model of the basilar membrane. An asymptotic treatment of the partial differential equations of motion – including the space-delayed terms represented by the feed-forward and feed-backward outer hair cells forces – shows some properties of the nonlinear vibrational behavior of the basilar membrane which may pave the way to understanding some experimentally observed nonlinear features.

1 Introduction

In the cochlea of the inner ear a mechanism takes place that enhances the wave with a consequent sharpening of the mechanical and neural response. Steele *et al.* (1993) [1] and Geisler and Sang [2] have introduced the notion of a process that involves feed-forward along the cochlea. The displacement is sensed at each point along the partition and causes an additional force to act on the partition a distance Δs_1 ahead in the propagation direction. Steele and Baker (1993) [3] (see also [5]) showed how this mechanism works well in

Walter Lacarbonara
Department of Structural and Geotechnical Engineering,
Sapienza University of Rome, via Eudossiana 18. 00814 Rome
e-mail: `walter.lacarbonara@uniroma1.it`

Charles R. Steele
Department of Mechanical Engineering, Stanford University,
Durand Building, Palo Alto, CA, U.S.A. 94305-4035
e-mail: `chasst@stanford.edu`

G. Stépán et al. (Eds.): Dynamics Modeling & Interaction Cont., IUTAM BOOK SERIES 30, pp. 165–175.
springerlink.com

the simple model of a string under tension attached to an elastic substrate and subject to a distribution of sensors and actuators.

A number of studies has highlighted the presence of nonlinear features in the cochlea such as the two-tone distortion whereby tones f_1 and f_2 result in distortions $2f_1 - f_2$ and $f_2 - f_1$. To explain these features, Hall [6] introduced quadratic and cubic mechanical nonlinearities at the basilar membrane level. Further, a systematic experimental investigation of the nonlinear components in the apical turn of cochlea in living guinea pigs was documented in [8]. The change in the magnitude of the harmonic components with the sound pressure was shown to be highly nonlinear. In addition to mechanical or hydrodynamical nonlinearities of the cochlea, nonlinearities in the mechanical to neural transduction processes have also been cited as possible sources of distortion products.

In this work, after documenting detailed observations on the 3D geometry of the partition of the Organ of Corti where the active process takes place, for simplicity we represent the phenomenon within a 1D setting considering a pre-tensioned nonlinear string resting on an elastic substrate subject to feed-forward and feed-backward forces. We treat the equations of motion by using the method of multiple scales. We identify the conditions under which the wave equation exhibits a Hopf bifurcation that is responsible for the onset of traveling waves. We study how the traveling waves are affected by the nonlinearity in the vicinity of the bifurcation.

This work, by proposing to treat a space-delay nonlinear PDE via a modified perturbation procedure, may serve as a basis for investigating a whole class of dynamical processes in biomechanics and in human-machine interactions where there is often an important interplay of time and space delays.

2 The Remarkable Active Space-Delayed Process in the Organ of Corti

The organ of hearing consists of the external ear, middle ear, and the inner ear. The acoustical signals collected by the external ear act on the middle ear, driving the stapes to produce fluid motion in the cochlea, a coiled snail-shell-like structure with three fluid-filled canals: scala vestibuli (SV), scala media (SM), and scala tympani (ST). The middle canal (SM) is triangular in shape and contains the hearing organ known as the Organ of Corti (OC). The basilar membrane (BM) forms the base of the SM while Reissner's membrane is the dividing partition between the SV and SM. Fluid motion in the canals is coupled to the BM and causes it to vibrate, with a spatially distributed tuning starting from high frequencies towards the base up to low frequencies towards the apex. Deiters' cells (D) are attached to the BM and support three rows of outer hair cells (OHC) and phalangeal processes (PhP) as shown in the SEMs of the OC (Figures 1a-b with top and side views). The top view of the OC shows that, across the width of the BM, the three rows of OHCs are attached to the reticular lamina (RL) on top of which the stereocilia (S) act as the sensing and promoting agents of the transduction process. Figures 1c-d portray a high-fidelity reconstruction of the top view and 3D view of the region of interest of the OC. The discovery of electromotility of the OHCs changed the view of the OHCs function from passive elements to active wave amplifiers. OHCs change their length when an electric current flows through them. Hyperpolarization causes the cell to elongate while depolarization results in shortening.

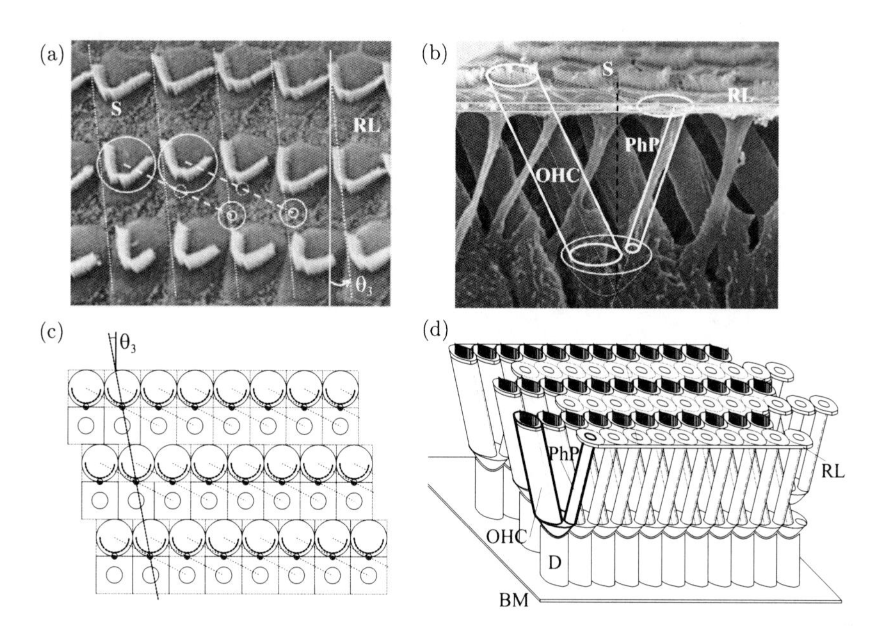

Fig. 1 SEMs of the Organ of Corti: (a) top view with the stereocilia (S) on the reticular lamina (RL) (courtesy of http://scienceblogs.com); (b) side view with the outer hair cells (OHC), the phalangeal processes (PhP), and the Deiters' cells (D) on the basilar membrane (BM) (courtesy of http://images.wellcome.ac.uk); (c) and (d) high-fidelity reconstructions of the top view and 3D geometry.

Figure 2a shows a schematic longitudinal view of the OC while Fig. 2b is a schematic cross section of the cochlear duct. The fluid pressure on the BM is transformed through the stiff pillar cell arch (AO) into a shearing motion between the RL and tectorial membrane (TM). A downward pressure causes a shearing force on the stereocilia that activates a process through which the OHCs are hyperpolarized. Because of the piezoeletric property of the OHCs, their incremental elongation results in an additional downward pressure on the BM. Consequently, a distributed "push-forward/pull-backward" active mechanism, whose model is indicated in Fig. 2a, is generated as suggested by [1]. The force applied by the OHCs on the BM is assumed to be proportional to the total force acting on the BM. To clarify the mechanism, suppose that the OHC whose apex is at s expands, then the Deiters' cell at the distance $s+\Delta s_1$ will be pushed down, while the extension of the connected phalangeal process will pull up the Deiters' cell whose apex is at $s-\Delta s_2$. The gains from the OHC push and the phalangeal process pull are α_1 and α_2, respectively. A detailed description of the push-pull mechanism can be found in [5] where it is asserted that, for long wavelengths, the push-pull effects are negligible (when $\alpha_1 = \alpha_2$), while for very short wavelengths, the fluid viscosity in the cochlea dominates. The push-pull is effective only in a band pass of wavelengths. It is assumed that the OHC force production does not roll-off with frequency and thus α_1 and α_2 are constant with frequency. Although this is controversial, there is evidence that suggests that the OHCs can generate forces for frequencies up to 100 kH, as measured in the guinea pig [7].

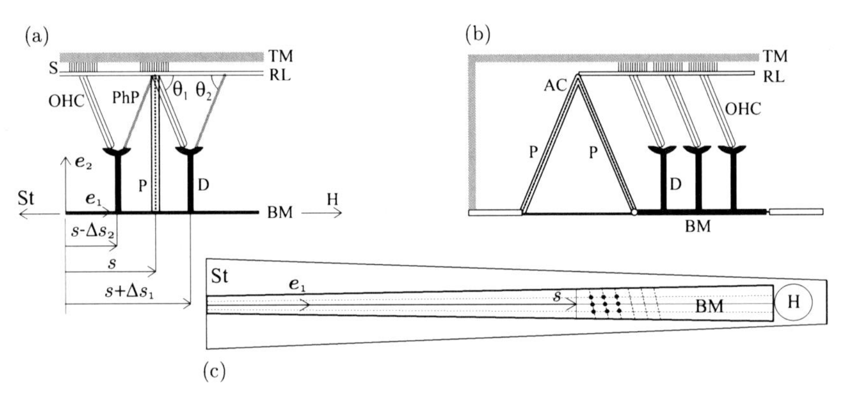

Fig. 2 (a) Schematic longitudinal view of the organ of Corti. St stands for stapes, H is the helicotrema. TM is the tectorial membrane and P is the pillar cell. (b) Schematic cross section with the arches of Corti (AC) and the pillar cells (P). (c) Schematic drawing of the cochlea where the coordinate s represents distances from St. The lines of the three skewed filled dots across the width of the BM denotes the actuation points of the active process.

2.1 A One-Dimensional Mechanical Model with Active Process

In this section we formulate the archetypal mechanical model of a nonlinear string on an elastic substrate subject to feed-forward and feed-backward forces. The string on the elastic substrate is taken with constant properties since the objective of our study is to use the simplest continuous system to examine the effects of the active spatially distributed process. The string is considered stress-free in the configuration $\mathcal{B}'$ (e.g., the straight configuration along the horizontal direction collinear with $\boldsymbol{e}_1$) and we let s be the arclength coordinate along a reference line of the string (e.g., the centerline); we further let $\boldsymbol{x}^{\circ}(s)$ be the position vector of the reference configuration. In this configuration the string is subject to a uniform tension N^{o} which causes the string to be in the prestressed configuration $\mathcal{B}^{\mathrm{o}}$ described by $\boldsymbol{r}^{\mathrm{o}}(s)$. The string stretch is given by $\boldsymbol{r}_s^{\mathrm{o}}(s) = \nu^{\mathrm{o}}(s)\boldsymbol{a}^{\mathrm{o}}(s)$ where subscript s denotes differentiation with respect to s and $\boldsymbol{a}^{\mathrm{o}}(s)$ is a unit vector in the tangent direction, here collinear with the fixed horizontal direction $\boldsymbol{e}_1$ (see Fig. 2).

In the current configuration $\mathcal{B}$, the string is described by the position vector $\boldsymbol{r}(s,t)$ and the total stretch is $\boldsymbol{r}_s(s,t) = \breve{\nu}(s,t)\boldsymbol{a}(s,t)$ where $\boldsymbol{a}(s,t)$ is the unit vector in the current tangent direction. It can be shown that $\breve{\nu}(s,t) = \nu^{\mathrm{o}}(s)\nu(s,t)$ where $\nu(s,t)$ is the incremental stretch arising during the motion from $\mathcal{B}^{\mathrm{o}}$ to $\mathcal{B}$. If we consider planar motions of the string, we let $\boldsymbol{r}(s,t) = \boldsymbol{r}^{\mathrm{o}}(s) + \boldsymbol{u}(s,t)$ with $\boldsymbol{u}(s,t) = u(s,t)\boldsymbol{e}_1 + v(s,t)\boldsymbol{e}_2$ and $\boldsymbol{r}^{\mathrm{o}} = s\boldsymbol{e}_1$. Thus,

$$\nu(s,t) = \sqrt{(\nu^{\mathrm{o}} + u_s)^2 + v_s^2}/\nu^{\mathrm{o}} \tag{1}$$

Moreover, $\boldsymbol{a}(s,t) = \cos\theta(s,t)\boldsymbol{e}_1 + \sin\theta(s,t)\boldsymbol{e}_2$ with $\cos\theta = (\nu^{\mathrm{o}} + u_s)/\breve{\nu}$ and $\sin\theta = v_s/\breve{\nu}$.

We let $\breve{\boldsymbol{n}}(s,t) = [N^{\mathrm{o}} + N(s,t)]\boldsymbol{a}(s,t)$ denote the actual string tension, expressed as the summation of the initial tension N^{o} and the incremental tension $N(s,t)$. The balance of linear momentum delivers the following equation of motion:

$$\partial_s \breve{\boldsymbol{n}} + \breve{\boldsymbol{f}} = \rho A \boldsymbol{r}_{tt} \tag{2}$$

Since $\mathcal{B}^{\mathrm{o}}$ is an equilibrium configuration with zero distributed forces, it is $N_s^{\mathrm{o}} = 0$ which implies that the pretension is uniform along the string. Thus the equation of motion can be written as

$$\partial_s(N\boldsymbol{a}) + \partial_s[N^{\mathrm{o}}(\boldsymbol{a} - \boldsymbol{e}_1)] + \boldsymbol{f} = \rho A \boldsymbol{r}_{tt} \tag{3}$$

where $\boldsymbol{f}$ is the incremental force per unit reference length and the adopted nonlinearly viscoelastic constitutive equation for N is $N = \hat{N}(\nu, \nu_t, s)$. We project equation of motion (3) along the tangential direction $\boldsymbol{a}$ and transverse direction $\boldsymbol{b}$ by considering $\boldsymbol{a}_s = \theta_s\boldsymbol{b}$, $\theta = \arctan(v_s/(\nu^{\mathrm{o}} + u_s))$. To this end, we assume that the force per unit reference length of the elastic substrate is $-k^{\mathrm{E}}v\boldsymbol{b}$ and the feed-forward and feed-backward forces at s act in the actual transverse direction $\boldsymbol{b}$, according to $\boldsymbol{f} = [\alpha_1 k^{\mathrm{E}}(s - \Delta s_1)v(s - \Delta s_1) + \alpha_2 k^{\mathrm{E}}(s + \Delta s_2)v(s + \Delta s_2)]\boldsymbol{b}$. Thus the componential equations of motion (in dimensional form) are

$$\begin{aligned} \partial_s N &= \rho A(u_{tt}\cos\theta - v_{tt}\sin\theta), \\ (N^{\mathrm{o}} + N)\partial_s\theta - kv^{\mathrm{E}} + \alpha_1 k^{\mathrm{E}} v(s - \Delta s_1) &+ \alpha_2 k^{\mathrm{E}} v(s + \Delta s_2) \\ &= \rho A(-u_{tt}\sin\theta + v_{tt}\cos\theta). \end{aligned} \tag{4}$$

2.2 Linear Analysis

Let us consider small-amplitude motions so that $\theta_s = v_{ss}$ whereby the linearized equations of motion reduce to

$$N^{\mathrm{o}} v_{ss} - k^{\mathrm{E}} v + \alpha_1 k^{\mathrm{E}} v(s - \Delta s_1) + \alpha_2 k^{\mathrm{E}} v(s + \Delta s_2) - c v_t = \rho A v_{tt} \tag{5}$$

where $c\, v_t$ indicates the overall dissipative force due to the string (membrane) and the surrounding fluid viscosity. We let

$$v(s,t) = V e^{i(ks - \omega t)} \tag{6}$$

where $i := \sqrt{-1}$, ω is the frequency, and k is the wave number. Substituting (6) into (5) yields the (dimensional) linear dispersion relation. When the dissipation and the feed-backward/feed-forward effects are neglected, the dispersion relation leads to $\omega^2 = k^2 N^{\mathrm{o}} + k^{\mathrm{E}}/\rho A$ from which we obtain the cut-off frequency as $\omega_0^2 := k^{\mathrm{E}}/\rho A$. We thus introduce the following nondimensional variables and parameters: $\sigma_1^* := k_0 \Delta s_1$, $\sigma_2^* := k_0 \Delta s_2$, $k^* := k/k_0$, $\omega^* := \omega/\omega_0$, $\zeta := c/(2\sqrt{\rho A k^{\mathrm{E}}})$ where $k_0^2 := k^{\mathrm{E}}/N^{\mathrm{o}}$ (note that $[k_0] = [L]^{-1}$). Parameters σ_1^* and σ_2^* play the role of nondimensional space delays. The nondimensional dispersion relation thus becomes

$$\omega^2 - 2i\omega\zeta = k^2 + 1 - \alpha_1 e^{ik\sigma_1} - \alpha_2 e^{-ik\sigma_2} \tag{7}$$

where, for notational slenderness, we dropped the stars in the nondimensional variables and parameters. In (7) we let $k = k_{\mathrm{R}} + i k_{\mathrm{I}}$, separate real and imaginary parts and obtain the following two transcendental equations:

$$\alpha e^{-k_{\mathrm{I}}\sigma} \cos\sigma_{\mathrm{R}} - k_{\mathrm{R}}^2 + k_{\mathrm{I}}^2 + \omega^2 - 1 = 0, \quad \alpha e^{-\sigma k_{\mathrm{I}}} \sin\sigma_{\mathrm{R}} - 2k_{\mathrm{I}} k_{\mathrm{R}} - 2\zeta\omega = 0 \tag{8}$$

where $\sigma_{\mathrm{R}} := \sigma k_{\mathrm{R}}$ and $\alpha = \alpha_1$, $\alpha_2 = 0$. In Fig. 3 we show variation of k_{I} with ω when $\alpha = 0.11$, $\zeta = 0.01$, for three values of σ, namely, $\sigma = (0.5, 0, -0.5)$.

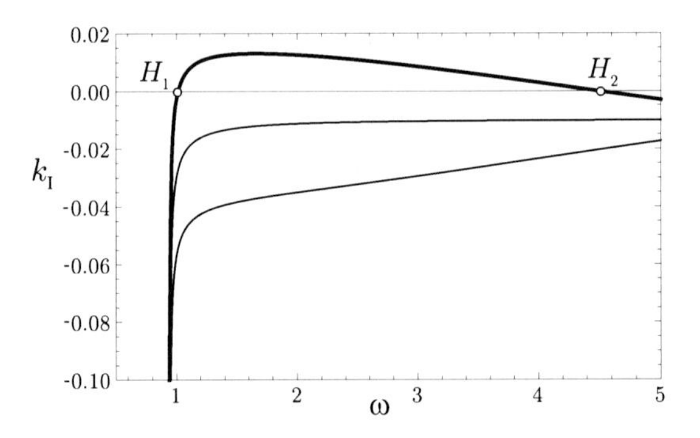

Fig. 3 Imaginary part of the wave number vs. ω when $\alpha = 0.11$ and $\zeta = 0.01$

The critical condition is attained when $k_{\mathrm{I}} = 0$. We thus let $k_{\mathrm{I}} = 0$ in (8), square and add $\sin\sigma_{\mathrm{R}}$ and $\cos\sigma_{\mathrm{R}}$ obtaining

$$k_{\mathrm{R}} = \sqrt{\omega^2 - 1 \pm \sqrt{\alpha^2 - (2\zeta\omega)^2}}, \quad \sigma = 1/k_{\mathrm{R}} \arctan\left(\frac{2\zeta\omega}{1 + k_{\mathrm{R}}^2 - \omega^2}\right) \tag{9}$$

The two wave numbers coalesce into $k_{\mathrm{R}} = \sqrt{\omega^2 - 1}$ when

$$\alpha = 2\zeta\omega, \quad \sigma = \tfrac{\pi}{2}/\sqrt{\omega^2 - 1}, \ \omega > 1 \tag{10}$$

We can find the locus in the plane (α, σ) substituting $\omega = \alpha/(2\zeta)$ thus obtaining

$$\sigma = \tfrac{\pi}{2}/\sqrt{(\alpha/(2\zeta))^2 - 1}, \ \alpha > 2\zeta \tag{11}$$

When $k_{\mathrm{I}} > 0$, there is an exponential growth of the wave in space. *This exponential growth is bounded by two possible distinct physical mechanisms: (i) a gain saturation due to the piezoelectricity-induced force saturation in the OHCs and (ii) mechanical nonlinearities in the basilar membrane (such as the stretching in the archetypal string model).*

The force saturation has been documented in previous studies by Steele and co-workers [4]. The saturation obviously induces a bounding effect on the wave since above a threshold displacement the feed-forward effects are turned off and the wave decays. However, during the decay, below the threshold, the feed-forward amplifying mechanism is turned on again and hence the response settles into a steady-state traveling wave.

The same occurs as a consequence of the stretching in the string/membrane. The latter mechanism may in fact be activated at lower displacement amplitudes without requiring the force saturation mechanism to take place. Since the features of the saturation have already been discussed in previous works employing a numerical approach, here we pause to investigate the other nonlinear mechanism (the mechanical nonlinearity) by employing the method of multiple scales.

3 Nonlinear Analysis via Asymptotic Approach

The nondimensional displacements are $u^* := k_0 u$ and $v^* := k_0 v$. The nondimensional force per unit reference length – due to the elastic substrate and feed-forward and feed-backward effects – becomes $\boldsymbol{f} = [-v^* + \alpha_1 v^*(s^* - \sigma_1^*) + \alpha_2 v^*(s + \sigma_2^*)]\boldsymbol{b}$ under the assumption that the elastic substrate stiffness is uniform. The nondimensional componential equations of motion become

$$\Lambda \nu_s = u_{tt} \cos\theta - v_{tt} \sin\theta, \tag{12}$$

$$[1 + \Lambda(\nu - 1)]\theta_s - v + \alpha_1 v(s - \sigma_1) + \alpha_2 v(s + \sigma_2) = -u_{tt} \sin\theta + v_{tt} \cos\theta \tag{13}$$

where we have dropped the stars and $\Lambda := EA/N^{\circ}$ represents the ratio between the elastic stiffness and the geometric stiffness of the string.

To obtain the perturbation expansion, we set $\nu^{\circ} = 1$, from which we obtain the incremental stretch and curvature as $\nu = \sqrt{(1 + u_s)^2 + v_s^2}$ and $\theta_s = [v_{ss}(1 + u_s) - v_s u_{ss}]/\nu^2$.The longitudinal motion is quasistatically induced by the transverse motion, which leads to the following ordering: $v = O(\varepsilon)$

and $u = O(\varepsilon^2)$ where $\varepsilon << 1$ is a small nondimensional number. We assume two space and time scales, namely, $s_0 = s,\ s_2 = \varepsilon^2 s,\ t_0 = t,\ t_2 = \varepsilon^2 t$. Therefore,

$$\partial_s = \frac{\partial}{\partial s_0} + \varepsilon^2 \frac{\partial}{\partial s_2} + \ldots, \qquad \partial_t = \frac{\partial}{\partial t_0} + \varepsilon^2 \frac{\partial}{\partial t_2} + \ldots, \qquad \text{etc.} \tag{14}$$

Then, the ansatz on the displacement components is

$$\begin{aligned} u(s,t) &= \varepsilon^2 u_2(s_0, s_2, t_0, t_2) + \ldots, \\ v(s,t) &= \varepsilon v_1(s_0, s_2, t_0, t_2) + \varepsilon^3 v_3(s_0, s_2, t_0, t_2) + \ldots \end{aligned} \tag{15}$$

Substituting (15) into (12) and (13), differentiating with respect to ε, and collecting terms of like powers of ε yields

O(ε) problem:

$$\frac{\partial^2 v_1}{\partial {t_0}^2} + 2\zeta \frac{\partial v_1}{\partial t_0} - \frac{\partial^2 v_1}{\partial {s_0}^2} + v_1 - \alpha_1 v_1(s_0 - \sigma_1) - \alpha_2 v_1(s_0 + \sigma_2) = 0 \tag{16}$$

O(ε^2) problem:

$$\frac{\partial^2 u_2}{\partial {t_0}^2} - \Lambda \frac{\partial^2 u_2}{\partial {s_0}^2} = -\frac{\partial v_1}{\partial s_0}\frac{\partial^2 v_1}{\partial {t_0}^2} + \Lambda \frac{\partial v_1}{\partial s_0}\frac{\partial^2 v_1}{\partial {s_0}^2} \tag{17}$$

O(ε^3) problem:

$$\begin{aligned} &\frac{\partial^2 v_3}{\partial {t_0}^2} - \frac{\partial^2 v_3}{\partial {s_0}^2} + v_3 + 2\zeta \frac{\partial v_3}{\partial t_0} - \alpha_1 v_3(s_0 - \sigma_1) - \alpha_2 v_3(s_0 + \sigma_2) \\ &\quad = -2\zeta \frac{\partial v_1}{\partial t_2} - 2\frac{\partial^2 v_1}{\partial t_0 \partial t_2} + \frac{\partial^2 u_2}{\partial t_0^2}\frac{\partial v_1}{\partial s_0} - \frac{\partial^2 u_2}{\partial s_0^2}\frac{\partial v_1}{\partial s_0} + \tfrac{1}{2}\frac{\partial^2 v_1}{\partial t_0^2}\left(\frac{\partial v_1}{\partial s_0}\right)^2 \\ &\quad + 2\frac{\partial^2 v_1}{\partial s_0 \partial s_2} + (\Lambda - 1)\frac{\partial^2 v_1}{\partial s_0^2}\frac{\partial u_2}{\partial s_0} + (\tfrac{1}{2}\Lambda - 1)\frac{\partial^2 v_1}{\partial s_0^2}\left(\frac{\partial v_1}{\partial s_0}\right)^2 \end{aligned} \tag{18}$$

The first-order problem admits a progressive wave since the space delays, gain factor, and damping satisfy the critical condition. Thus, the solution reads

$$v_1 = A(s_2, t_2)e^{i(ks_0 - \omega t_0)} + \bar{A}(s_2, t_2)e^{-i(ks_0 - \omega t_0)} \tag{19}$$

where $k \in \Re^+$ is the wave number and ω is the frequency (above cut-off) that satisfies the critical condition (9). The overbar denotes the complex conjugate. We substitute (19) into (17) and obtain, by integration,

$$\frac{\partial u_2}{\partial s_0} = -i\tfrac{1}{4}kA^2 e^{2i(ks_0 - \omega t_0)} + i\tfrac{1}{4}k\bar{A}^2 e^{-2i(ks_0 - \omega t_0)} \tag{20}$$

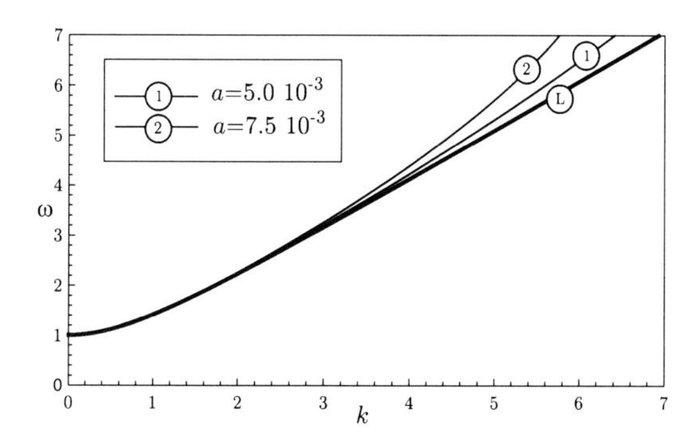

Fig. 4 Dispersion relations obtained by the linear model, denoted L, and the nonlinear model, labelled 1 ($a = 5 \cdot 10^{-3}$) and 2 ($a = 7.5 \cdot 10^{-3}$) when $\Lambda = 5 \cdot 10^2$.

We then substitute (19) and (20) into (18) and obtain an inhomogeneous problem whose solvability is enforced by Fredholm's Alternative Theorem. The resulting solvability condition reads

$$2i\omega(\dot{A} + C_{\mathrm{g}}A') - 2\zeta\dot{A} + \Gamma|A|^2 A = 0 \tag{21}$$

where the dot indicates differentiation with respect to t_2 and the prime with respect to s_2, respectively. Moreover, $\Gamma = \frac{1}{2}k^2[\omega^2 - k^2(2\Lambda - 1)]$ is the effective wave nonlinearity coefficient and $C_{\mathrm{g}} := \partial\omega/\partial k = k/\omega$ is the group velocity (recall that the linear dispersion relation is $\omega = \sqrt{k^2+1}$).

To determine the solutions of (21), we let the complex-valued function $A(s_2, t_2)$ be expressed in polar form as $A(s_2, t_2) = 1/2a(s_2, t_2)\exp[i\phi(s_2, t_2)]$. We substitute the above polar form into (21), separate real and imaginary parts, and obtain

$$\dot{a} + C_{\mathrm{g}}a' - (\zeta/\omega)a\dot{\phi} = 0, \quad a(\dot{\phi} + C_{\mathrm{g}}\phi') + (\zeta/\omega)\dot{a} - (\Gamma/8\omega)a^3 = 0 \tag{22}$$

We seek monochromatic waves with constant amplitude, that is, $\dot{a} = 0 = a'$. Therefore, (22) yields $\dot{\phi} = 0$ and, in turn,

$$\phi = \frac{\Gamma}{8\omega C_{\mathrm{g}}}a^2 s_2 + \phi_0 \tag{23}$$

Consequently, the progressive wave, to within lowest order, is expressed as

$$v(s,t) = \varepsilon a \cos\left[\left(k + \varepsilon^2 \frac{\Gamma}{8\omega C_{\mathrm{g}}}a^2\right)s - \omega t\right] + O(\varepsilon^3) \tag{24}$$

We observe that the wave number is corrected by the nonlinearity according to

$$k^{\mathrm{N}} = k + \varepsilon^2 \tfrac{1}{16}k[\omega^2 - k^2(2\Lambda - 1)]a^2 \tag{25}$$

where the correction consists of a term that increases monotonically with the square of the amplitude and with Λ (the elasticity of the string relative to the initial tension). We thus conclude that the effect of the nonlinearity, by decreasing the wave number, lengthens the waves at higher frequencies as shown in Fig. 4. The wave propagation properties thus studied relate to frequency ranges near the Hopf bifurcation points H_1 and H_2 (see Fig. 3). Clearly, for the low-frequency range (near H_1), the nonlinear corrections are negligible for the presumed amplitude ranges and nonlinearity strengths while they become appreciable at moderately higher frequencies near H_2 (about one order of magnitude higher than the cut-off frequency) already at relatively low amplitudes.

4 Concluding Remarks

We have shown, within the context of an archetypal model represented by a nonlinear string resting on an elastic substrate and subject to feed-forward and feed-backward forces, that the space-delayed active mechanism of the outer hair cells, by counteracting the viscous forces, explains the existence of waves in the cochlea that can propagate at certain frequencies and wave numbers. The critical conditions at which the waves propagate without attenuation are identified as Hopf bifurcations. The conducted asymptotic analysis indicates that the (hardening) nonlinearity, although being representative of a nonlinearity whose source may well be different, is responsible for a lengthening of the waves at higher frequencies. The nonlinearity may also explain other observed nonlinear cochlear behaviors such as distortion products and clicks.

Acknowledgments. This work was partially supported by a FY 2009 Grant from La Sapienza AST (Italy) and Grant No. R01-DC007910 from NIDCD of NIH (USA).

References

1. Steele, C.R., Baker, G., Tolomeo, J., Zetes, D.: Electro-mechanical models of the outer hair cells. In: Proc. of Biophysics of Hair Cell Sensory Systems, Groningen (1993)
2. Geisler, C.D., Sang, C.: A cochlear model using feed-forward outer-hair-cell forces. Hear. Res. 86, 132–146 (1995)
3. Steele, C.R., Baker, G.: Hydroelastic waves in the cochlea. In: Hajela, P., McIntosh Jr., S.C. (eds.) Advances in Aerospace Sciences, pp. 443–451 (1993)
4. Lim, K.-M., Steele, C.R.: Response suppression and transient behavior in a nonlinear active cochlear model with feed-forward. Int. J. Solids Struct. 40, 5097–5107 (2003)

5. Yoon, Y., Puria, S., Steele, C.R.: A cochlear model using the time-averaged Lagrangian and the push-pull mechanism in the organ of Corti. J. Mech. Mater. Struct. 4, 977–986 (2009)
6. Hall, J.L.: Two-tone distortion products in a nonlinear model of the basilar membrane. J. Acoust. Soc. Am. 56, 1818–2828 (1974)
7. Hemmert, W., Zenner, H.P., Gummer, A.W.: Characteristics of the travelling wave in the low-frequency region of a temporal-bone preparation of the guinea-pig cochlea. Hear. Res. 142, 184–202 (2000)
8. Khanna, S.M., Haob, L.F.: Nonlinear vibrations in the apex of guinea-pig cochlea. Int. J. Solids Struct. 38, 1919–1933 (2001)

Sensitivity Investigation of Three-Cylinder Model of Human Knee Joint

I. Bíró, B.M. Csizmadia, G. Krakovits, and A. Véha

Abstract. The contact surfaces of wide-spread applied human knee joint prostheses can be described with simple geometrical elements. The relative motion realized by knee joint is ensured by complicated condyle surfaces. For this reason the implanted prostheses comply with requirements limited and it causes additional load on the diseased bony tissue. In order to describe the motion of knee joint at first the authors constrained coordinate-systems on the basis of anatomical landmarks to the femur and tibia and then they joined a three-cylindrical mechanism as mechanical model to the axes of coordinate-systems. In the second phase 'the' authors determined the six independent kinematical parameters of tibia compared to the fixed femur during flexion and extension. The experimental examinations were carried out on cadaver knees. The positioning was tracked by optical positioning appliance. Needed parameters can be obtained from the recorded data determined by the applied kinematical model.

Considering the irregular shapes of femur and tibia the anatomical coordinate systems can be joined with 1-2 mm and 2-4 degree position deflections. The aim of this paper is the determination of the effects of position deflections on kinematical parameters.

Keywords: knee joint, kinematical model, optical positioning, accuracy, sensitivity investigation.

1 Introduction

In recent decades the number of kinematical models of anatomical joints has increased [4-7]. In case of certain models so many researchers have measured and described joint motion with less than six degrees of freedom. Obviously the

I. Bíró · A. Véha
University of Szeged, Faculty of Engineering, Szeged, Hungary

B.M. Csizmadia
Szent István University, Faculty of Mechanical Engineering, Gödöllő, Hungary

G. Krakovits
Szent János Hospital, Budapest, Hungary

G. Stépán et al. (Eds.): Dynamics Modeling & Interaction Cont., IUTAM BOOK SERIES 30, pp. 177–184.
springerlink.com

treatment of six degree of freedom models is the most difficult. The motion of human knee joint can be described by the following components: The flexion-extension is defined around the medio-lateral axis, internal-external rotation around the tibial axis and the abduction-adduction around the anterior-posterior (floating) axis. The medio-lateral translation is measured along the medio-lateral axis, proximal-distal translation along the tibial axis and antero-posterior translation along the mutually perpendicular floating axis [1-3].

Description of motion components in such a way is a little bit subjective. In order to mark out describe the motion components precisely it is needed to join coordinate-systems to the femur and tibia consequently.

In recent decades different methods have been developed in order to measure kinematical parameters of human knee joint. In these methods it is measured and processed how the motion of markers fastened to femur and tibia referring to each other. In vitro-mechanical investigations there are mainly phantom or simulated computer models or cadaver motion experiments.

In spite of that the newly introduced techniques developed in an enormous numbers in the last decades e.g. the radiology, fluoroscopy, three-dimensional CT, MRI, stereophotogrammetry, ultrasound, etc., most of the results of which were unreliable, inconsistent with other published data. The range of the tibia out and in-rotation along the flexion-extension motion of the knee had been established by different authors as between *5* up to *17* degrees, moreover the character of this diagram is variable [1-7, 9, 10, 13, 15, 16]. On the basis of difference of the published results it is quite difficult to establish the exact character concerning the motion of knee joint. Some sensitivity analysis can be found among published results. The effect of different position of axis on change of kinematical parameters of human knee joint is at present a debated question.

2 Method

Experiments were carried on cadaver knees for the analysis of motion of the knee using an experimental test rig built by the research team [11, 12, 14]. To the presented sensitivity investigation it is necessary to determine anatomical landmarks on femur and tibia, furthermore, the coordinate-systems joining to the determined anatomical landmarks.

Needed anatomical landmarks on femur are: centre of the femoral head (*fh*); medial and lateral epicondyles (P_{mcl}, P_{lcl}). On tibia: apex of the head of the fibula (*hf*); prominence of the tibial tuberosity (*tt*); distal apex of the lateral and medial malleolus (*kb, bb*). Current international standards and conventions (e.g. from the International Society of Biomechanics) were used by the authors [16].

Some of anatomical landmarks (*kb, bb, fh*) on the whole cadaver body were determined. Other landmarks (P_{lcl}, P_{mcl}, *hf, tt*) can be determined on the whole cadaver body or on the fixed resection in the equipment. Polaris infrared optical positioning system (Fig.1) was used for the determination of the position of the anatomical points during the experiments.

Above mentioned anatomical landmarks are not similar to a peak rather like a small surface. For this reason the optical positioning of anatomical landmarks can be achieved with *1-2 mm* position deflection.

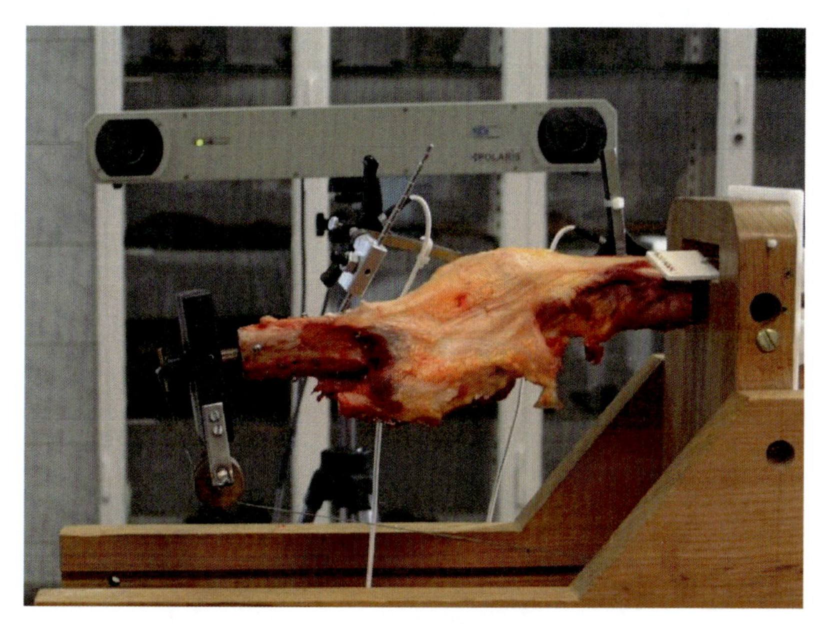

Fig. 1 Experimental equipment with the Polaris system

As a result of different vector operations and coordinate transformations using data recorded by Polaris system during motion investigation we can follow the change of position of anatomical coordinate-system joined to tibia in anatomical coordinate-system joined to femur. If the anatomical landmarks on femur and tibia are determined with position deflection the kinematical parameters of knee joint will be modified.

3 Three-Cylindrical Mechanism

Similar spatial mechanical models are often used in theory and practice of mechanisms. The above mentioned kinematical parameters can be obtained by the aid of a three-cylindrical mechanism model (Fig.2) put in between the above defined anatomical coordinate-systems.

In case of similar spatial structures the so-called Denavit-Hartenberg (*HD*) coordinates can be applied (Fig.4). The advantage of the application of *HD* coordinates is: the transformation matrix contains – instead of six – four (Θ_i, d_i, l_i, α_i) variable physical quantities joining to geometrical characters of bodies and their constraint.

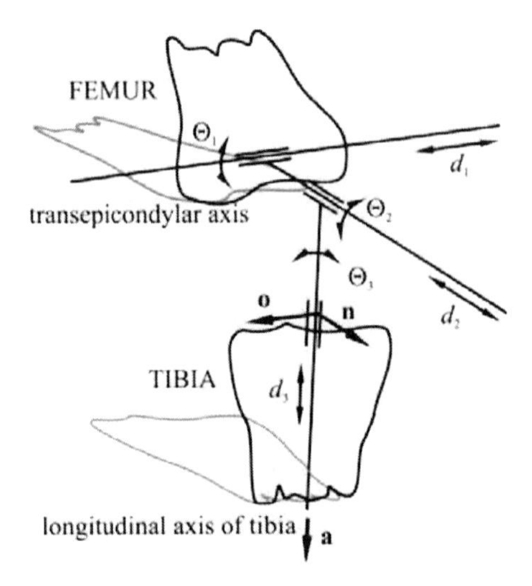

Fig. 2 Kinematical parameters of human knee joint in three-cylindrical mechanism model

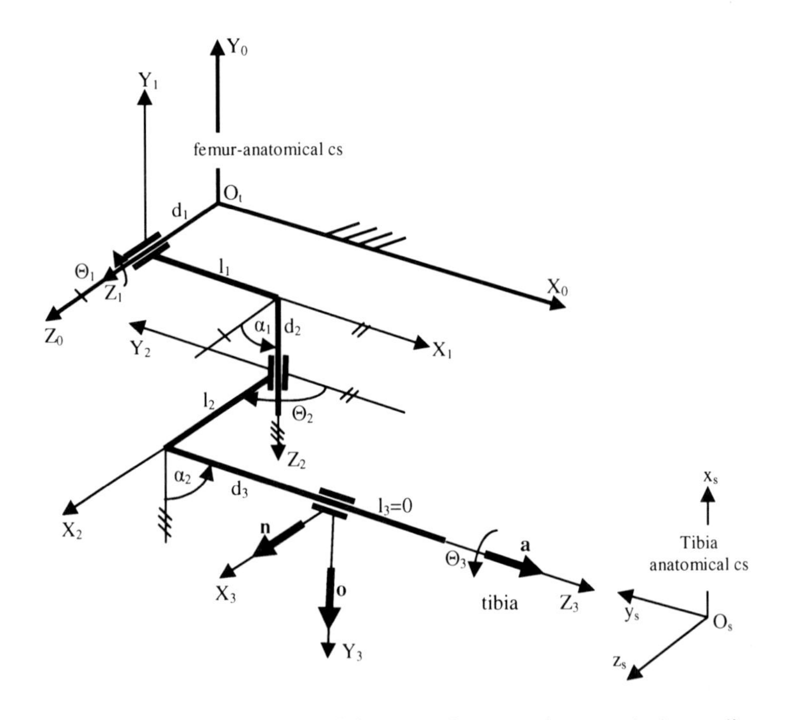

Fig. 3 Three-cylindrical mechanism model between the created anatomical coordinate-systems (extended position)

In Fig.4 the *HD* coordinates can be seen in extended position of the leg. In the mechanism α_i, l_i, $(i=1,2,3)$ can be adjustable optionally according to the special geometry of knee joint. On the basis of the published recommendations the following data supply proper approaching: $\alpha_1=\alpha_2=90^o$, $\alpha_3=0^o$, $l_1=l_2=l_3=0$.

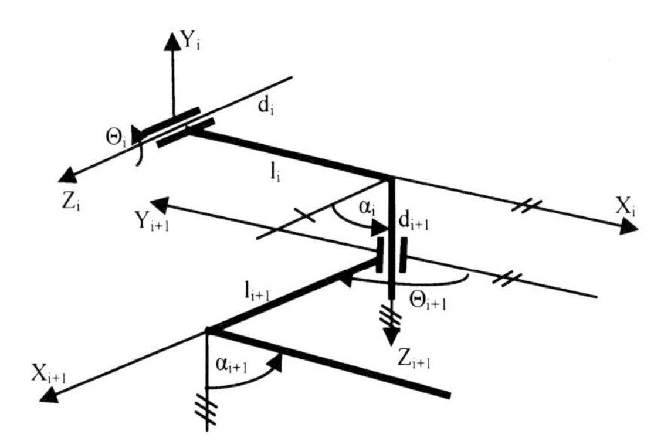

Fig. 4 Connection of ith and i+1th bodies and joined coordinate-systems

The application of the model enables the calculation of the following quantities (Fig.2,3):

Θ_1 – flexion, in drawn position *0* degree,
Θ_2 – abduction/adduction, in drawn position *90* degree,
Θ_3 – rotation of the tibia, in drawn position *0* degree,
d_1, d_2, d_3 – moving on accordant axes.

On the basis of approaching $\alpha_1=\alpha_2=90^o$, $\alpha_3=0^o$, $l_1=l_2=l_3=0$ for the kinematical chain of Fig.3 the following matrix equation can be written down where n, o, a are unit vectors of coordinate-system $X_3Y_3Z_3$ in coordinate-system $X_0Y_0Z_0$ and P_x, P_y, P_z are coordinates of origin of coordinate-system $X_3Y_3Z_3$ in coordinate-system $X_0Y_0Z_0$.

$$\begin{bmatrix} \cos\Theta_1 & 0 & \sin\Theta_1 & 0 \\ \sin\Theta_1 & 0 & -\cos\Theta_1 & 0 \\ 0 & 1 & 0 & d_1 \\ 0 & 0 & 0 & 1 \end{bmatrix} * \begin{bmatrix} \cos\Theta_2 & 0 & \sin\Theta_2 & 0 \\ \sin\Theta_2 & 0 & -\cos\Theta_2 & 0 \\ 0 & 1 & 0 & d_2 \\ 0 & 0 & 0 & 1 \end{bmatrix} * \begin{bmatrix} \cos\Theta_3 & -\sin\Theta_3 & 0 & 0 \\ \sin\Theta_3 & \cos\Theta_3 & 0 & 0 \\ 0 & 0 & 1 & d_3 \\ 0 & 0 & 0 & 1 \end{bmatrix} = \begin{bmatrix} n_x & o_x & a_x & P_x \\ n_y & o_y & a_y & P_y \\ n_z & o_z & a_z & P_z \\ 0 & 0 & 0 & 1 \end{bmatrix}$$

Roots of equation system are: Θ_1, Θ_2, Θ_3, d_1, d_2, d_3, which determine precisely the position of tibia compared to femur.

4 Sensitivity Investigation

The obtained kinematical parameters and diagrams partly depend on the method of measurement, the anatomical specialties of the cadaver, but principally on the determination of the applied coordinate-systems constrained to femur and tibia. Considering the irregular shapes of femur and tibia the anatomical landmarks can be measured with position deflection, for this reason the anatomical coordinate-systems joined to anatomical landmarks have position deflection as well. The optical positioning of anatomical landmarks was made by a pointer. The position deflection of the centre of femoral head, apex of the head of the fibula and prominence of the tibial tuberosity cause small angular deflection because these anatomical landmarks are located relatively far from the origins of coordinate-systems.

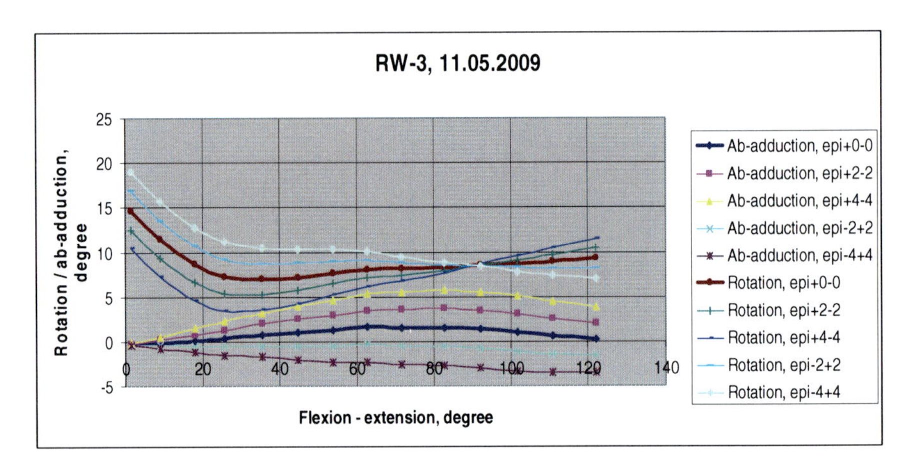

Fig. 5 Effect of modification of position of epicondyles on kinematical functions

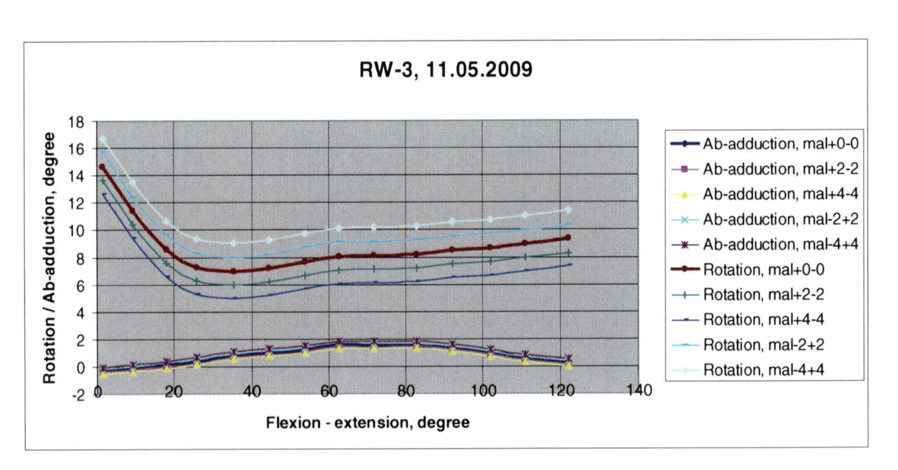

Fig. 6 Effect of modification of position of apices of the lateral and medial malleolus on kinematical functions

The origins of coordinate-systems are determined between epicondyles and apices of the lateral and medial malleolus, therefore the effect of position deflection on position and orientation of coordinate-systems is quite significant. As first step of sensitivity investigation the positions of epicondyles were modified in vertical direction (cadaver lying on his back, extended position) step by step *(±2 mm)*. In the second phase the positions of apices of the lateral and medial malleolus were modified step by step (*±2 mm*) approximately in the vertical plane. Difference from basic function (thickened curve) can be seen in Fig.5-6. (The basic function contains some mistakes because it was calculated from measured data.)

Katona et al.[17] carried out more motion investigation using the same cadaver knee joint. Before every single investigation they always recorded the position of anatomical landmarks on femur and tibia by the aid of Polaris system. On comparing the experimental and calculated results to each other the reasons of differences can be explained, thus.

5 Conclusion

As a result of sensitivity investigation the followings are established:

- The flexion-rotation and flexion-ab/adduction diagrams depend strongly on position mistakes of determined coordinate-systems, for this reason it is important to determine the anatomical landmarks with a great accuracy depending on technical possibilities.
- The cause of modification of diagrams is first of all the position deflections of epicondyles in vertical direction: Its effect can be seen over *60* degree in flexed position. In this part the diagrams are approximately linear, their gradient variable depending on position mistakes of epicondyles. The shapes of ab/adduction diagrams are strongly deformed.

Summing up it can be established that the using of anatomical coordinate-systems enables the comparison and generalization of results of different motion investigation of human knee joints. On the basis of above mentioned it can be decided which parts of the obtained results are acceptable and certainly which parts and its conclusions need cautious treatment. By more precise prescription of taking up of anatomical coordinate-system the comparison of results can be intensified.

References

[1] Grood, E.S., Suntay, W.J.: A joint coordinate system for the clinical description of three-dimensional motions: application to the knee. J. biomech. Engng. 105, 136–144 (1983)

[2] Pennock, G.R., Clark, K.J.: An anatomybased coordinate system for the description of the kinematic displacements in the human knee. J. Biomechanics 23(12), 1209–1218 (1990)

[3] Blankevoort, L., Huiskes, R., Lange, A.: Helical axes of passive knee joint motions. J. Biomechanics 23(12), 1219–1229 (1990)

[4] Hollister, A.M., et al.: The axes of rotation of the knee. Clinical Orthopaedics and Related Research (290), 259 (1993)

[5] Churchill, D.L., et al.: The transepicondylar axis approximates the optimal flexion axis of the knee. Clinical Orthopaedics and Related Research (356), 111–118 (1998)
[6] Bull, A.M.J, Amis, A.A.: Knee joint motion: description and measurement. Proc. Instn. Mech. Engrs. 212 Part H (1998)
[7] Iwaki, H., Pinskerova, V., Freeman, M.A.R.: Tibiofemoral movement 1: the shapes and relative movements of the femur and tibia in the unloaded cadaver knee. The Journal of Bone and Joint Surgery [Br] 82-B, 1189–1195 (2000)
[8] Mosta, E., Axea, J., Rubasha, H., Lia, G.: Sensitivity of the knee joint kinematics calculation to selection of flexion axes. Journal of Biomechanics 37, 1743–1748 (2004)
[9] Wilson, D.R., Feikes, J.D., Zavatsky, A.B., O'Connor, J.J.: The components of passive knee movement are coupled to flexion angle. Journal of Biomechanics 33, 465–473 (2000)
[10] Eckhoff, D.G., et al.: Three-dimensional mechanics, kinematics, and morphology of the knee viewed in virtual reality. Journal of Bone & Joint Surgery-American Volume 87(Suppl. 2), 71 (2005)
[11] Zoltán, S.: Mérőberendezés térdízület mozgásvizsgálatához. GÉP LVII, 37–40 évfolyam 1. szám (2006)
[12] Katona, G., Csizmadia, M.B., Fekete, G.: Kísérleti vizsgálatok térd mechanikai modelljének megalkotásához, 2. In: Magyar Biomechanikai konferencia, Debrecen, június 30 – július 01, pp. 49–50 (2006)
[13] Eckhoff, D., et al.: Difference between the epicondylar and cylindrical axis of the knee. Clinical Orthopaedics and Related Research 461, 238–244 (2007)
[14] Bíró, I., Csizmadia, M.B., Katona, G.: New approximation of kinematical analysis of human knee joint, pp. 330–338. Bulletin of the Szent István University (2008), ISSN 1586-4502
[15] Aglietti, P., et al.: Rotational Position of Femoral and Tibial Components in TKA Using the Femoral Transepicondylar Axis. Clinical Orthopaedics and Related Research 466(11), 2751–2755 (2008)
[16] Victor, J.: An Experimental Model for Kinematic Analysis of the Knee. The Journal of Bone and Joint Surgery (American) 91, 150–163 (2009), doi:10.2106/JBJS.I.00498
[17] Katona, G., Csizmadia, M.B., Bíró, I., Andrónyi, K., Krakovits, G.: Cadaver térdek mozgásánakértékelése anatómiai koordináta-rendszerek felhasználásával. In: Proceedings of 4th HungarianConference on Biomechanics, Pécs (May 2010)
[18] http://www.ulb.ac.be/project/vakhum

On the Performance Index Optimization of a Rheological Dynamical System via Numerical Active Control

Paweł Olejnik and Jan Awrejcewicz

Abstract. Methods of active control can be adopted in optimization of fast impulsive response systems occurring in body interacting biomechanics. Active control of some mechanical or biological structures is not a new topic but can be still explored and successively used. The work focuses on application of one controlling force to minimize a relative compression of human chest cave that has been caused by some impacting action of an elastic external force. A virtual actuator controlling deformation in the analyzed rheological dynamical system of three degree of freedom acts between humans back and a supporting it fixed wall. Reduction of internal displacements in the thorax has been estimated solving the linear quadratic regulator (LQR) optimization problem. Solution of the system dynamics, Riccati's equation and the objective function's $J(t_0, t_f)$ minimization have been conducted in a numerical way with the use of Python programming language. Time histories of the controlled and non-controlled system responses, evaluation of the response's shape after changing coefficients of the control method as well as dependency of the objective function's estimation on the proportional gain vector are presented and discussed.

1 Introduction

It is obvious that todays modern high modern technology cannot exists without modeling and computer simulations. These tools bring an important information about specific properties of the model being under investigation that could be checked and validated in the final design of many products. Usually, it is one of the earliest stages of products design, its meaning and observations are seriously taken into consideration. Most of computer simulations in the field of biomechanics of human, animals or flora are shared out between some finite elements methods [8] and classical rigid

Paweł Olejnik · Jan Awrejcewicz
Department of Automation and Biomechanics, Technical University of Łódź, 1/15 Stefanowski St., 90-924 Łódź, Poland
e-mail: {olejnikp, awrejcew}@p.lodz.pl

G. Stépán et al. (Eds.): Dynamics Modeling & Interaction Cont., IUTAM BOOK SERIES 30, pp. 185–195.
springerlink.com

and multi-body mechanics. Active control of building structures was a motivation for the approach shown in the paper, therefore a few representative examples of the second branch can be found in [4].

Investigations on tension or deformation of a fully shape described parts of our surrounding nature lie in the scope of interest of the first field of science. This paper follows the second way to illustrate the possibility of global action on the analyzed structure, where dynamics of the bodies represented by point masses is reconstructed. It gives some valuable advantages like, for example, less complexity of the mathematical description, low costs of testing of the prototype before setting it to the construction, etc. There appear, as usually, some drawbacks resulting, for example, from omission of internal structure deformations of the investigated bodies, and others. Disregarding of wear phenomena or influence of changes of temperature fields may serve as instances of these both concepts.

Numerical methods are today popular in biodynamics and, in particular, in dynamical analysis of communication accidents [5]. Generally, frontal impacts are considered to be the most common vehicle collision and causing many injuries [7]. When a human body is exposed to an impact load, soft tissues of the internal organs can sustain large stress and strain rate. To investigate the mechanical responses of the internal organs, sometimes complex modeling of the organs is required. Homogeneous and linear elasticity material properties are assigned to each part of the model, whereas the human cartilages and bones may have different material properties. In order to have more realistic representation, more complex tissue material properties should be applied [3].

Model development converges often to some advance and, by a mathematical description, some complex analyzes including passive or active control of some weak points. This work focuses on the linear quadratic regulator (LQR) method that is known to have very good robustness properties. These properties are independent on choosing of the weighting matrices in the objective function estimation (the cost functional), so if a control system belongs to the class, these robustness properties are definitely assured. It has been confirmed by some applications of the method in [1, 2, 9].

Observe that in the existing literature any use of linear quadratic regulation of impulsively acting load on a rigid body elastic and damped biomechanical model of human organs is reported, hence our considerations are the first attempt of checking of such possibility. Methodology concerns on a dynamical modeling of human chest cave as an elastic connection of its main components considered separately as a mechanical rigid body having a point-focused mass. Such assumption does some averaging in the behavior of thorax but is very useful for the examination of its controlled realization.

2 Dynamical Modeling of the Analyzed Problem

Equilibrium of forces in the gravitational field leads to a system of three second order differential equations and a one of first order (because of the massless point of coupling at x_4 (see Fig. 1 b), valid for the rheological properties of the inner part of the thorax). One could said that we deal with a three and a half degree-of-freedom mechanical model.

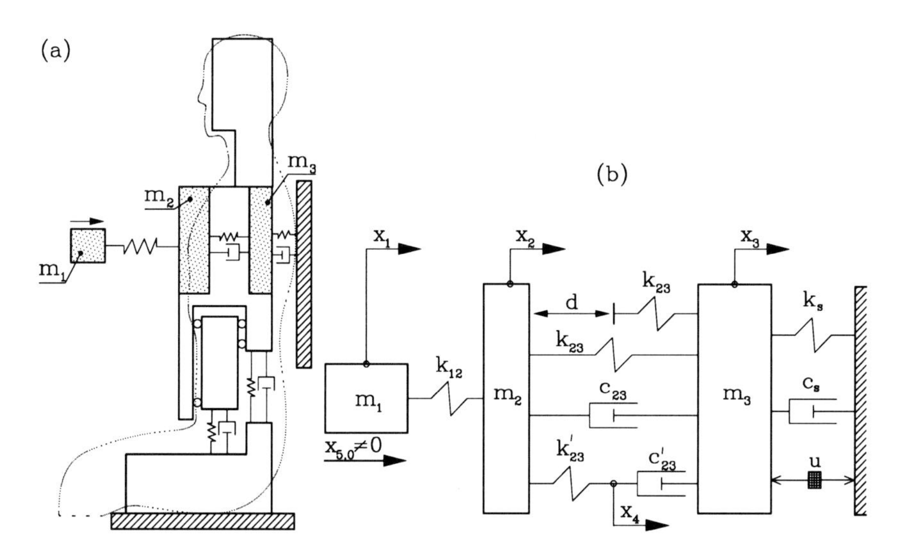

Fig. 1 A simplified scheme of a sitting human body (a) and the marked out investigated region of the chest. Particular redraw of the third degree of freedom of the dynamical system (b) with mass m_1 elastically impacting the chest (in the posterior surface).

The system of equations has been written (for further numerical purpose) in the form of seven first order differential equations

$$\dot{x}_i = x_{i+4}, \quad (i = 1,2,3)\,, \tag{1}$$

$$\dot{x}_4 = x_7 + \frac{\bar{k}_{23}}{\bar{c}_{23}}(x_2 - x_4)\,, \tag{2}$$

$$\dot{x}_5 = \frac{1}{m_1}\big(k_{12}(x_2 - x_1)\big)\,, \tag{3}$$

$$\dot{x}_6 = \frac{1}{m_2}\big(k_{12}(x_1 - x_2) - k_{23}(x_2 - x_3) - \bar{k}_{23}(x_2 - x_4) - c_{23}(x_6 - x_7)\big)\,, \tag{4}$$

$$\dot{x}_7 = \frac{1}{m_3}\big(k_{23}x_2 - (k_{23} + k_s)x_3 + c_{23}x_6 - (c_{23} + \bar{c}_{23} + c_s)x_7 + \bar{c}_{23}x_8 - u\big)\,, \tag{5}$$

where: $m_1 \dots m_3$ denote the separated point masses of the model, $\bar{k} = [k_{12}, k_{23}, k'_{23}, k_s]$, $\bar{c} = [c_{23}, c'_{23}, c_s]$ are the vectors of system stiffness and damping (respectively), $\bar{x}_d = [x_1 \dots x_4]$ is the vector of system displacements in each direction, $\bar{x}_v = [x_5 \dots x_8]$ is the vector of system velocities, but with regard to the introduced rheological description of the model the massless point reduces dimension of the system to 7. Thus, object of control (the plant) is described by an odd-dimension system state vector, means $\bar{x} = [\bar{x}_d, x_5 \dots x_7]$.

Rheological properties of the mechanical rigid bodies (here, in point focused masses) connection model are introduced in two places: firstly, when a relative displacement $x_r = x_2 - x_3$ (the controlled reference distance) between front and back side of the thorax ranges over $d = 3.8$ cm then k_{23} doubles its value, and secondly, when a relative velocity $v_r = x_6 - x_7$ becomes negative then c_{23} doubles its value as well. It means that stiffness of the rheological coupling increases discontinuously with regard to a bigger than d compression of chest cave x_r and that damping ability of the coupling varies in time as the thorax is under compression or depression. Such discontinuity in the system's stiffness and damping vectors is very interesting and will need a special attention during estimation of controlling force.

3 The Control Methodology

This paper takes into a numerical investigation possible modification of standard LQR optimization of proportional system control. A qualitative change in system dynamics comes from a discontinuous changes of system parameters. The case of such behavior results from the assumed biomechanical system of human thorax of which some material properties (in reality too) introduce to its dynamics the discontinuities of stiffness k_{23} and damping c_{23}. The way of their evaluation has been explained above in Sec.2.

For the particular case, the analyzed problem has been mathematically described in [6]. The state-space representation of the dynamical system takes the form

$$\begin{aligned}
\frac{d\bar{x}}{dt} &= \mathbf{A}\bar{x} + \mathbf{B}\bar{u} = \\
&= \begin{bmatrix} 0 & 0 & 0 & 0 & 1 & 0 & 0 \\ 0 & 0 & 0 & 0 & 0 & 1 & 0 \\ 0 & 0 & 0 & 0 & 0 & 0 & 1 \\ 0 & \frac{\bar{k}_{23}}{\bar{c}_{23}} & 0 & \frac{-\bar{k}_{23}}{\bar{c}_{23}} & 0 & 0 & 1 \\ \frac{-k_{12}}{m_1} & \frac{k_{12}}{m_1} & 0 & 0 & 0 & 0 & 0 \\ \frac{k_{12}}{m_2} & \frac{-a_{62}}{m_2} & \frac{k_{23}}{m_2} & \frac{\bar{k}_{23}}{m_2} & 0 & \frac{-c_{23}}{m_2} & \frac{c_{23}}{m_2} \\ 0 & \frac{-a_{72}}{m_3} & \frac{-a_{73}}{m_3} & \frac{-\bar{k}_{23}}{m_3} & 0 & \frac{c_{23}}{m_3} & \frac{-a_{77}}{m_3} \end{bmatrix} \bar{x} + \begin{bmatrix} 0 \\ 0 \\ 0 \\ 0 \\ 0 \\ 0 \\ \frac{-1}{m_3} \end{bmatrix} \bar{u}\,, &(6) \\
\bar{y} &= \mathbf{C}\bar{x} + \mathbf{D}\bar{u} = \\
&= \begin{bmatrix} 0 \; 1 \; -1 \; 0 \; 0 \; 0 \; 0 \end{bmatrix} \bar{x}\,, &(7) \\
\bar{x}(t_0) &= [0, 0, 0, 0, x_5(0) \neq 0, 0]\,, &(8)
\end{aligned}$$

where: the state vector $\bar{x} = [x_1 \dots x_7]^T$, control input vector $\bar{u} = u$, state-space representation matrices: **A** - system matrix, **B** - input matrix, **C** - output matrix, **D** - input transforming matrix and constants: $a_{62} = k_{12} + a_{72}$, $a_{72} = k_{23} + \bar{k}_{23}$, $a_{73} = k_{23} + k_s$, $a_{77} = c_{23} + c_s$. Nonzero initial velocity $x_5(0)$ of the impacting mass (the impactor's mass is a part of the whole system) states the external excitation.

Our task focuses on searching for the control force $u(t)$ that at some weighting matrices would satisfactorily minimize the objective function J in time $t \in [t_0;t_f]$:

$$J(t_0,t_f) = \frac{1}{2}\int_{t_0}^{t_f} \begin{bmatrix} \bar{x}(t) \\ \bar{u}(t) \end{bmatrix}^T \begin{bmatrix} \mathbf{Q} & 0 \\ 0 & \mathbf{R} \end{bmatrix} \begin{bmatrix} \bar{x}(t) \\ \bar{u}(t) \end{bmatrix} dt =$$

$$= \frac{1}{2}\int_{t_0}^{t_f} \begin{bmatrix} x_1 \\ \vdots \\ x_7 \\ u \end{bmatrix}^T \begin{bmatrix} q_1 & & & \\ & \ddots & & \\ & & q_7 & \\ & & & r \end{bmatrix} \begin{bmatrix} x_1 \\ \vdots \\ x_7 \\ u \end{bmatrix} dt =$$

$$= \frac{1}{2}\int_{t_0}^{t_f} \left(\sum_{i=1}^{n=7} (q_i x_i^2(t)) + r u^2(t) \right) dt \,, \tag{9}$$

where weighting matrices of the LQR control method are: quality matrix $\mathbf{Q}$ has nonzero elements only on its main diagonal, reaction matrix $\mathbf{R} = r$ is reduced to a single constant. Computation of integral J will be done by means of the numerical trapeze integration procedure. One needs to note, that $\bar{x}(t)$ represents a state-space vector of dynamical changes of the controlled system (if $u(t)$ is present in Eq. 5 then k_s and c_s are deleted).

Control law for the minimal realization of J in Eq. 9 is as follows:

$$u(t) = -r\mathbf{B}^T \mathbf{K}_x \bar{x}_f(t) \,. \tag{10}$$

Equation 10 introduces one new matrix $\mathbf{K}_x$ of dimension (7×7) called the Riccati matrix. This matrix is symmetrical along the main diagonal, so we get 28 unknown elements ($\xi_{21} = \xi_{12}$, $\xi_{31} = \xi_{13}$ and so on) that need to be estimated. Observe that the sought control law $u(t)$ is governed by a proportional relation to the state vector solution $\bar{x}_f(t)$ of the free system. It is confirmed here that the best method of estimation of the K_x matrix and thereby estimation of $u(t)$ is utilization of a proper convergent numerical procedure. This procedure solves the following matrix equation

$$\left(\dot{\mathbf{K}}_x + \mathbf{K}_x\mathbf{A} + \mathbf{A}^T\mathbf{K}_x - \mathbf{K}_x\mathbf{B}\frac{1}{r}\mathbf{B}^T\mathbf{K}_x + \mathbf{Q} \right) \bar{x}_f(t) = 0 \,, \tag{11}$$

Resolving Eq. 11 with respect to all provided forms of matrices brings 28 first order differential equations on each independent element of the symmetric matrix $\mathbf{K}_x$.

Exemplary expansion of the first equation on $\xi_{1,1}$ element of $\mathbf{K}_x$ holds

$$\frac{d\xi_{1,1}}{dt} = 2k_{12}\left(\frac{\xi_{1,5}}{m_1} - \frac{\xi_{1,6}}{m_2} \right) + \frac{\xi_{1,7}^2}{m_3^2 r} - q_1 \,. \tag{12}$$

Eq. 10 can be used now for calculation of the control law

$$u(t) = \bar{f}_x \cdot \bar{x}_f(t) = \left[\frac{\xi_{i,7}}{rm_3} \right] \cdot \bar{x}_f(t) = \sum_{i=1}^{7} \frac{\xi_{i,7} x_{f,i}}{rm_3} \,, \tag{13}$$

where $\bar{f}_x = -r\mathbf{B}^T\mathbf{K}_x$ is the proportional gain of the control feedback loop. Because of the specification of the analyzed rheological dynamical system $\bar{f}_x$ will switch between two states as it has been described below.

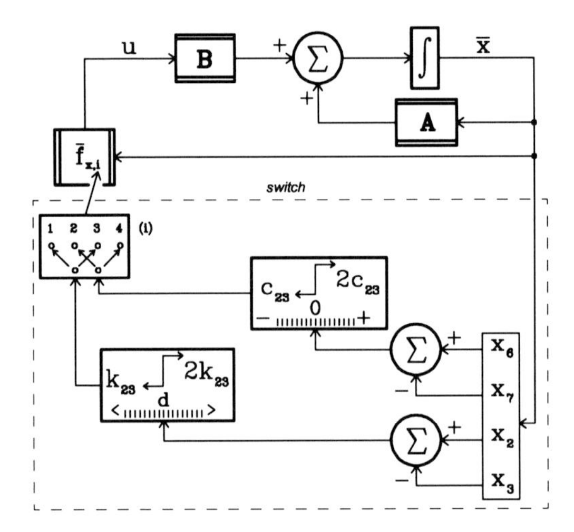

Fig. 2 A block diagram of the standard LQR system control with an additional switching function selecting the proper i-th constant gain vector of the state-feedback loop.

A block diagram of the considered system with presence of the proportional control from the full vector of state is schemed in Fig. 2. Before the scalar product of time-dependent vector of state and the vector of constant gain $(\bar{x} \cdot \bar{f}_{x,i})$ is calculated, one needs to regard to rheology of the bio-inspired model and to select the proper vector of proportional gain. Because of two 2-state switching blocks there are four combinations like $[k_{23}, c_{23}]$, $[k_{23}, 2c_{23}]$, $[2k_{23}, c_{23}]$ and $[2k_{23}, 2c_{23}]$. As the numerical solution of the free system dynamics confirms, x_r is always greater than d. Therefore, Riccati matrix equation associated with the controlled system have to be solved only twice (this time, the two last combinations are unnecessary), and during numerical solution to the control problem, as the relative velocity v_r varies in time, the first element of damping vector $\bar{c}$ will switch between two values c_{23} and $2c_{23}$ generating some unexpected disturbances of gain.

4 Numerical Simulation

The following set of system parameters is assumed in the numerical model: $m_1 = 1.6$, $m_2 = 0.45$, $m_3 = 27$ kg, $d = 3.8$ cm, $k_{12} = 281$, $k_{23} = 26.3$, $\bar{k}_{23} = 13.2$, $k_s = 10 \cdot 10^3$ N/m, $c_{23} = 1.23$, $\bar{c}_{23} = 0.18$, $c_s = 0.11 \cdot 10^3$ Ns/m and initial conditions: $\bar{x}_f(t_0 = 0) = [0, 0, 0, 0, 13.9, 0, 0, 0]$.

At first sight let us examine the switching nature of time histories of the first component of $\bar{f}_x$ that are matched by two scatter plots in Fig. 3.

The resulting shape of the component's characteristics that is taken to tune the control system is matched by solid line on which direction of a sudden switchings is pointed by some arrows.

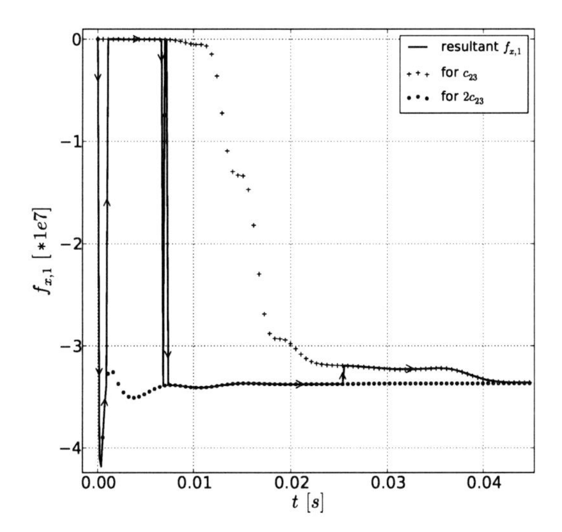

Fig. 3 Changes of first component of the control gain vector. Line with arrows exposes its switching nature that is the effect of the rheological properties of connection inside the chest model. Scatter plots represent two parallel transient $f_{x,1}$ time histories for c_{23} and $2c_{23}$.

Figure 3 presents first 250 iterations for $t_f = 0.045$ s of the first element of vector $\bar{f}_x$ and as it has been confirmed, its 60-th iteration at (0.0108 s) provides this minimal value J_{min} of the objective function J. It refers to the first value of $\bar{f}_{x,1} = -3.40e7$ in Tab. 1.

Table 1 Elements of $j = 60$ iteration of $\bar{f}_x$ dependent on sign of the relative velocity v_r, the optimal tunning ($h = 1.8e-4$, numerical integration step) ensures $J_{min} = 1.39e7$ (see Fig. 5).

i		1 ↘	2	3	4	5	6	7
if $c_1 = c_{23}$,	$\bar{f}_{x,i} =$	$[-3.40e7,$	$3.66e7,$	$-3.89e6,$	$1.34e6,$	$-2.81e3,$	$1.57e5,$	$-1.53e5]$
if $c_1 = 2c_{23}$,	$\bar{f}_{x,i} =$	$[-8.35e6,$	$9.30e6,$	$6.40e5,$	$-3.12e5,$	$8.28e3,$	$4.40e3,$	$-1.28e4]$

Proportional control law is realized by a virtual actuating impulsive mechanism characterized by a rapid force response having a shape of $u(t)$, the time-dependency visible in Fig. 4. There is simultaneously visible actual relative displacement x_c of the controlled system supported with the actuating mechanism shown in Fig. 1.

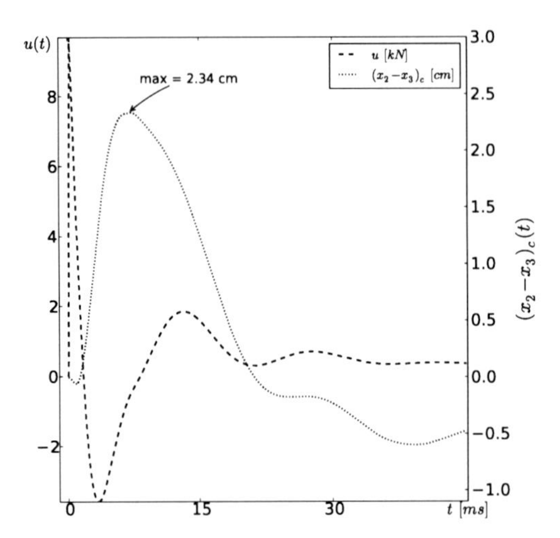

Fig. 4 Time history of control force applied on the anterior surface of the thorax model on background of the controlled system relative displacement x_c [cm]. For a better picture t_f on the time history is shortened to 40 ms

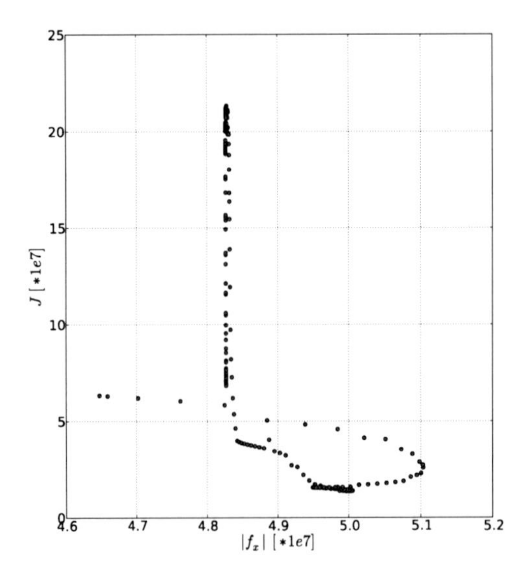

Fig. 5 Dependency of the objective function $J_f(t_f = 0.3)$ on the length of i-th ($i \in [5:250]$) iteration of the gain vector $|f_x|[i] = \sqrt{\sum_{k=1}^{7} f_{x,(k)}^2}[i]$.

Before $x_c(t)$ reaches its maximal value peak at 2.34 cm the control force $u(t)$ exposes very rapid changes of amplitude from about 10 to -4 kN and that is the most important part of the response taking about 10 ms. In practice, this time interval is probably longer and making it longer here could be possible after introduction of some time delays either in response of the mechanism or in a reaction of the shock-excited body. Nevertheless, one confirms that the method of active control can be used for finding a shape of force characteristics that has to be realized by the actuator.

One can find in Fig. 5 (visible here a collection of 246 points) that $J_{min} = 1.39e7$. This confirms that there has been chosen a gain vector of the LQR method that will guarantee the optimal (but not the minimal x_r, see Fig. 7) control response of the system. It is necessary to stress that this stage of control procedure tuning has to be preceded by as precise as possible selection of coefficients of matrix Q and the scalar value r.

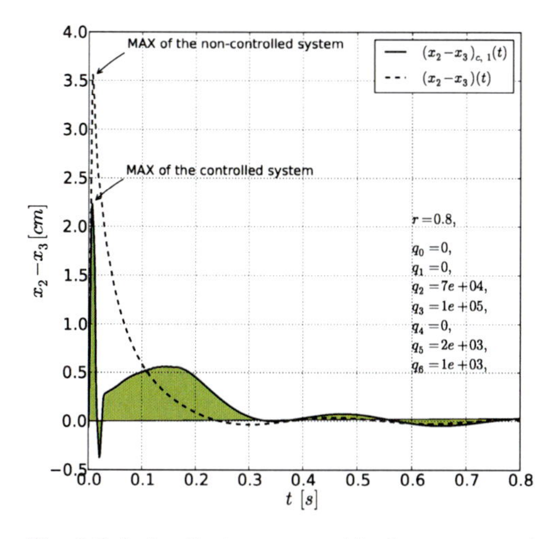

Fig. 6 Relative displacement $x_c(t)$ of mases m_2 and m_3 of the thorax before (dashed line) and after (solid line) the final estimation of control parameters.

The final result of control obtained with attention to J_{min} is presented in Fig. 6. The set of listed in its area parameters of the control procedure has been estimated on the basis of observation of minimal amplitude of $x_c(t)$. The last comparison of results is done in Fig.7. It shows that the control strategy still has some margin, because a decreasing of the r tuning control parameter gains a smaller maximal peak value of x_c. It reflects in a smaller compression of the chest cave but gains the second mode of the transitory vibration of which amplitude reaches almost 1.5 cm. Such evolution of the response makes the system not optimally controlled (as it is done in Fig. 6), but quite different effect of the control is achieved, the maximal compression is reduced to about 1.75 cm, and it does state about 50% reduction of deformation in comparison to the free system response.

A few responses of the controlled system plotted in Fig. 7 visualize gaining of the second mode's amplitude. Such undesirable time history of control variable having two big amplitudes in a short interval of time causes appearance of some jerks of the thorax.

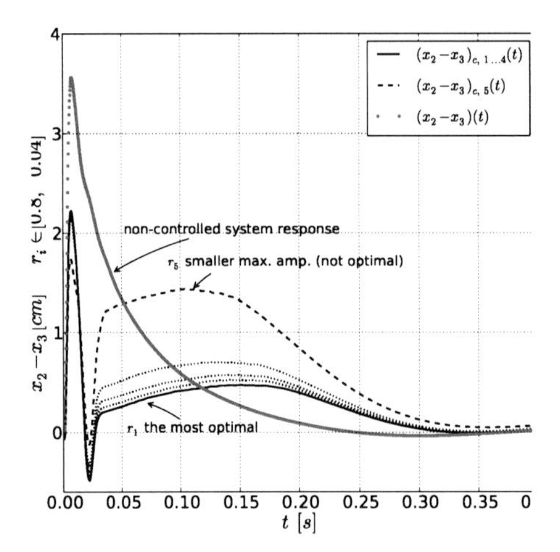

Fig. 7 Overtuning of r control parameter confirms that the most optimal controlled system response is found for $r_1 = 0.8$.

5 Concluding Remarks

In the presented example of interesting biomechanical application there has been obtained visible reduction of maximal amplitude of the output signal. Dependently on requirements the output response of the controlled system can be optimized with respect to minimum value of the performance index of the LQR method or even with respect to the minimal amplitude.

Thinking about practical implementation of the estimated shape of the controlling force to compensate an impact-induced compression, one would to design a very fast reaction forcing actuator that could be located on back side of the human thorax.

Acknowledgements. This paper has been financially supported by *Mistrz Programme* of the Foundation for Polish Science in 2010-2012.

References

1. Alavinsab, A., Moharami, H., Khajepour, A.: Active control of structures using energy-based LQR method. Computer-Aided Civ. and Infrastruct. Engineering 21, 605–611 (2006)
2. Bernussou, J., Peres, P.L.D., Geromel, J.C.: A linear programming oriented procedure for quadratic stabilization of uncertain systems. Syst. Control Lett. 13, 65–72 (1989)
3. Harrigan, T.P., Hamilton, J.J.: Necessary and sufficient conditions for global stability and uniqueness in finite element simulations of adaptive bone. International J. Solids and Structures 31(1), 97–107 (1994)
4. Jalihal, P., Utku, S.: Active control in passively base isolated buildings subjected to low power excitations. Computers and Structures 66(2–3), 211–224 (1998)

5. Noureddine, A., Digges, K.H., Bedewi, N.: An evaluation of deformation based chest injury using a hybrid III finite model. International J. Crashworthiness 1, 181–189 (1996)
6. Olejnik, P., Awrejcewicz, J.: Reduction of deformation in a spring-mass realisation of human chest occurred after action of impact. J. KONES Powertrain and Transp. 17(1) (2010)
7. Pietrabissa, R., Quaglini, V., Villa, T.: Experimental Methods in Testing of Tissues and Implants. Meccanica 37, 477–488 (2002)
8. Plank, G., et al.: Finite element modeling and analysis of thorax restrain system interaction. In: 14-th ESV Conf. 94-S1-0-16 (1994)
9. Uchida, K., Shimemura, E., Kubo, T., Abe, N.: The linear quadratic optimal control approach to feedback control design for systems with delay. Automatica 24, 773–780 (1988)

Characterization of a Least Effort User-Centered Trajectory for Sit-to-Stand Assistance

Viviane Pasqui, Ludovic Saint-Bauzel, and Olivier Sigaud

Abstract. Sit-to-stand transfer is a prerequisite for locomotion and induces a lot of effort from elderly or disabled people. In the context of a project based on a locomotion and sit-to-stand assistance robotics device, we present a methodology to tune the trajectory of active handles so that the verticalisation effort of the user is minimised. The methodology is user-centered in the sense that the robot will generate a specific trajectory for each particular user.

1 Introduction

A robotic device to assist the locomotion function of elderly and disabled people may improve a lot their autonomy, the efficacy of rehabilitation efforts and, more generally, may improve their quality of life. But assisting locomotion is of less interest if it does not help the user in transfering himself or herself from the sitted position to the standing position. As a matter of fact, very few robotics systems are endowed with both a locomotion assistance and a verticalisation assistance capability [7], [1], [11]. However the sit-to-stand motion induces a lot of effort from the patient. Thus it is of the most importance to design the robotic assistance device so that this effort is minimised.

In this paper, we present a preliminary study of a user-centered methodology that is designed to generate an assistance trajectory for robotic handles that minimises the effort from the user depending on his/her own sit-to-stand transfert strategy.

The generated trajectories are tailored to a particular user using a set of five parameters:

- P_I and P_F, the initial and final positions of the trajectory, are chosen by the user,

Viviane Pasqui · Ludovic Saint-Bauzel · Olivier Sigaud
Institut des Systèmes Intelligents et de Robotique - CNRS UMR 7222
Université Pierre et Marie Curie, Pyramide Tour 55 - Boîte Courrier 173,
4 Place Jussieu, 75252 Paris CEDEX 5, France
e-mail: {pasqui, saintbauzel}@isir.upmc.fr, olivier.sigaud@upmc.fr

G. Stépán et al. (Eds.): Dynamics Modeling & Interaction Cont., IUTAM BOOK SERIES 30, pp. 197–204.
springerlink.com

- T_F, the sit-to-stand transfert time, is set by the experimenter,
- (dev_1, dev_2) describe the shape of the trajectory, they must be different for each user. Our method tunes dev_1 and dev_2 so as to minimise the effort.

The paper is organised as follows. In Section 2, we present Monimad, the robotic assistance device that we use, as well as our method to generate trajectories. In Section 3, we describe our method to choose among five trajectories the one that is most adequate for a particular user, and then our method to tune dev_1 and dev_2 depending on the user effort. In Section 4, we present the preliminary empirical results that we obtained with the method. These results are discussed in Section 5 before we conclude.

2 Robotic Device

Monimad is an original assistive device, combining sit-to-stand transfer and walking aid for elderly and disabled people. It allows mobility rehabilitation and assistance, safe walking (postural stabilization) and safe sit-to-stand transfer [11]. The designed robotic system is basically a two degrees of freedom arms mechanism mounted on an active mobile platform (Fig. 1). The lower part (mobile platform) consists of two electric motors actuating the wheels. The upper part of the mechanism (articulated arms) consists of two plane parallelograms to maintain the handles horizontally, actuated by linear actuators. In addition, they are independent in order to restore lateral balance. The end effectors consists of handles equipped with a six axis forces/torques sensor that are used in the experiments presented below to evaluate the relevance of the sit-to-stand trajectory. A preliminary analysis of the

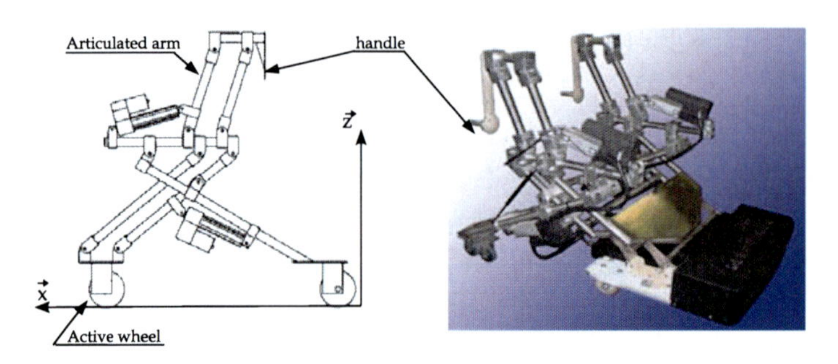

Fig. 1 Description of Monimad a Robotic assistant for sit-to-stand and walking

movement in elderly sit-to-stand clinical trials, with a specific measurement system [11], has allowed to describe the trajectory of the hands during assisted movements. First, the handles must pull slowly the patient to an antepulsion configuration (see Fig. 2). Then, they go from this down position to the up position, used for walking. For each patient, several sit-to-stand transfer trajectories were recorded, some

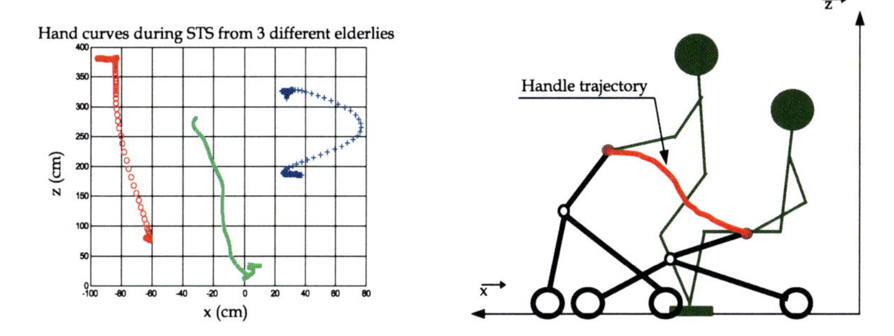

Fig. 2 Experimental sit-to-stand transfer trajectories

examples of these trajectories are given in Fig. 2. The analysis of these transfer trajectories shows that the global shape of the trajectory is a "s-like" curve and is not directly related to the age or height of the patient but seems to be correlated with its own personal strategy to stand up or sit-down. This seems to reflect invariants of the trajectory generation [14]. The trajectory of the handles has to be similar to the general curve presented in Fig. 3. The term "trajectory" refers here to Cartesian-space

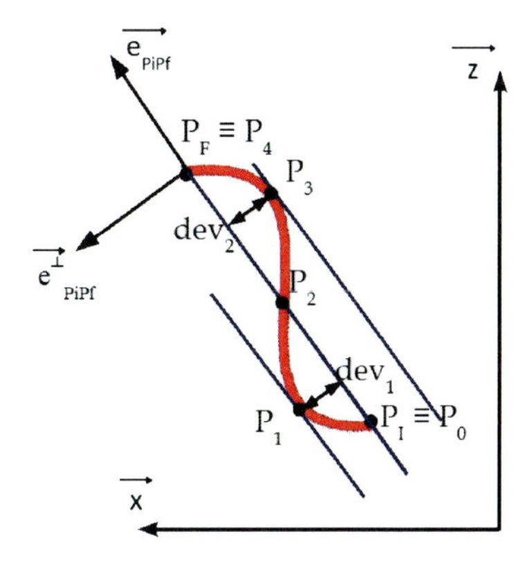

Fig. 3 The handles trajectory in cartesian space. This kind of curves look like a third order polynom in the $(P_i, \mathbf{e}_{P_iP_f}, \mathbf{e}^{\perp}_{P_iP_f})$ frame. The input of these curves are the initial and final handles positions (P_i and P_f) and the curvatures of both pieces of the curve (dev_1 and dev_2).

planning of the handle movement. A "natural" trajectory is requested for comfortable human movement assisted by robotic devices. From our point of view, "natural" means that the trajectory path must be compatible with hand movements when the sit-to-stand transfer is assisted by someone else. It must also be smooth and generate a continuous motion of the hand. As proposed in [15], smoothness can be quantified as a function of jerk, which is the time derivative of the acceleration.

The method to obtain such a trajectory consists in decomposing its characteristics into a physiological part and a mechanical part. The minimum jerk criterion is a physiological constraint for smoothness and is only related to trajectory quality. Thus, the curvilinear abscissa is used to describe the law of motion satisfying the minimum jerk criterion.

Let $s(t)$ represents the distance that the handles have moved along the curve at instant t. The curvilinear abscissa $s(t)$ defined by Eq. 1 (see [15]), guarantees the smoothness of the handles trajectory.

$$s(t) = s(T_i) + (s(T_f) - s(T_i))(10(\frac{t}{T_f - T_i})^3 - 15(\frac{t}{T_f - T_i})^4 + 6(\frac{t}{T_f - T_i})^5) \quad (1)$$

Where T_i is the initial time and T_f is the final one.

The geometrical path describing the hand are not time dependent and may be expressed in terms of Euclidean coordinates. An assisted sit-to-stand transfer trajectory follows a path similar to the one shown in Fig. 3.

This kind of curve will be defined by a third order polynomial in the $(P_i, \mathbf{e}_{P_iP_f}, \mathbf{e}^{\perp}_{P_iP_f})$ plane (with : $\mathbf{e}^{\perp}_{P_iP_f} = \mathbf{z} \wedge \mathbf{e}_{P_iP_f}$). Considering the point P_j on the handle path curve: $\mathbf{P_iP_j} = U_j \mathbf{e}_{P_iP_f} + V_j \mathbf{e}^{\perp}_{P_iP_f}$, with : $V_j = \sum_{i=0}^{3} \alpha_i U_j^i$. The knowledge of the coordinates of the points P_k $(k = 0, \ldots, 4$, Fig.3) leads to the equations below:

$$\begin{cases} \alpha_0 = 0 \\ \alpha_1 U_1 + \alpha_2 U_1^2 + \alpha_3 U_1^3 = V_1 \ (= dev_1) & (A) \\ \alpha_1 U_2 + \alpha_2 U_2^2 + \alpha_3 U_2^3 = 0 & (B) \\ \alpha_1 U_3 + \alpha_2 U_3^2 + \alpha_3 U_3^3 = V_3 \ (= dev_2) & (C) \\ \alpha_1 U_4 + \alpha_2 U_4^2 + \alpha_3 U_4^3 = 0 \ (where\ U_4 = \overline{P_iP_f}) & (D) \\ \left(\frac{dV}{dU}\right)_{U_1} = \alpha_1 + 2\alpha_2 U_1 + 3\alpha_3 U_1^2 = 0 & (E) \\ \left(\frac{dV}{dU}\right)_{U_3} = \alpha_1 + 2\alpha_2 U_3 + 3\alpha_3 U_3^2 = 0 & (F) \end{cases}$$

A solution of this system is obtained in two steps. First, solving the linear system $\{(A), (B), (C)\}$, α_i coefficients can be expressed in term of U_1, U_2, U_3. In a second step, U_1, U_2, U_3 values are the solutions of the minimisation problem based on equations $\{(D), (E), (F)\}$ defined as:

$$min((D)^2 + (E)^2 + (F)^2) \text{ under } \begin{cases} -U_1 < 0 \\ U_1 - U_2 < 0 \\ U_2 - U_3 < 0 \\ U_3 - U_4 < 0 \end{cases} \quad (2)$$

The starting point of the optimization procedure, is taken such that:

$$\begin{cases} -U_1 < 2\% \text{ of the length curve } (s(T_f) - s(T_i)) \\ U_1 - U_2 < 2\% \text{ of the length curve} \\ U_2 - U_3 < 2\% \text{ of the length curve} \\ U_3 - U_4 < 2\% \text{ of the length curve} \end{cases} \quad (3)$$

We have observed that the resulting polynomial is always very close to the experimental curve. The handle trajectory in spatiotemporal space is defined by U and V coordinates as time functions, such that

$$(\frac{dU}{ds})^2 + (\frac{d}{ds}(\sum_{i=1}^{3} \alpha_i U^i))^2 = 1$$

Finally, the trajectory is a time function of Cartesian coordinates generating a smooth movement. Estimated coordinates of the point P are computed with the following recursive scheme:

$$\begin{cases} U(k+1) = \frac{(t(k+1)-t(k))(4\dot{s}_M \cdot t(k) \cdot (T_f - t(k)))}{(T_f^2 \sqrt{1+(\sum_{i=1}^{3} i\alpha_i U(k)^{i-1})^2})} + U(k) \\ V(k+1) = \sum_{i=1}^{3} \alpha_i U(k+1)^i \end{cases} \quad (4)$$

And $\dot{s}_M$ is computed with the handle path length using simultaneously the curvilinear abscissa and the Euclidean coordinates:

$$\dot{s}_M = \frac{3(\sum_{i=0}^{N} \sqrt{(U_{i+1} - U_i)^2 + (V_{i+1} - V_i)^2})}{2T_f} \quad (5)$$

In order to observe the posture of the user, a six axis force/torque sensor is placed under the user's feet to measure the ground contact forces and the forces applied to the handles. As a consequence, the anticipatory postural adjustement (APA) determination can be used to initiate the sit-to-stand motion of the robot handles. Both ground and hand forces, treated with fuzzy logic laws, give information on the sit-to-stand transfer phase and on the stability state of the user [16].

3 Experimental Study

The goal of our experiment is to tune the dev_1 and dev_2 parameters of the trajectory so as to ensure that the user makes as few effort as possible. Our methology consists of three steps:

- First, we recorded the global effort of the users along a set of five trajectories that have been shown through clinical experiments to provide a satisfactory basis for sit-to-stand assistance trajectory generation [14]. The active handles are controlled in position, than the user has no other choice then following the imposed trajectory. The difference between the natural trajectory the subject would use if helped by a human assistant and the imposed trajectory generates an additional effort from the user. Thus we assume that the most natural trajectory is one that generates the least global effort from the user.
- In a second step, we control the active handles with some impedance, so that the user can locally alter the handles trajectory. We assume that the subject will alter the handles trajectory so as to make it more comfortable, provided that the corresponding additionnal effort is not too high.

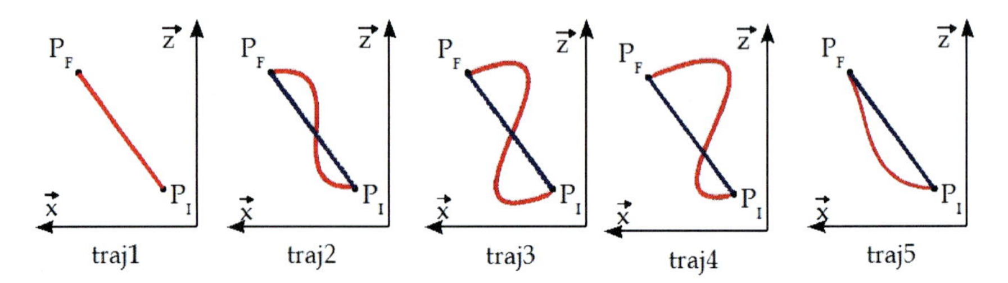

Fig. 4 Trajectories for experiments

The altered trajectory corresponds to modified dev_1 and dev_2 parameters that can be extracted from this second step [17].

- In the third step, we record again the total effort of the user with a position controlled handles trajectory, where the trajectory is generated according to the dev_1 and dev_2 values determined in the second step. We check that the global effort is smaller than initially.

4 Results

The experiments were performed with 8 subjects. All were healthy male volunteers between 25 and 30 years old (average: height :175,3 $\pm$ 5,36 cm, weight : 66,2 $\pm$ 9,91 kg). They were instructed to stand-up as naturally as possible. Every subject performed 10 sit-to-stand tranfers. The data corresponding to one subject has been discarded as inconsistent.

Among the subjects, three select traj5 in Fig. 4, two select traj2 and, traj1 and traj3 are selected by one subject.

These trajectories are the least deformed when the verticalisation is made with the active handles controled with an impedance law. That is that among the proposed trajectories, the one that is followed most closely is the one where the sum of efforts measured at the hands and at the feet are the least. The new parameters dev_1, dev_2 are determined from the average of ten sit-to-stand trajectories recorded, when the active handles are controled with an impedance law.

Now these trajectory are used again with the active handles controlled in position. Fig. 5 shows, for one subject, that the resulting trajectory is not deformed a lot, so validating the hypothesis according to which the most comfortable trajectory is the one which brings the least effort.

5 Discussion

The fact that different subjects select diverse trajectories among our basis of five trajectories reveals the importance of a user-centered approach.

Indeed, given that each subject has his own sit-to-stand transfer strategy, any methodology that would propose an unique trajectory as optimal for all the subjects

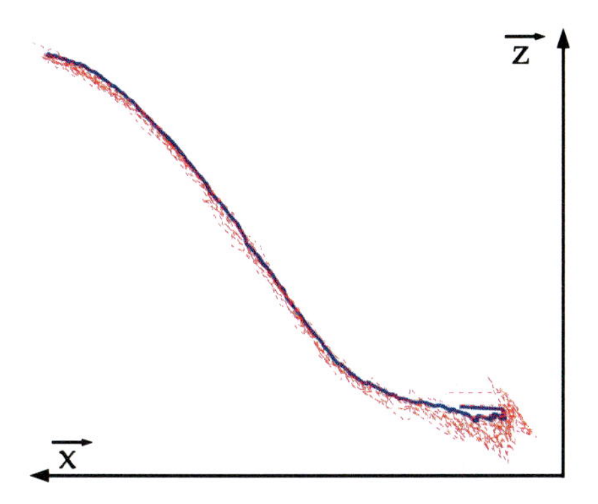

Fig. 5 The solid thick trajectory is made with the active handles controlled in position and the dashed thin ones are recorded when the active handles are controlled with an impedance law

would fail to take their specificities into account and would generate unnecessary effort and discomfort for some of these subjects.

Second, the fact that our method effectively results in a global decrease of effort for all subjects is encouraging. However, we cannot claim for any of the subject that we have found the trajectory that would minimize his effort. To further optimize the dev_1 and dev_2 parameters, we will call in the future upon gradient descent approaches that are guaranteed to reach at least a local optimum, provided enough data.

6 Conclusion and Future Work

In this paper, we have presented a preliminary methodology that validated the feasability of minimising the effort of users in the context of robot-assisted sit-to-stand transfert. The method shapes the active handles trajectory just by tuning two parameters, dev_1 and dev_2. Immediate future work will consist in calling upon gradient descent tools to further optimize the parameters. Then, the next step will consist in incorporating the method in the broader context of our robotic-assisted locomotion project and, in parallel, in evaluating its benefits with elderly and disabled patients.

Acknowledgements. This work is supported by the French ANR (Agence Nationale de la Recherche) TecSan program under grant No ANR-08-TECS-009-02 (MIRAS Project).

References

1. Chuy, O., Hirata, Y., Wang, Z., Kosuge, K.: Approach in assisting a sit-to-stand movement using robotic walking support system. In: International conference on intelligent robots and systems, Beijing, China, pp. 4343–4348 (2006)

2. Lacey, G., Dawson-Howe, K.M.: The Application of Robotics to a Mobility Aid for the Elderly Blind. In: Robotics and Auton. Systems, vol. 23, pp. 245–252 (1998)
3. Alwan, M., Rajendran, P.J., Ledoux, A., Huang, C., Wasson, G., Sheth, P.: Stability Margin Monitoring in Steering-Controlled Intelligent Walkers for the Elderly. In: AAAI Fall 2005 Symposium (EMBC). AAAI Press, Menlo Park (2005)
4. Graf, B., Hans, M., Rolf, D.S.: Care-O-Bot II- Development of a next generation robotic home assistant. In: Autonomous Robots, vol. 16, pp. 193–205 (2004)
5. Dubowsky, S., Genot, F., Godding, S., Kozono, H., Skwersky, A., Yu, H., Yu, L.S.: PAMM- A robotic aid to the elderly for mobility assistance and monitoring: a "'helping-hand"' for the elderly. In: IEEE International Conference on Robotics and Automation, pp. 570–576 (2000)
6. Wolfe, R.R., Jordan, D., Wolfe, M.L.: The WalkAbout: A new solution for preventing falls in the elderly and disabled. Arch. Phys. Med. Rehabil. 85(12), 2067–2069 (2004)
7. Lee, C.Y., Seo, K.H., Kim, C.H., Oh, S.K., Lee. J.J.: A system for gait rehabilitation: mobile manipulator approach. In: Proc.of IEEE of Int. Conf. on Robotics and Automation, pp. 3254–3259 (2002)
8. Nemoto, Y., Egawa, S., Fujie, M.: Power Assist Control Developed for Walking Support. Journal of Robotics and Mechatronics 11(6), 473–476 (1999)
9. Kuzelicki, J., Zefran, M., Burger, H., Bajdand, T.: Synthesis of standing-up trajectories using dynamic optimization, gait and posture (2005)
10. Wisneski, K., Johnson, M.: Quantifying kinematics of purposeful movements to real, imagined, or absent functional objects: Implications for modelling trajectories for robot-assisted ADL tasks. Journal of NeuroEngineering and Rehabilitation 4(7) (2007)
11. Médéric, P., Lozada, J., Pasqui, V., Plumet, F., Bidaud, P., Guinot, J.C.: An optimized design for an intelligent walking aid. In: CLAWAR 2003, Catania, Italy, pp. 53–60 (2003)
12. Médéric, P.: Conception et commande d'un système robotique d'assistance à la verticalisation et à la déambulation. PhD thesis, Université Paris (December 6, 2006)
13. Médéric, P., Pasqui, V., Plumet, F., Bidaud, P.: Design of a walking-aid and sit to stand transfer assisting device for Elderly people. In: 15th CISM-IFToMM Symposium on robotic design, Dynamic and control, St Hubert, Canada (2004)
14. Médéric, P., Pasqui, V., Plumet, F., Bidaud, P.: Elderly people sit-to-stand transfert experimental analysis. In: 8th Conference on climbing and walking robots, CLAWAR, Madrid, Spain (2005)
15. Flash, T., Hogan, N.: The coordination of arm movements: an experimentally confirmed mathematical model. The journal of Neuroscience 5(7), 1688–1703 (1985)
16. Saint-Bauzel, L., Monteil, I., Pasqui, V.: A reactive robotized interface for lower limb Rehabilitation: Clinical results. In: IEEE Int. Conference on Robotics and Automation, pp. 283–295 (2007)
17. Pasqui, V., Bidaud, P.: Bio-mimetic trajectory generation for guided arm movement during assisted sit-to-stand transfer. In: 9th International Symposium on Climbing and walking Robots and Associated Technologies, September 11–14, Royal Military Academy, Brussels (2006)
18. Khalil, W., Dombre, E.: Modélisation identification et commande des robots. In: Khalil, W., Dombre, E. (eds.) Commande en effort, 2nd edn., Janvier, pp. 397–415 (1999)

Micro-electromechanical Systems

This section is devoted to papers dealing with modeling and vibration control of micro-electromechanical systems. The work of A. Miklós introduces a multibody model of longitudinal elastic rods for doubly clamped MEMS resonators. In the paper of V. Puzyrev the mechanisms to control elastic wave propagation in circular and hollow piezoceramic cylinders are studied. The problems of design and actuation as well as structural health monitoring of composite shells are addressed in the papers by P. Kedziora and A. Muc. The paper by M. Chwal gives an insight into vibration control of defects in carbon nanotubes. Finally, the work by S. Ramakrishnan study the dynamics of an array of nonlinear micromechanical oscillators subjected to a combination of deterministic and white noise excitations.

Multibody System Model of MEMS Resonator

Ákos Miklós and Zsolt Szabó

Abstract. The present work proposes a multibody model for doubly clamped MEMS resonators. First, frequently used models are listed from literature. Next, the equation of motion is derived for the model, which can also be written in dimensionless form. Then the static behaviour of the model is checked, and finally, the dynamics of the system is investigated. With the obtained model it is possible to identify nonlinear behaviour arising from the geometric nonlinearity of the system.

1 Introduction

As an effect of the advances in production of Micro-Electro-Mechanical Systems (MEMS) nowadays MEMS resonators are widely used. Typical areas of use are Radio Frequency Filters (RF Filters) and other signal processing applications such as gyroscopes. Further, MEMS devices prove in many cases optimal solution because of its compact dimensions, small mass and large strength to mass ratio. Because of this last feature large displacements are enabled in MEMS devices, thus nonlinear effects can be showed, even if the material does not reach its linear elastic limit. A typical example for that is the doubly-clamped MEMS resonator, which has a significant longitudinal tension in case of large lateral displacement, so the lateral stiffness is growing with the displacement. Another nonlinear effect is, that in most cases the excitation force is an electrostatic force, which is a nonlinear function of the displacement. The resonator and a base electrode form a capacitor, and the force acting on the two parts depends on the distance between them. This electrostatic driving force has a softening effect on the lateral stiffness,

Ákos Miklós
Department of Applied Mechanics, Budapest University of Technology and Economics, Műegyetem rkp. 5., 1111 Budapest, Hungary
e-mail: miklosa@mm.bme.hu

Zsolt Szabó
Department of Applied Mechanics, Budapest University of Technology and Economics, Műegyetem rkp. 5., 1111 Budapest, Hungary
e-mail: szazs@mm.bme.hu

G. Stépán et al. (Eds.): Dynamics Modeling & Interaction Cont., IUTAM BOOK SERIES 30, pp. 207–214.
springerlink.com

and in case of strong excitation it can lead to the Pull-in effect, which is an important limitation for MEMS resonators [1].

For the modelling of MEMS resonators various approaches are known. Lifshitz and Cross used a one degree-of-freedom model to show nonlinear behaviour and to study resonator arrays [1]. Mestrom *et. al.* showed the softening effect of the electrostatic excitation by simulating a one degree-of-freedom model and the results were compared with measurements [2]. Doubly-clamped MEMS resonators are often modelled by beams. Han *et. al.* investigated the four widely used engineering theory for beams: the Euler-Bernoulli, Rayleigh, Shear and Timoshenko theories, but none of them handles the longitudinal stretch of the beam [3]. Ahmadian *et. al.* derived a complete nonlinear finite element model for electrostatic driven MEMS resonators [4]. Rubinstein used a multibody model for flexible robotic arms, which takes into account large displacements, but longitudinal stretch is also neglected in this model [5].

The current work presents a multibody model of longitudinal elastic rods to study doubly clamped MEMS resonators. The bending stiffness is modelled by torsion springs in the joints between the adjacent rods. After deriving the equations of motion we study the static behaviour of the model, finally, nonlinear effects are investigated in the case of harmonic excitation.

2 Multibody Model

To model the possible large displacements of a doubly-clamped beam the longitudinal stretch of the beam cannot be neglected. To do that we substitute the beam with longitudinal elastic rods, that are connected by joints. In the joints the bending stiffness is modelled by torsion springs (see Fig. 1.).

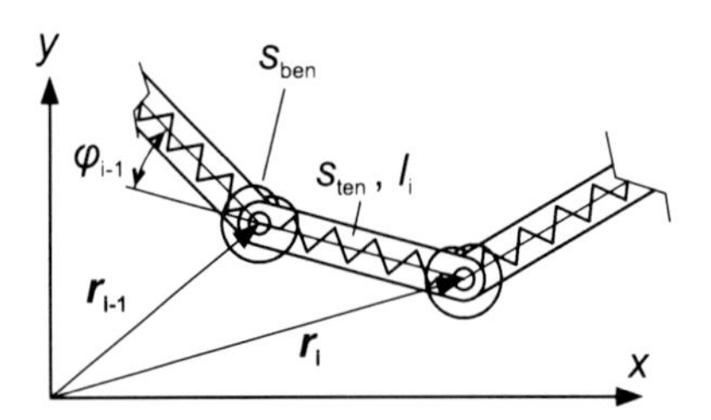

Fig. 1 Multibody model of longitudinal elastic rods

We choose the coordinates of the position vectors of the joints as generalized coordinates. The design of MEMS resonators is in most cases such that one of the second moment of inertia of the cross section is much higher than the other, so the vibration can be treated in plane. The longitudinal stiffness of the rods is denoted by s_{ten}, and the torsional stiffness of the torsion springs in the joints is represented by s_{ben}. By considering that the stretched lengths of the rods are l_i, the area and the second moment of inertia of its cross-section is A and J respectively, and the

Young-modulus of the material is E, the longitudinal and bending stiffness can be calculated as follows:

$$s_{\text{ten}} = \frac{AE}{l_i}; \; s_{\text{ben}} = \frac{J_z E}{l_i}. \tag{1}$$

To simplify the calculation we split the beam into rods having equal lengths l and same cross-sections A. Furthermore we assume, that the displacements inside the rods are linear functions of the local parameter p, which varies between the limits 0 and 1, and for the limits the function values are the displacements of the joints at both ends of the rod, respectively. So the kinetic energy of the whole system can be obtained as

$$T = \sum_{i=1}^{N} \int_0^1 \left(p\dot{x}_i + (1-p)\dot{x}_{i-1}\right)^2 + \left(p\dot{y}_i + (1-p)\dot{y}_{i-1}\right)^2 m\,dp\,, \tag{2}$$

where $m = \rho A l$ is the mass of the individual rods, $\dot{x}_i$ and $\dot{y}_i$ are the velocities associated with the choosen generalized coordinates and N is the number of the moving rods. In addition, the potential energy stored by the springs is

$$U = \frac{1}{2} s_{\text{ten}} \sum_{i=1}^{N} \Delta l_i^2 + \frac{1}{2} s_{\text{ben}} \sum_{i=0}^{N} \Delta \varphi_i^2\,, \tag{3}$$

where Δl_i is the stretch of the rods, and $\Delta\varphi_i$ denote the angle difference between the adjacent rods. $\Delta\varphi_i$ can be obtained from cross product of the direction vectors of the adjacent rods, so Δl_i and also $\Delta\varphi_i$ can be derived from the generalized coordinates as

$$\Delta l_i = \sqrt{\left(x_i - x_{i-1}\right)^2 + \left(y_i - y_{i-1}\right)^2} - l = l_i - l\,, \tag{4}$$

$$\Delta\varphi_i = \arcsin\left(\frac{\left(\mathbf{r}_i - \mathbf{r}_{i-1}\right)\times\left(\mathbf{r}_{i+1} - \mathbf{r}_i\right)}{l_i\, l_{i+1}}\right), \tag{5}$$

where l_i is the length of the stretched rods. After we derived the form of the kinetic and potential energy we can obtain the equations of motion for the system. The system of equations of motion can be written in matrix form as follows:

$$\mathbf{M}\ddot{\mathbf{x}} + \mathbf{S}_{\text{x,ten}}\left(\mathbf{x},\mathbf{y}\right) + \mathbf{S}_{\text{x,ben}}\left(\mathbf{x},\mathbf{y}\right) = \mathbf{f}_{\text{x}}\left(t\right), \tag{6}$$

$$\mathbf{M}\ddot{\mathbf{y}} + \mathbf{S}_{\text{y,ten}}\left(\mathbf{x},\mathbf{y}\right) + \mathbf{S}_{\text{y,ben}}\left(\mathbf{x},\mathbf{y}\right) = \mathbf{f}_{\text{y}}\left(t\right), \tag{7}$$

where $\mathbf{M}$ is a tridiagonal matrix with constant coefficients, $\mathbf{S}_{\text{x}}$ and $\mathbf{S}_{\text{y}}$ are pentadiagonal matrices. All matrices are linear functions of m, s_{ten} and s_{ben}, respectively. Because of the chosen generalized coordinates the equations for x and y directions can be decoupled. The equations of motion can be linearized around the static equilibrium position, and in this case the eigenfrequencies and eigenmodes of the system are in the 1% range of those obtained with the Euler-Bernoulli theory. The first four eigenmodes are shown in Fig. 2.

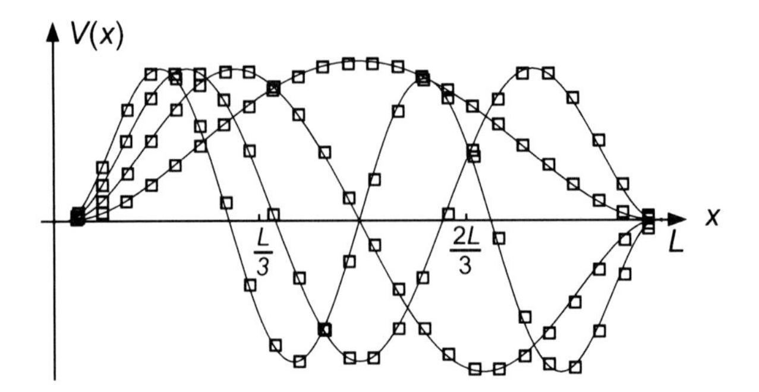

Fig. 2 Eigenmodes of the linearized system with multibody model using 25 rods (symbols) and Euler-Bernoulli theory (solid lines)

3 Dimensionless Equation of Motion

Since our model does not handle the damping of the material of the resonator and the surrounding medium, a simple modal damping is added to the equations of motion. Furthermore it is useful to investigate the dependence of the equations of motion on the parameters of the system, and to create the dimensionless equations

$$\hat{\mathbf{M}}\ddot{\mathbf{X}} + D\dot{\mathbf{X}} + \hat{\mathbf{S}}_{x,\text{ten}}(\mathbf{X},\mathbf{Y}) + \varepsilon\hat{\mathbf{S}}_{x,\text{ben}}(\mathbf{X},\mathbf{Y}) = f_0\mathbf{f}_x(\tau), \tag{8}$$

$$\hat{\mathbf{M}}\ddot{\mathbf{Y}} + D\dot{\mathbf{Y}} + \hat{\mathbf{S}}_{y,\text{ten}}(\mathbf{X},\mathbf{Y}) + \varepsilon\hat{\mathbf{S}}_{y,\text{ben}}(\mathbf{X},\mathbf{Y}) = f_0\mathbf{f}_y(\tau), \tag{9}$$

with

$$X_i = \frac{x_i}{L};\ Y_i = \frac{y_i}{L};\ \alpha = \sqrt{6\frac{s_{\text{ten}}}{m}};\ \beta = \sqrt{6\frac{s_{\text{ben}}}{m}};\ D = 6\frac{k}{m\alpha};\ \varepsilon = \frac{s_{\text{ben}}}{s_{\text{ten}}};$$

$$f_0 = \frac{6}{L\alpha^2 m};\ \tau = \alpha t, \tag{10}$$

where X_i and Y_i are the dimensionless position coordinates scaled by the total length L of the beam, α and β are angular velocities computed from longitudinal and bending stiffnesses respectively, but they do not have real physical meaning for the vibration of the beam. Symbol τ is the dimensionless time parameter, D is the dimensionless damping, ε is the dimensionless bending stiffness to longitudinal stiffness ratio and f_0 is the factor of the driving force.

4 Static Behaviour

The investigated resonator in the present work has the following dimensions and parameters

$$L = 80\,\mu\text{m};\ h = 1\,\mu\text{m};\ b = 3\,\mu\text{m};\ \rho = 2300\,\text{kg/m}^3;\ E = 130\,\text{GPa}\,,$$

where h is the height of the cross-section and b is the width of the cross-section normal to the plane of the motion. To check the dimensionless equations of motion a clamped-free beam is taken, and a force couple is applied on the free end as shown in Fig. 3. The beam is sectioned into 10 elastic rods, and the boundary conditions are applied by fixing the rod on the clamped end of the beam. The magnitude of the elements of the force couple applied on the ending rod is varied by the integer parameter n

$$F_n = 2\pi \cdot f_0 \frac{n}{10} \frac{JE}{L^2}, \tag{11}$$

which expression also implies that in case of $n = 10$ the beam is curved into a full circle.

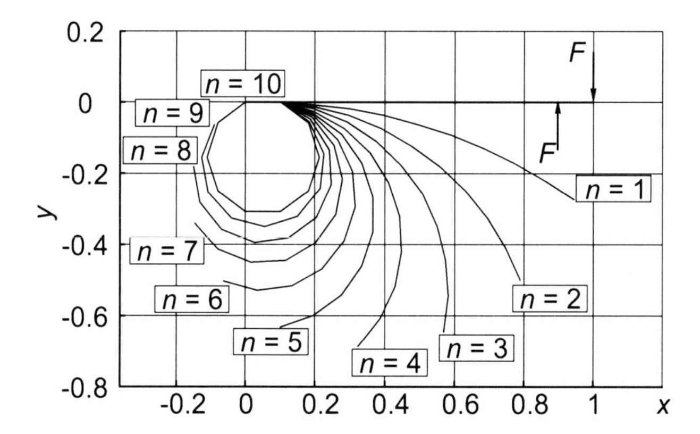

Fig. 3 Large deformations of a clamped-free beam loaded by force couple

The nonlinear behaviour of the clamped-clamped beam can be also shown. A ten-element model is loaded with constant force in all moving joints. The acting force in all joints is defined as follows:

$$F_n = 10^{-5} \cdot f_0 \frac{n}{5}, \tag{12}$$

where the distributed force components are scaled by the factor 10^{-5} in order to fit to the micro scale model. In Fig.4 the stiffening behaviour can be seen.

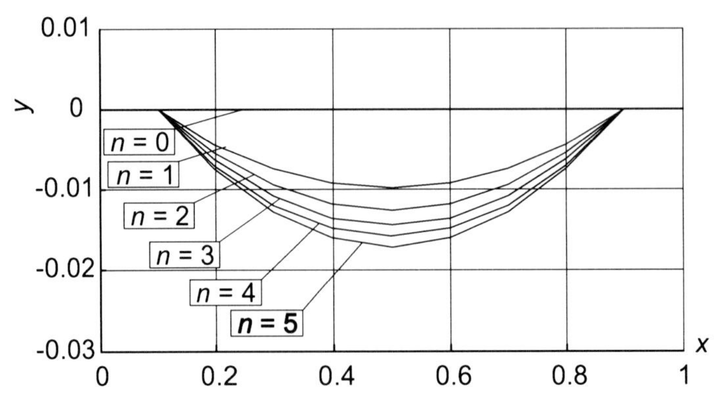

Fig. 4 Large deformations of a clamped-clamped beam loaded by constant force in all joints

5 Dynamic Behaviour

We also investigated the dynamic behaviour of the beam. The doubly clamped resonator was excited by equal harmonic forces.

$$F(\tau) = 10^{-5} \cdot f_0 \sin\left(\frac{\omega}{\alpha}\tau\right), \tag{13}$$

in all moving joints, and ω was changed between $4 \cdot 10^6$ 1/s and $4 \cdot 10^7$ 1/s in $1 \cdot 10^6$ 1/s steps. The frequency sweep was made first in the increasing direction then in the decreasing direction. Near to the peak of the resonance curve the hysteresis-like effect of the hardening nonlinear behaviour can be seen (Fig. 5.).

The maximum amplitudes of all joints of the beam are shown in Fig. 6. It can be seen, that the shapes of the resonating beam for the symmetric exciting force are also symmetric.

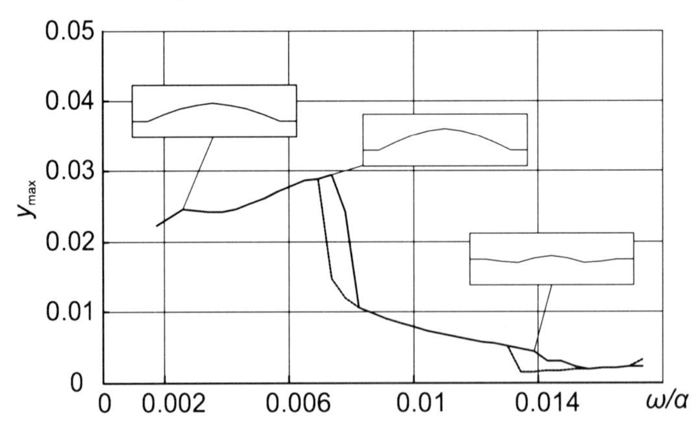

Fig. 5 Maximal deflection of the middle joint of a doubly clamped beam excited by harmonic force and shapes of the resonating beam

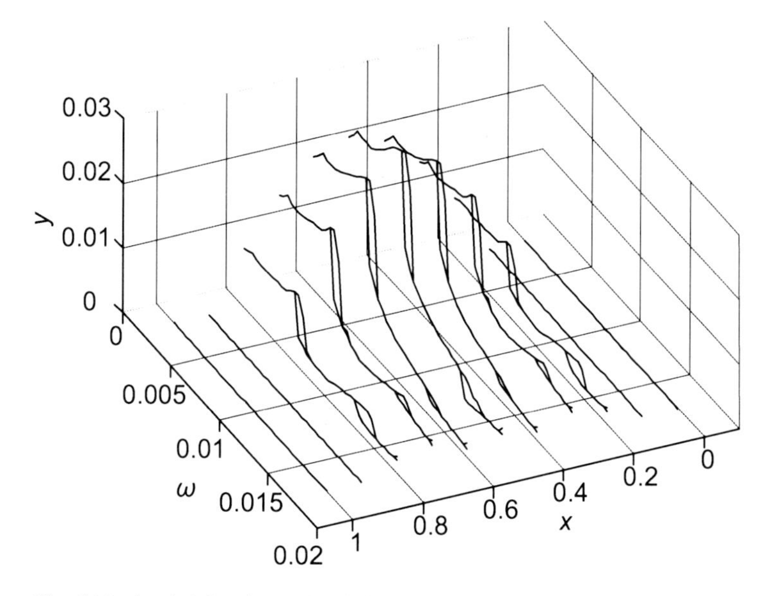

Fig. 6 Maximal deflections of all joints of the harmonic excited beam resonator

6 Conclusions

The present work introduced a multibody model of longitudinal elastic rods for doubly clamped MEMS resonators. Taking longitudinal elasticity of the beam into account enabled us to observe the stiffening geometric nonlinear behaviour. The model uses well known parameters of the MEMS resonator, so modelling of real devices is simple and accurate. Only the value of the damping cannot be calculated from known parameters, and it has to be adjusted. The benefits of the multibody model are that we obtain information about the full movement of the beam, since the computation is still simple. The model gives with 10 elements good results. The proposed multiple degree-of-freedom model may be able to identify the pull-in effect, which is an important limitation of MEMS resonator applications. Therefore the nonlinear electrostatic exciting force also has to be taken into account.

References

1. Lifshitz, R., Cross, M.C.: Nonlinear dynamics of nanomechanical and micromechanical resonators. In: Schuster, H.G. (ed.) Preprint of contribution to Review of Nonlinear Dynamics and Complexity. Wiley, Chichester (2008)
2. Mestrom, R.M.C., Fey, R.H.B., van Beekb, J.T.M., Phan, K.L., Nijmeijer, H.: Modelling the dynamics of a MEMS resonator: Simulations and experiments. Sensors and Actuators A 142, 306–315 (2008)

3. Seon, M.H., Haym, B., Timothy, W.: Dynamics of transversely vibrating beams using four engineering theories. Journal of Sound and Vibration 225(5), 935–988 (1999)
4. Ahmadian, M.T., Borhan, H., Esmailzadeh, E.: Dynamic analysis of geometrically nonlinear and electrostatically actuated micro-beams. Communications in Nonlinear Science and Numerical Simulation 14, 1627–1645 (2009)
5. Rubinstein, D.: Dynamics of a flexible beam and a system of rigid rods, with fully inverse (one-sided) boundary conditions. Comput. Methods Appl. Mech. Engrg. 175, 87–97 (1999)

Control of Elastic Wave Propagation in Piezoceramic Cylinders of Sector Cross Section

Puzyrev Vladimir and Storozhev Valeriy

Abstract. In the present paper the problem of elastic waves propagation in piezoceramic cylindrical waveguides of noncircular cross section is studied for the cylinders of circular and hollow cross section with sector cut of arbitrary angular measure. Boundary surfaces of sector cut are covered by non-extensible membranes and cylindrical surfaces can have arbitrary boundary conditions. The equations of motion of the piezoelectric cylinder are analytically integrated and dispersion equations that exactly satisfy the boundary conditions are obtained. A selected set of numerical results including dispersion spectrums, distributions of cutoff frequencies and phase velocities for cylinders with various cross section geometry and boundary conditions is presented, and main effects of their transformation are discussed. Results are in good agreement with the results obtained by other researches for the special case of circular and hollow cylinders. The model can be used to control the dispersion spectrums and wave characteristics by changing the geometry and mechanical properties of the waveguides.

1 Introduction

In past years, the problem of elastic wave propagation in infinite noncircular cylinders has received a great deal of attention. The researches carried out are of interest for the search of new fundamental laws and specific characteristics of axially anisotropic cylinders. A large amount of literature devoted to this topic can be found in [1–5]. Comparatively, little attention has been given to the propagation of electroelastic waves in noncircular piezoceramic cylinders, and to the imaginary and complex branches of all modes. Waveguide properties of anisotropic piezoelectric cylinders have been studied in the cases of homogeneous circular and hollow cylinders [6]. Piezoelectric and thermoelastic solid cylindrical waveguides of arbitrary cross section are studied in [7, 8] using the Fourier expansion collocation method.

Puzyrev Vladimir · Storozhev Valeriy
Donetsk National University, Mathematical Department, 24 Universitetskaya Str., Donetsk 83001, Ukraine
e-mail: vladimir.puzyrev@gmail.com, stvi@i.ua

G. Stépán et al. (Eds.): Dynamics Modeling & Interaction Cont., IUTAM BOOK SERIES 30, pp. 215–222.
springerlink.com

In the present paper we study the mechanisms to control elastic wave propagation in circular and hollow piezoceramic cylinders by making sector cut of arbitrary angular measure in cross section. The method is based on exact analytical integration of wave equations by using wave potentials. The frequency equations obtained from boundary conditions in a form of functional determinants of the fourth or eight order are analyzed numerically. Dispersion spectrums and distributions of cutoff frequencies are presented, and main effects of their transformation by variation of sector cut angular measure and ratio of outer and inner radii are discussed.

2 Problem Formulation

2.1 System Geometry and Equations of Motion

Consider piezoceramic cylindrical waveguides of cross section that are constant along the x_3 axis of a rectangular Cartesian coordinate system. Waveguides of the first type (C-type) have circular cross section with sector cut of any angular measure; waveguides of the second type (H-type) are hollow cylinders with sector cut in cross section. The angular measure of waveguides is denoted by α, the radius of C-type cylinder, inner and outer radii of H-type cylinder are denoted by R, R_1 and R_2 respectively. The prepolarization axis of the piezoceramic material has longitudinal orientation.

The linear deformation theory of piezoelectric solids includes three-dimensional stress equations of motion, quasistatic approximation of Maxwell's field equations and constitutive relations for a linearly electroelastic medium

$$\sigma_{ij,j} - \rho \ddot{u}_i = 0, \quad D_{i,i} = 0, \tag{1}$$

$$\sigma_{ij} = c_{ijkl} K_{kl} - e_{ijk} E_k, \quad D_i = e_{ikl} K_{kl} + \varepsilon_{ik} E_k, \quad K_{ij} = (u_{i,j} + u_{j,i})/2. \tag{2}$$

Here σ_{ij}, K_{kl}, u_i, E_i, and D_i are, respectively, the components of stress, strain, displacement, electric field intensity, and electric displacement; c_{ijkl}, e_{ijk}, and ε_{ik} denote the elastic, piezoelectric, and dielectric constants respectively, ρ is the mass density. The electric field intensity E is related to the electric potential φ by $E = -grad\varphi$.

We seek solutions of the problem in the form $u_j = \mathrm{Re}(u_j^{(0)} \cdot e^{-i(\omega t - kx_3)})$, $\varphi = \mathrm{Re}(\varphi^{(0)} \cdot e^{-i(\omega t - kx_3)})$ for harmonic electroelastic waves. Here ω is the angular frequency and k is the wavenumber. Displacement equations of motion are obtained in the following form of a system of partial differential equations

$$\left\| L_{ij} \right\| \mathbf{f}^{(0)} = 0, \tag{3}$$

where $\mathbf{f}^{(0)} = (u_1^{(0)}, u_2^{(0)}, u_3^{(0)}, \varphi^{(0)})^T$.

2.2 *Boundary Conditions*

The cases of stress-free and displacement-fixed cylindrical boundaries are considered in the paper. In the first case the cylindrical surface (or both inner and outer surfaces for waveguide of H-type) is stress free and covered by thin short-circuited electrodes. Thus, the following stress and electric potential components are zero:

$$(\sigma_{rr})_\Gamma = (\sigma_{r\theta})_\Gamma = (\sigma_{rz})_\Gamma = (\varphi)_\Gamma = 0 . \tag{4}$$

In the second case the cylindrical surface is displacement-fixed, and the following components are zero:

$$(u_r)_\Gamma = (u_\theta)_\Gamma = (u_z)_\Gamma = (\varphi)_\Gamma = 0 . \tag{5}$$

The sector cut surfaces of the waveguides studied in the present paper are covered by non-extensible membranes:

$$(\sigma_{\theta\theta})_{\Gamma_\pm} = (u_r)_{\Gamma_\pm} = (u_z)_{\Gamma_\pm} = (\varphi)_{\Gamma_\pm} = 0 . \tag{6}$$

3 Problem Solution

The method of generalized wave potentials is used

$$u_1^{(0)} = \partial_1\varphi_1 + \partial_2\varphi_4 , \quad u_2^{(0)} = \partial_2\varphi_1 - \partial_1\varphi_4 , \quad u_3^{(0)} \equiv \varphi_2 , \quad \varphi^{(0)} \equiv \varphi_3 . \tag{7}$$

It leads to a system of differential equations for the wave potentials φ_1, φ_2, φ_3 and a separate equation for φ_4

$$\tilde{L}_{ij}\varphi_j = 0 \quad (i, j = \overline{1,3}) , \quad \tilde{L}_{44}\varphi_4 = 0 . \tag{8}$$

We seek the solutions of the Eqs. (8) in the form

$$\varphi_p = \sum_{j=1}^{3} \beta_{pj}\chi_j(x_1, x_2) \qquad p = (\overline{1,3}) , \quad \varphi_4 = \chi_4(x_1, x_2) ; \tag{9}$$

where functions χ_j are determined from the metaharmonic equations $D^2\chi_j + \gamma_j^2\chi_j = 0$, and γ_j^2 $(j = \overline{1,3})$ are the roots of a bicubic algebraic equation. After substituting Eqs. (9) in Eqs. (7) the displacement and electric potential components are obtained in the following form

$$u_1^{(0)} = \sum_{j=1}^{3} \beta_{1j}\partial_1\chi_j(x_1, x_2) + \partial_2\chi_4(x_1, x_2) ,$$

$$u_2^{(0)} = \sum_{j=1}^{3} \beta_{1j} \partial_2 \chi_j(x_1, x_2) - \partial_1 \chi_4(x_1, x_2),$$

$$u_3^{(0)} = \sum_{j=1}^{3} \beta_{2j} \chi_j(x_1, x_2), \quad \varphi^{(0)} = \sum_{j=1}^{3} \beta_{3j} \chi_j(x_1, x_2). \tag{10}$$

Then the expressions for u_α, $\sigma_{\alpha\beta}$ and D_α in cylindrical coordinate system are obtained and the frequency equations are developed from boundary conditions. A set of normal waves considered here splits into two subsets: symmetric (S) waves that satisfy the conditions $u_r(r,-\theta) = u_r(r,\theta)$, $u_\theta(r,-\theta) = -u_\theta(r,\theta)$, $u_z(r,-\theta) = u_z(r,\theta)$, $\varphi(r,-\theta) = \varphi(r,\theta)$, and antisymmetric (A) waves that satisfy the conditions $u_r(r,-\theta) = -u_r(r,\theta)$, $u_\theta(r,-\theta) = u_\theta(r,\theta)$, $u_z(r,-\theta) = -u_z(r,\theta)$, $\varphi(r,-\theta) = -\varphi(r,\theta)$.

According to the geometry and boundary conditions the functions χ_j for waveguides of C-type are chosen in the following form for the symmetric case:

$$\chi_j(r,\theta) = \sum_{n=0}^{\infty} A_{jn} J_{\lambda_n}(\gamma_j r) \cos \lambda_n \theta \quad (j = \overline{1,3}),$$

$$\chi_4(r,\theta) = \sum_{n=0}^{\infty} A_{4n} J_{\lambda_n}(\gamma_4 r) \sin \lambda_n \theta; \tag{11}$$

where $\lambda_n = (2n+1)\pi/2\alpha$, $J_n(r)$ is the nth-order Bessel function of the first kind. For H-type waveguides the functions χ_j for the symmetric case are of the form:

$$\chi_j(r,\theta) = \sum_{n=0}^{\infty} \left(A_{jn}^{(1)} J_{\lambda_n}(\gamma_j r) + A_{jn}^{(2)} Y_{\lambda_n}(\gamma_j r) \right) \cos \lambda_n \theta \quad (j = \overline{1,3}),$$

$$\chi_4(r,\theta) = \sum_{n=0}^{\infty} \left(A_{4n}^{(1)} J_{\lambda_n}(\gamma_4 r) + A_{4n}^{(2)} Y_{\lambda_n}(\gamma_4 r) \right) \sin \lambda_n \theta; \tag{12}$$

where $Y_n(r)$ is the nth-order Bessel function of the second kind. For A-waves the functions χ_j have similar form.

The boundary conditions (6) on the sector cut surfaces $\Gamma_\pm$ are satisfied identically. Application of the boundary conditions (4) or (5) on cylindrical surface of C-type waveguide results in four linear, homogeneous, algebraic equations in the unknown wavenumber k and normalized frequency Ω for each value of

$n = \overline{0,\infty}$. A nontrivial solution of them exists only if the determinant of the system matrix vanishes:

$$F_n(\Omega,k) = \begin{vmatrix} \Delta_r^{(1,n)} & \Delta_r^{(2,n)} & \Delta_r^{(3,n)} & \Delta_r^{(4,n)} \\ \Delta_\theta^{(1,n)} & \Delta_\theta^{(2,n)} & \Delta_\theta^{(3,n)} & \Delta_\theta^{(4,n)} \\ \Delta_z^{(1,n)} & \Delta_z^{(2,n)} & \Delta_z^{(3,n)} & \Delta_z^{(4,n)} \\ \Delta_\varphi^{(1,n)} & \Delta_\varphi^{(2,n)} & \Delta_\varphi^{(3,n)} & 0 \end{vmatrix} = 0, \tag{13}$$

where $\Delta_r^{(j,n)} = \beta_{1j}\left(\lambda_n R^{-1} J_{\lambda_n}(\gamma_j R) - \gamma_j J_{\lambda_{n+1}}(\gamma_j R)\right)$, $\Delta_r^{(4,n)} = \lambda_n R^{-1} J_{\lambda_n}(\gamma_4 R)$, … $\Delta_\varphi^{(j,n)} = \beta_{3j} J_{\lambda_n}(\gamma_j R)$.

The dispersion equations for H-type waveguides are obtained analogously in the form of eight-order determinants. Each of these equations describes an independent partial spectrum with the certain value of generalized wavenumber n, and the full spectrum is a superposition of all partial spectrums.

4 Numerical Results

A few numerical examples are given below to demonstrate the approach. The main question of the paper is how does the inner radius (or its absence in the case of C-waveguides) and the angular measure of sector cut affect the characteristics of cylinders under consideration. The elastic, piezoelectric, and dielectric constants of the $BaCaTiO_3$ material are taken from [9]. Dispersion curves are presented in the form of normalized frequency Ω versus dimensionless wavenumber k. Numerical calculations of the real and imaginary branches have been carried out by the bisection method. A computer code developed for spectrum calculation applies the method to the consecutive horizontal lines (each with its own value of Ω) and vertical lines (each with its own $\mathrm{Re}(k)$ or $\mathrm{Im}(k)$ value).

Dispersion curves of symmetric waves for the C-type waveguide of angular measure $\alpha = \pi/4$ with displacement-fixed cylindrical surface and $n = 0$ are shown in Fig. 1. Curves of antisymmetric waves are found to be not qualitatively different from the symmetric motions, although they have another mode order in few cases. No mode has zero cutoff frequency, and a backward wave occurs in the first mode. Nonpropagating modes with purely imaginary values of k come close to each other in inflection points and form almost vertical lines in the left part of the spectrum.

Partial dispersion spectrums of symmetric waves for the H-type waveguides of angular measure $\alpha = \pi/4$, $n = 0$, displacement-fixed cylindrical surfaces and two different ratios of inner to outer radius are illustrated in Fig. 2. These curves are plotted analogously to those for the circular waveguide and a comparison can be made. As seen in Fig. 2, ratio of inner-outer radii affects significantly on spectrum's quantitative characteristics. Modes of the spectrum for the waveguide with

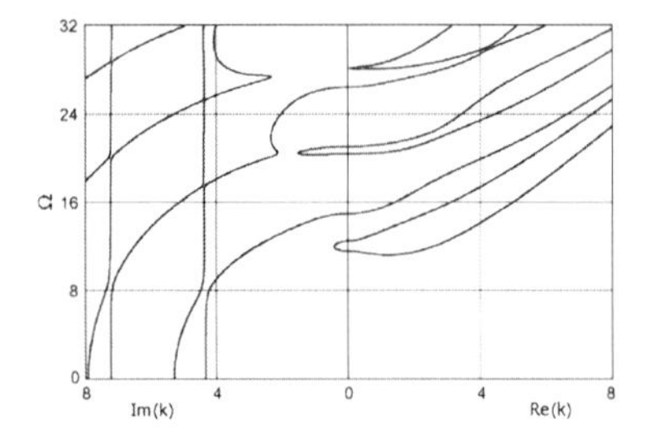

Fig. 1 Dispersion curves for C-waveguide with displacement-fixed cylindrical surface, $\alpha = \pi/4$, $n = 0$.

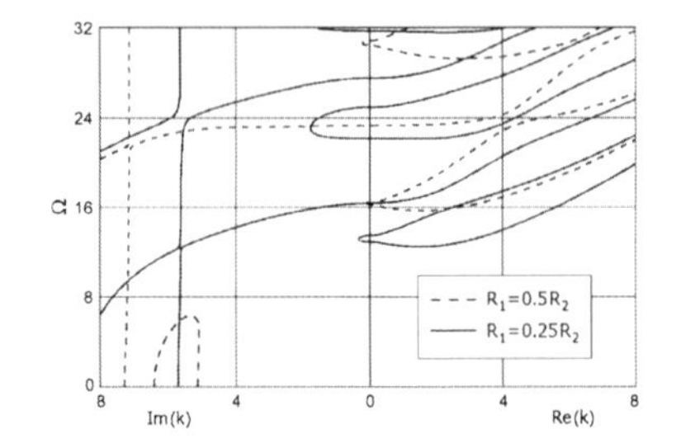

Fig. 2 Dispersion curves for H-waveguides with displacement-fixed cylindrical surfaces, $\alpha = \pi/4$, $n = 0$.

smaller inner radius $R_1 = 0.25R_2$ (solid lines) have smaller cutoff frequencies than modes of spectrum for the waveguide with $R_1 = 0.5R_2$ (dotted lines). In the last case the spectrum has a purely imaginary loop, which is absent in the case of $R_1 = 0.25R_2$ (the loop disappears when R_1 becomes smaller than $0.42R_2$). The first and second modes of the spectrum for the waveguide with $R_1 = 0.25R_2$ are similar to the two lowest modes of spectrum for C-waveguide (Fig. 1).

Figure 3 shows the cutoff frequencies of the three lowest modes for the two H-waveguides with inner radius $R_1 = 0.25R_2$ (solid lines) and $R_1 = 0.5R_2$ (dotted lines). Cutoff frequencies of shear elastic modes (S) have the entirely radial particle displacement. The displacement for the modes of electroelastic longitudinal type (L) is entirely axial. It can be seen very clearly from this plot that the geometry of waveguide's cross section has a considerable influence on its cutoff frequencies. Not only the values of cutoff frequencies, but also the order of S- and L-modes is sensitive to the angular measure of the cross section. Also, the waveguide with inner radius $R_1 = 0.25R_2$ has lower cutoff frequencies than the waveguide with $R_1 = 0.5R_2$.

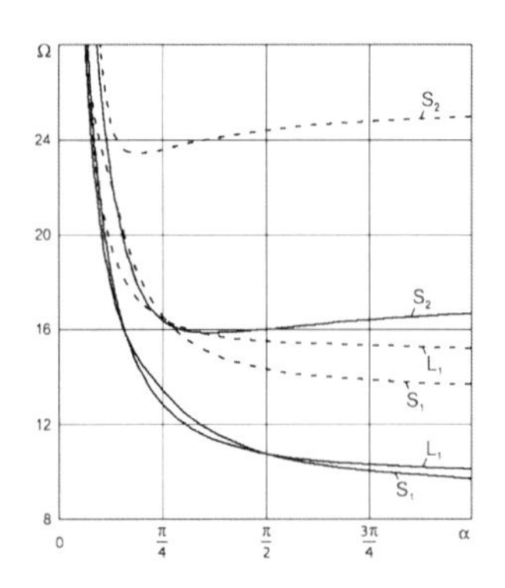

Fig. 3 Cutoff frequencies for H-waveguides with displacement-fixed surfaces, $n = 0$.

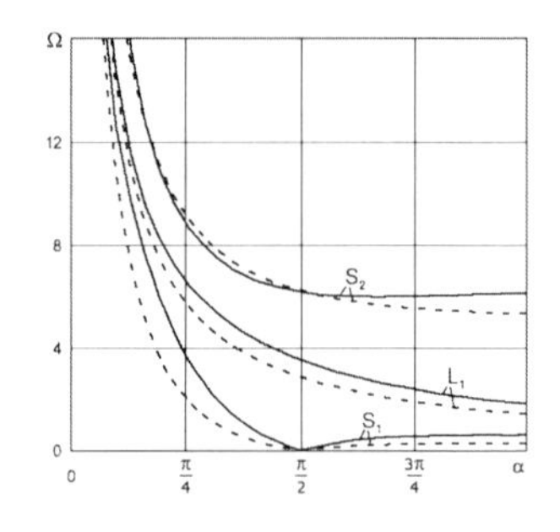

Fig. 4 Cutoff frequencies for H-waveguides with stress-free surfaces, $n = 0$.

Cutoff frequencies of the three lowest modes for the two H-waveguides with the same ratios of inner-outer radii ($R_1 = 0.25R_2$ denoted by solid lines, and $R_1 = 0.5R_2$ denoted by dotted lines) and stress-free cylindrical surfaces are presented in Fig. 4. It can be found from the comparison of Fig. 3 and Fig. 4 that the boundary conditions on the cylindrical surfaces of waveguide have considerable influence on the cutoff frequencies distributions. The first mode of a semicircle waveguide ($\alpha = \pi/2$) has zero cutoff frequency and further increasing in α leads to an increase of the cutoff frequency of this mode. The influence of inner radius is less noticeable than for the cylinders with displacement-fixed cylindrical surfaces (Fig. 3).

Results of the similar qualitative characteristics are obtained for several simulations with different piezoelectric materials, boundary conditions on the cylindrical surfaces and angular measure of cross section. Also, the comparison is made with results from literature obtained by other researches. In particular, the curves for semicircular H-waveguide are found to be identical to those presented in [6]. The high-frequency limit of phase velocities of the first mode waves in waveguides with stress-free cylindrical surfaces is the Rayleigh surface wave velocity. In the case of displacement-fixed waveguides the high-frequency limit for the phase velocity of normal waves is the velocity of shear elastic waves.

5 Conclusion

A method for studying normal waves in axially polarized piezoceramic cylinders with sector cut of arbitrary angular measure and arbitrary boundary conditions on the cylindrical surfaces is presented. The frequency equation of the system in a form of a transcendental function of circular frequency and wavenumber is developed through the boundary conditions at the cylinder's surfaces. Numerical analysis of the dispersion spectrums in circular and hollow waveguides of sector cross section is carried out. Real and imaginary branches for circular and hollow waveguides are presented and compared; changes in distributions of cutoff frequencies are described. The variation of cross section geometry is considered to be the mechanism to control the dispersion characteristics of a waveguide.

References

[1] Damljanovic, V., Weaver, R.L.: Propagating and evanescent elastic waves in cylindrical waveguides of arbitrary cross section. J. Acoust. Soc. Am. 115, 1572–1581 (2004)

[2] Elmaimouni, L., Lefebvre, J.E., Zhang, V., Gryba, T.: A polynomial approach to the analysis of guided waves in anisotropic cylinders of infinite length. Wave Motion 42, 177–189 (2005)

[3] Honarvar, F., Enjilela, E., Sinclair, A.N., Mirnezami, S.A.: Wave propagation in transversely isotropic cylinders. Int. J. Solids Struct. 44, 5236–5246 (2007)

[4] Ying, Z.G., Wang, Y., Ni, Y.Q., Ko, J.M.: Stochastic response analysis of piezoelectric axisymmetric hollow cylinders. Journal of Sound and Vibration 321, 735–761 (2009)

[5] Graff, K.F.: Wave motion in elastic solids. Dover Publications, New York (1991)

[6] Shulga, N.A.: Propagation of harmonic waves in anisotropic piezoelectric cylinders. Homogeneous piezoceramics waveguides. Int. Appl. Mech. 38, 933–953 (2002)
[7] Paul, H.S., Venkatesan, M.: Propagation in a piezoelectric solid cylinder of arbitrary cross section. J. Acoust. Soc. Am. 82, 2013–2020 (1987)
[8] Ponnusamy, P.: Wave propagation in a generalized thermoelastic solid cylinder of arbitrary cross-section. Int. J. Solids Struct. 44, 5336–5348 (2007)
[9] Berlincourt, D.A., Curran, D.R., Jaffe, H.: Piezoelectric and piezomagnetic mateials and their functions in transducers. In: Mason, W.P. (ed.) Physical Acoustics 1A, Academic, New York (1964)

SHM of Composite Cylindrical Multilayered Shells with Delaminations

A. Muc

Abstract. In this study, effective computational procedures are introduced and used to characterize the dynamic behavior of cylindrical panels with circular cross-section having the single interlaminar delamination between laminate layers. Based on the computed results it is possible to determine the effect of delamination on the overall structural dynamic behavior. Those results are used to quantify the difference between the results of the relevant parameters in the cases of perfect and defected structures. Usually, the wave propagation can be observed with the use of piezoelectric sensors. Therefore, in the next step of our analysis we modeled delaminated structures with a finite number of PZT sensors to consider also their influence on the structural dynamic response. The numerical analysis have been conducted with the use of 3D finite elements. A lot of numerical results allow us to understand better the influence of various parameters on the form of wave propagation in cylindrical multilayered shells.

1 Vibrations of Cylindrical Shells

Anisotropic laminated cylinders or circular cylindrical shells are widely used in various engineering applications. Because of the algebraic complexity analytical solutions for waves in anisotropic cylinders are very difficult to obtain. Generally speaking, exact solutions based on elasticity are difficult to find, and most of them deal with plate that extend to infinity. In fact, no solutions in closed form are known to exist for vibrations of plates or shells with unrestricted dimensional ratios and with traction free boundary conditions. Special numerical methods therefore need to be developed and then used.

There have been many works on wave propagation problems related to composite shells. Mirsky [1] and Nowinski [2] solved for axially symmetric waves in orthotropic shells. Chou and Achenbach [3] provided a three-dimensional solution for orthotropic shell as well. Yuan and Hsieh [4] proposed an analytical method for the investigation of free harmonic wave propagation in laminated shells. Nayfeh [5] discussed scattering of horizontally polarized elastic waves from

A. Muc
Institute of Machine Design, Cracow University of Technology, Kraków, Poland

G. Stépán et al. (Eds.): Dynamics Modeling & Interaction Cont., IUTAM BOOK SERIES 30, pp. 223–230.
springerlink.com

multilayered anisotropic cylinders embedded in isotropic solids. The numerical description of the waves traveling into waveguides and slender structures has also raised many interests – an information about those problems are discussed in Refs [6].

The fundamental relations can be developed by applying Hamilton's principle, both in 3D or 2D formulations. To visualize the effect of anisotropy on wave propagation six representations of wave surfaces are used: velocity, phase slowness, phase wave surfaces, group velocity, group slowness and group wave surfaces.

2 Wave Scattering by Cracks

The study involving the monitoring, detection and arrest of the growth of flaws, such as cracks, constitutes what is universally termed as Structural Health Monitoring. SHM has four levels: 1, confirming the presence of damage, 2, determination of the size, location and orientation of the damage, 3, assessing the severity of the damage, 4, controlling the growth of damage. For cylindrical shells such an analysis is mainly conducted in two ways: a) numerically with the use of a numerical-analytical method or a strip element method, b) experimentally using smart (piezoelectric) patches. In general, two typical modes of failure are investigated, i.e. intralaminar cracks arising due to a damage of an individual ply in a laminate or interlaminar cracks due to debonding of individual plies.

Level 1 can be demonstrated experimentally using both PZT as well as magnetorestrictive sensors/actuators. The indication of the damage can be obtained by the analysis of the difference between the values of the Open Circuit Voltage across a sensor before and after the delamination. The difference is called as the Damage Induced Voltage and needs to be of the order of millivolts for meaningful measurements. Numerically the presence of the interlaminar damage can be studied by the analysis of eigenfrequencies for shells with and without delaminations.

By passing a voltage across the actuator it is possible to induce a strain in a cylindrical shell which changes stresses and hence the voltage across the PZT sensors. To establish the severity of the damage it is necessary to determine the fracture parameters. It may be done by measuring the changes of the voltage across the sensor if plots of the SIF or SERR with the sensor voltage for various damage configurations are available. Using a point of line dynamic excitation one can compare signals in the form of elastic waves for damaged and undamaged structures. Those values are computed with the use of the above-mentioned variants of the finite element analysis and may be represented as displacements or fracture parameters.

It is worth to mention that for composite multilayered structures the wave propagation is dependent not only on the form of a damage but also on other factors characterizing material properties of composites as well as laminate configurations (stacking sequences and fiber orientations in each individual plies constituting a laminate).

3 Formulation of the Problem

Consider a cylindrical shell with throughout circumference delamination, which has mid-surface radius R, thickness t, length L and mass density ρ, as shown in Fig. 1. h_1 and h_2 are the thicknesses of the upper and lower sublaminates, respectively. The nondimensional parameter $\hat{h}_1= h_1/t$ is used to describe the delamination thickness. The axial coordinate is x, the circumferential coordinate is y, and the thickness coordinate normal to the shell surface is z. R is the radius of the cylindrical shell segment and the circumferential coordinate is replaced by $y = R\beta$. The displacement field for an intact cylindrical shell is described with the use of a first-order, shear deformation theory. The displacement components of middle surface and the rotations for delaminated shell are different for each three parts of the shell, where the subscript I corresponds to the part of the shell with no delaminations, whereas II and III corresponds to two sublaminates.

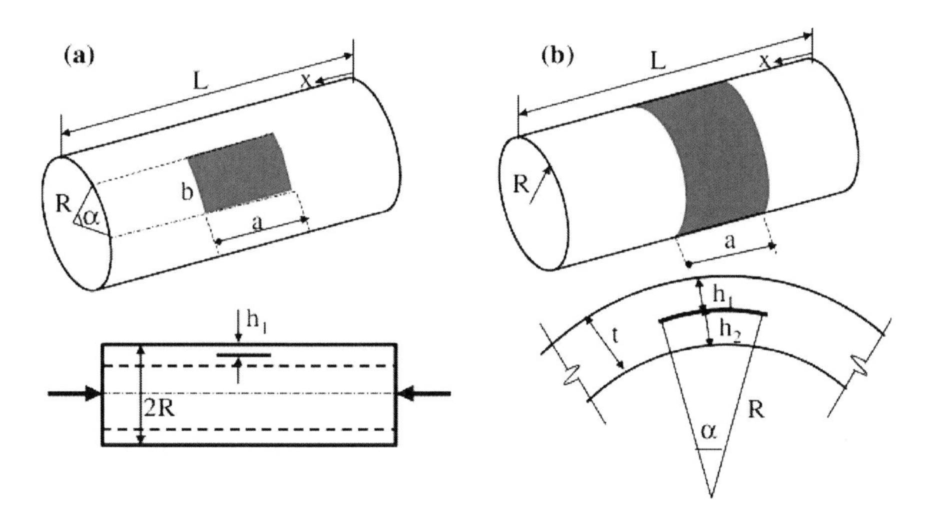

Fig. 1 Geometry of a cylindrical shell with a delamination: a) a local delamination, b) an axisymmetric delamination.

Consider a cylindrical shell with throughout circumference delamination, which has mid-surface radius R, thickness t, length L and mass density ρ, as shown in Fig. 1. h_1 and h_2 are the thicknesses of the upper and lower sublaminates, respectively. The non-dimensional parameter $\hat{h}_1= h_1/t$ is used to describe the delamination thickness. The axial coordinate is x, the circumferential coordinate is y, and the thickness coordinate normal to the shell surface is z. R is the radius of the cylindrical shell segment and the circumferential coordinate is replaced by $y = R\beta$. The displacement field for an intact cylindrical shell is described with the use of a first-order, shear deformation theory. The displacement components of middle surface and the rotations for delaminated shell are different for each three parts of the shell, where the subscript I corresponds to the part of the shell with no delaminations, whereas II and III corresponds to two sublaminates and are characterized in the following way:

$$\Omega_{II} = \{L/2 - a/2 \leq x \leq L/2 + a/2\} \times \{0 \leq y \leq 2\pi R\} \times \{-h_1/2 \leq z \leq h_1/2\}$$

$$\Omega_{III} = \{L/2 - a/2 \leq x \leq L/2 + a/2\} \times \{0 \leq y \leq 2\pi R\} \times \{-h_2/2 \leq z \leq h_2/2\} \quad (1)$$

In order to enforce compatibility along the interface, constraint equations can be imposed. The constraint equations will tie the upper and lower sublaminates in both displacements and rotations along the interface so that both sublaminates together deform like an intact single laminate. For the delaminated cylindrical shell, the total potential energy Π can be written as:

$$\Pi = K - U = \sum_{r=I}^{III} (K_r - U_r) \quad (2)$$

where K represents the kinetic energy and U represents the strain energy for each of the shell region. Those terms are defined in the classical manner.

Let us assume that the unknown parameters of deformations $u^0_{xr}(x,y,t)$, $u^0_{yr}(x,y,t), u^0_{zr}(x,y,t), \psi_{xr}(x,y,t), \psi_{yr}(x,y,t)$, r=I,II,III (15 parameters characterizing shell deformations) are expressed in the form of double infinite trigonometric series. Substituting trigonometric series into the final form of Eq. (2) and performing the necessary calculation of differentiation and integration, we obtain the total potential energy expressed as the function of 15 uknowns. Using the Rayleigh–Ritz method for minimizing the total potential energy, one arrives at a system of dynamic governing equations by equating the derivatives of Π with respect to unknowns parameters to zero. Such a procedure is carried out almost automatically with the use of the symbolic package MATHEMATICA. Finally, the system of equations can be reduced to the following form:

$$\mathbf{M}\ddot{\mathbf{q}} + \mathbf{K}\mathbf{q} = \mathbf{0} \quad (3)$$

where **q** is the vector of unknowns, and **M**, **K** are the mass and the stiffness matrix, respectively. From the above equation the natural frequencies can be derived as the classical eigenvalues.The details of that approach are presented in Ref [7].

4 Free Vibration Analysis

In order to verify the present analytical and numerical model, the linear natural frequency of shell without delamination is calculated. The geometrical and material parameters of the shell are the same as in Refs [8]. ω and $\overline{\omega}$ are dimensionless and dimensional linear natural frequency of shell, respectively where the dimensionless quantity is defined as follows: $\overline{\omega} = L^2\sqrt{\rho/(A_{22}t)}\overline{\omega}$ and A_{22} is the membrane term of the stiffness matrix. The theoretical analysis has been conducted for the delamination in the form presented in Fig. 1b.

The numerical model of the analyzed problem (Fig.1) has been also build to verify the analytical relations and to estimate their accuracy. We have employed herein the FE package NISA II. For simplicity, assuming the symmetry of the problem 1/8 of the shell have been discretized with the use of 15200 to 28400 quadrilateral shell finite elements having six degrees of freedom at each of the nodes – three displacements and three rotations. In the delaminated zone the gap elements have been introduced in order to get an information about possible interpenetration of delaminated layers. If it occurs such a solution is not taken into account.

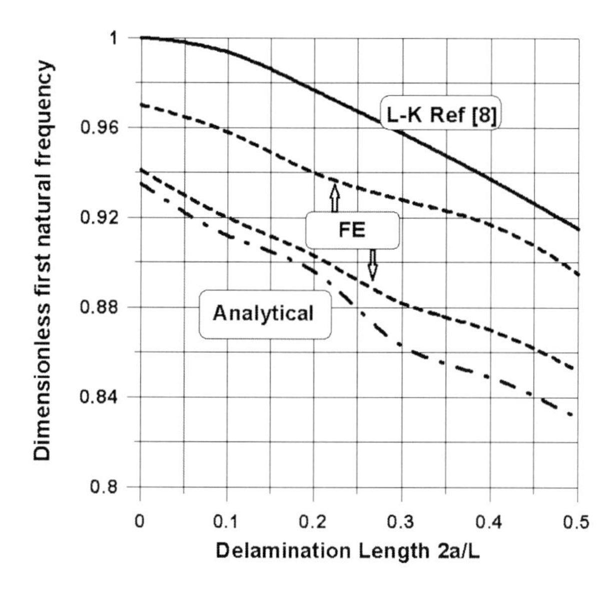

Fig. 2 Variations of the first natural frequency with the delamination length

The present results for the first natural frequencies of the shell without delamination are compared in Table 1 with those given by literature [8]. A good agreement between these two sets of values is observed.

Table 1 Comparison of natural frequencies

	Present		Literature [8]
Boundary conditions	Analytical	FEM	
Simply supported	7.583	7.634	7.653
Clamped	10.134	10.284	10.275

The effects of the delamination length parameter a, the delamination depth parameter h_1 and the material properties on the first natural frequencies of the shell with both ends clamped are presented in Fig.2.

They are also compared with the results presented in Ref [8]. The latter comparison has been carried out in order to estimate the accuracy of the Love-Kirchhoff model since such a hypothesis has been employed in the cited work. In this case, the geometrical parameters of the delaminated shell are L/R = 5/3, R/h = 30 and the stacking sequences of laminates are $[0^0/90^0/0^0]_{10}$. The elastic constants are following: E_1/E_2=40, G_{12}/E_2=0.5, ν_{12}=0.25.

It may be observed that the first natural frequencies decreases with the delamination length, however, the decrease is not so rapid as for the buckling loads for cylindrical shells with delaminations. Natural frequencies computed with the use of the Love-Kirchhoff hypothesis are higher than those determined with the use of the first order transverse shear deformation theory. The difference increases if the transverse shear effects are much more significant, i.e. as the G_{13}/E_2 ratio decreases. The analytical model is a very good approximation of the numerical results and in this way it may be treated as the benchmark for the finite element computations.

The method of the analysis can be also applied to the investigation of the influence of the rectangular delamination (Fig. 1a) on natural frequencies. Using such an approach it is possible to carry out parametric studies for multilayered laminated composite cylindrical shells.

5 Propagation of Elastic Waves

Let us consider the cylindrical shell having the identical properties as described in the previous section and loaded by a circular ring of radial loads. Assume that the ring load with time function:

$$F(\hat{t}) = \sin(\pi\hat{t}), 0 < \hat{t} < 2 \tag{4}$$

which is one cycle of sine function and $\hat{t}$ is a dimensionless time parameter. By applying the above kernel function (the dimensionless excitation function) the displacement response of the intact cylindrical shell and of the shell with the interlaminar delamination is evaluated with the use of the finite element method.

Figure 3 illustrates the distribution of radial displacements at the edge of delamination.

As it may be observed the interlaminar crack both shift the radial displacements and enlarge them in the comparison with displacements for the intact shell. First natural frequencies decreases with the delamination length, however, the decrease is not so rapid as for the buckling loads for cylindrical shells with delaminations. The analysis of the time response should be always cut off at the specified values of time since for higher times the initial response is disturbed by the waves representing the reflection from boundaries.

The similar analysis have been done for the radial point excitation of the shell (mathematically it is characterized by the 3D Dirac's delta distribution) to model the real experimental situation of the actuation of the shell by the PZT actuator

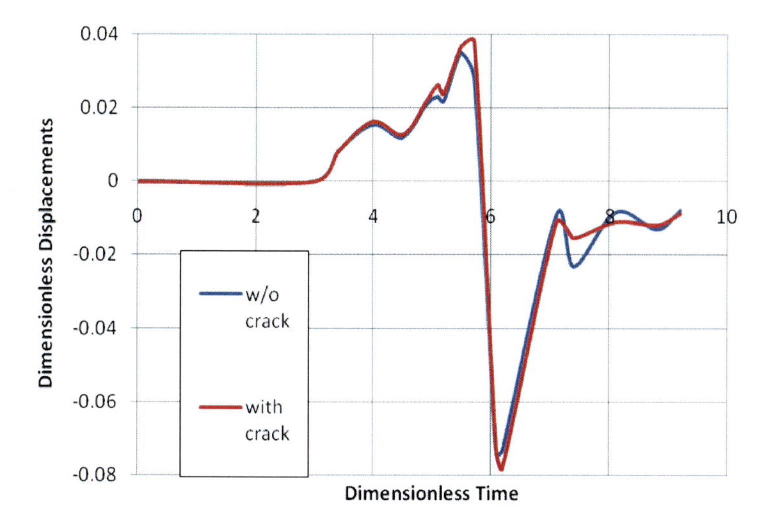

Fig. 3 Time history for radial displacements on the upper surface of cylindrical shell with and without interlaminar crack (delamination).

and then investigate the response at the chosen points. Those results will be demonstrated later together with the comparison of experimental data.

6 Conclusions

In the paper the analytical model accounting for transverse shear deformations has been presented in order to evaluate free vibrations of cylindrical multilayered shells with delaminations. It has been compared with numerical finite element analysis and with the results available in the literature both for the Love-Kirchhoff theory and the first order transverse shear deformation theory for cylindrical shells having variable length of axisymmetric delaminations. It has been observed that natural frequencies computed with the use of the Love-Kirchhoff hypothesis are higher than those determined with the use of the first order transverse shear deformation theory. The difference increases if the transverse shear effects are much more significant, i.e. as the G_{13}/E_2 ratio decreases. It has been demonstrated that the analytical model is a very good approximation of the numerical results.

Then, the finite element analysis of the cylindrical shell with axisymmetric delaminations and subjected to axisymmetric excitations have been conducted in order to compare the radial displacements for shells with a crack and without the crack. The interlaminar crack both shift the radial displacements and enlarge them in the comparison with displacements for the intact shell. Thus, the numerical results show evidently that those cracks can be detected experimentally with the use of the PZT patches.

Acknowledgments. The Polish Research Foundation PB 1174/B/T02/2009/36 is gratefully acknowledged for financial support.

References

[1] Mirsky, I.: Axisymmetric vibrations of orthotropic cylinders. J. Acoust. Soc. Am. 36, 2106–2112 (1964)
[2] Nowinski, J.L.: Propagation of longitudinal waves in circular cylindrical bars. J. Engng. Ind. 89, 408–412 (1967)
[3] Chou, F.H., Achenbach, J.D.: Three-dimensional vibrations of orthotropic cylinders. ASCE J. Engng. Mech. 98, 813–822 (1981)
[4] Nayfeh, A.H.: Wave propagation in layered anisotropic media with applications to composites. Elsevier, Amsterdam (1995)
[5] Yuan, F.G., Hsieh, C.C.: Three-dimensional wave propagation in composite cylindrical shells. Compos. Str. 42, 153–157 (1998)
[6] Ichchou, M.N., Mencik, J.-M., Zhou, W.: Wave finite elements for low and mid-frequency description of coupled structures with damage. Comput. Methods Appl. Mech. Engrg. 198, 1311–1326 (2009)
[7] Muc, A., Kędziora, P.: Free vibrations of cylindrical shells with delaminations. In: Shell Structures, Theory and Applications (SSTA 2009), pp. 187–190. Taylor & Francis Group (2010)
[8] Yang, J., Fu, Y.: Analysis of dynamic stability for composite laminated cylindrical shells with delaminations. Compos. Str. 78, 309–315 (2007)

Optimal Design of Eigenfrequencies for Composite Structures Having Piezoelectric Sensors or Actuators

A. Muc and P. Kędziora

Abstract. Modeling of composite structures having smart piezoelectric sensors or actuators are very similar to conventional composite layered structures, however, there is one difference reflected in the constitutive laws in the form of the electromechanical coupling. It affects also the additional complexities in the FE formulation. The present formulation of optimal design introduces boundaries of piezoelectric patches as new class of design variables. In addition classical design variables in the form of ply orientation angles of orthotropic layers are also taken into account. The design objective is the minimization (maximization) of natural frequencies. The standard Rayleigh-Ritz method is used, however, the accuracy of optimal design are verified with the aid of the FE package ABAQUS. Examples are presented to illustrate the performance of the proposed model.

1 Introduction

In recent years, the application of piezoelectric actuators for active control of low frequency structural noise and vibration problems has drawn considerable attention. One of the primary reasons for this growing interest is the ability to tailor geometric and material properties of the actuators (e.g., shape, size, location, orientation and polarization profile etc.) in order to prescribe specific actuator/substructure mode coupling characteristics. Investigations on the effect of shaping and varying the polarization profile of the actuator (thus varying the moment distribution function over the surface of the actuator) have shown that selective mode coupling is possible. However, a comprehensive study into the effect of shaping the actuator and its polarization profile is still lacking, especially for the two dimensional case. To design actuators for specific applications, existing actuator models need to be viewed in terms of the equivalent forces and moments they exert on the structure and also analyzed in terms of their wave-number (spatial frequency) components to predict the modal excitation characteristics.

Shape control of flexible structures using piezoelectric (PZT) materials has attracted much attention in many fields such as space and aeronautical engineering

A. Muc · P. Kędziora
Institute of Machine Design, Cracow University of Technology, Kraków, Poland

G. Stépán et al. (Eds.): Dynamics Modeling & Interaction Cont., IUTAM BOOK SERIES 30, pp. 231–238.
springerlink.com

[1, 2]. A main focus in static shape control using piezoelectric materials as the actuators is on finding the best control voltages so that the actuated shape is as close as possible to the desired one. Except for in some simple structures with simple boundary conditions, the shape control of structures is usually achieved using numerical methods (particularly the finite element method) and numerical optimization schemes. The investigations in this area are conducted in different directions. A great effort is put in the development of a consistent finite element model of structures integrated with piezoelectric transducers; see e.g. Ref [3-6]. The optimal design of actuator patterns as regards features such as a size, orientation and location has attracted the most attention as well as the analysis of the possibility of applications of different variants of optimization algorithms to those problems – see Refs [7-12]. A broad review of the used performance indexes that gauge the effectiveness of the optimization process is also presented by Jin et al. [13]. Optimal control voltages constitute a separate class of optimization problems for PZT actuators. It is especially introduced to save electrical energy. A broader discussion of the above group of problems can be found e.g. in Refs [14-17].

The formulation of the problem and the methods of the analysis are continuation of the works Muc and Kędziora [18, 19]. In addition, due to the lack of the space we shall not dwell on the description of the used optimization methods; the details can be found in Refs [18-20].

2 Formulation of the Problem

We recall briefly the 3D formulation of linear piezoelectricity that appeared in Ref. [21] and then discuss the simplifications resulting in 2D form of relations.

The 3D constitutive equations are derived by considering the electric entropy:

$$G = \sigma_{ij}\varepsilon_{ij} - D_i E_i, \quad i, j = 1,2,3 \tag{1}$$

where $\boldsymbol{D}$, $\boldsymbol{E}$ denotes the vector of the electric displacement and electric field, respectively. The electric field $\boldsymbol{E}$ is the gradient of the electric potential Φ_{el}. To solve dynamic problems the relations (1) should be supplemented by terms corresponding to the density of kinetic energy:

$$T = \rho \dot{u}^T \dot{u} / 2 \tag{2}$$

and to the work of external forces. $\boldsymbol{u}$ is a 3D displacement vector. In the classical manner, the constitutive relations of piezoelectricity can be obtained by considering Legendre transformation to eqn (1) and finally we find:

$$\sigma_{ij} = c_{ijkl}\varepsilon_{kl} - e_{mij}E_m, \quad D_m = e_{mij}\varepsilon_{ij} + \mu_{mn}E_n \tag{3}$$

In which the strain is defined by:

$$\varepsilon_{ij} = \frac{1}{2}\left(u_{i,j} + u_{j,i}\right) \tag{4}$$

and $[c]$ is the matrix of elastic moduli of size 6x6, $[e]$ is the matrix of piezoelectric coefficients of size 6x3, $[\mu]$ is the permittivity matrix of size 3x3. The standard variational calculus provides the following results:

$$\sigma_{ij,j} = 0, \quad D_{m,m} = 0 \tag{5}$$

where the body force and the electric charge density is omitted. They characterize the elastic equilibrium equations and one of the Maxwell equations.

For most practical problems piezoelectric materials are located on plated or shell laminated multilayered structures, so that the analysis is reduced to 2D one. For 2D analysis, we normally employ a kinematical hypothesis to model plate/shell deformations. Assuming the validity of the Love-Kirchhoff hypotheses the 3D components of displacements may be expressed in the following way:

$$\begin{gathered} u_i(x,y,z) = \Phi_i(x,y) + z\psi_i(x,y), \psi_\alpha(x,y) = -w_{,\alpha}, \alpha = 1,2, \\ \psi_3(x,y) = 0, \Phi_3(x,y) = w(x,y) \end{gathered} \tag{6}$$

where the comma after the symbol means the differentiation with respect to the variable α, and the subscripts 1, 2 denote x and y, respectively, whereas 3 – z. For the assumed loading conditions and the form of the stacking sequence (antisymmetric, angle-ply laminates) the in-plane displacements $\Phi_\alpha(x,y)$ are equal to zero since the in-plane and bending states are uncoupled. The above simplifications imply that the transverse displacement $w(x,y)$ becomes the first unknown variable. The electric potential Φ_{el} is the second unknown in the problem considered. However, the assumed 2D simplification results in the further requirements with respect to the electric potential distribution along the z coordinate. Usually, it is assumed that that function can be written in the following form: (see Wang et al. [22]):

$$\Phi_{el} = \left[1 - \left(\frac{2z - t_{PZT} - t}{t_{PZT}} \right)^2 \right] \varphi(x,y) \tag{7}$$

where the symbol $\varphi(x,y)$ denotes the value of the electric potential on the mid-surface of piezoelectric layer. Due to the above formulation it is much more convenient to express the physical relation (5b) in the equivalent integral form, i.e.:

$$\int_{t/2}^{t/2+t_{PZT}} div[D]dz = 0 \tag{8}$$

3 The Rayleigh-Ritz Quotient

The Rayleigh-Ritz method is based on the variational statements, i.e. the principle of minimum the Hamiltonian being the integral form of the sum of eqs (1), (2). In general, it is equivalent to the fundamental differential equations as well as to the so-called natural or static boundary conditions including force boundary conditions. In this approach, the approximation of the 2D displacement variable $w(x,y)$ is given by the appropriate Ritz infinite series approximation in the following form:

$$w(x,y) = a_{mn}\sin\left(\frac{n\pi x}{L_x}\right)\sin\left(\frac{m\pi y}{L_y}\right), \quad m,n = 1,3,5,... \tag{9}$$

where the a_{mn} denote undetermined parameters. The form of equations (9) corresponds to simply-supported boundary conditions. It is also possible to use other approximations. Inserting the series (9) into the Hamiltonian, from the stationary condition follows the system of the infinite set of non-linear algebraic equations. In order to ensure a convergence of the approximate solutions (9) to the true solution as the number of terms increases the infinite systems of equations are cut off to a finite number as the ratios of the functional of the Hamiltonian are less than 1% for the increasing number of terms a_{mn}. MATHEMATICA, a mathematics symbolic program was used to carry out the algebraic operation and to verify the convergence of the approximations (9).

4 Numerical Results

In this section, the geometric optimization of piezoelectric patches for optimal eigenfrequencies based on the proposed approach is carried out, and the simulation is implemented for a simply supported plate. Mechanical characteristics of the system are detailed in Table 1. The plate is made of layers having an identical thickness and the total thickness is equal to 0.0012 [m]. The laminate is assumed to be antisymmetric angle-ply so that fibre orientations are characterized by one variable representing fibre orientation θ. The length of the plate at the y direction b=0.1[m]. The piezoelectric actuators ($e_{13} = e_{23} = -260\text{x}10^{-12}$) have 0.0002[m] thickness each and are located on the opposite sides of plate.

Table 1 Rectangular plate and piezoelectric patch specifications.

Material	E_1 [GPa]	E_2 [GPa]	G_{12} [GPa]	ν_{12}	Density [kg/m^3]
Glass fibre/epoxy	55	16	7.6	0.28	1800
Piezoelectric	62.5	62.5	24	0.3	7500

In order to verify the correctness of the results computed with the use of the Rayleigh-Ritz method the analysis has been also conducted with the use of the FE package ABAQUS. In the latter case, both the laminated plate and the piezoelectric layers have been modeled with the use of appropriate 3D finite elements In general, we can conclude that the comparison with FE results shows a very good agreement and the discrepancy lies between 1% to 2%.

Distributions of the first natural frequencies (m=n=1) of the plate versus fibre orientations are plotted in Fig. 1 and for different host plate geometrical properties. It is assumed that the piezoelectric covers the whole plate area – so-called bimorph plate. The fibre orientation corresponding to the minimal normal deflection depends on the value of the ratio *a*/*b* and host plate material properties. There is no voltage applied to the top and bottom actuators.

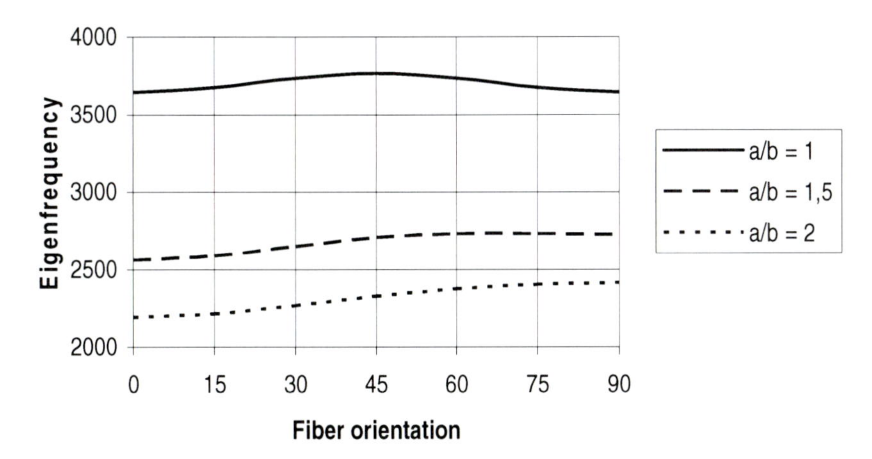

Fig. 1 Variations of the natural frequencies (*m*=*n*=1) with fibre orientations for bimorph plate, $E_z = 0$.

Figure 2 demonstrates the vibration mode for bimorph square plate. As it may be observed the mode is typical for such class of problems.

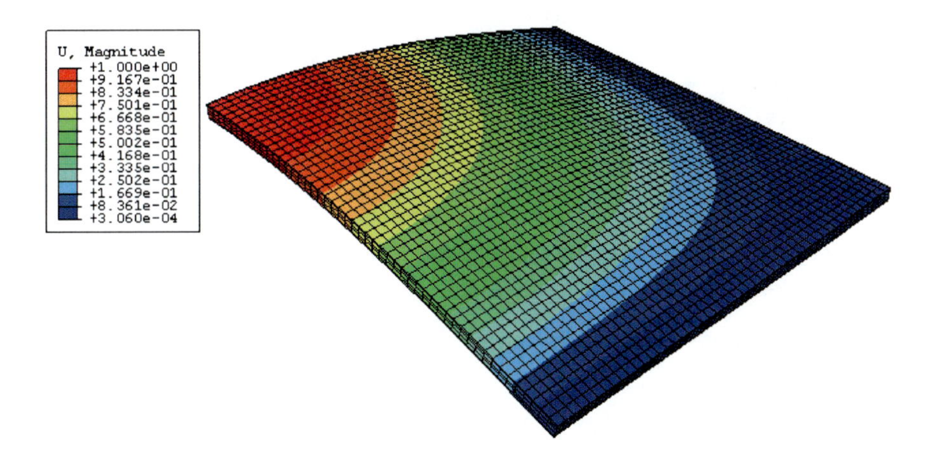

Fig. 2 Vibration mode for square bimorph plate (*m*=*n*=1).

As the area covered by the piezoelectric is reduced both the values of eigenfrequencies and their distributions with fibre orientations are changed in comparison with the results plotted in Fig.1. Their values and distributions are strongly affected by the position and shape of the piezoelectric layer – see Figs. 3, 4. However, for the cases considered the deformation mode is identical and also similar to that for the bimorph plate – Fig. 5. In the present analysis the ratio the area of the single rectangular actuator to the area of the composite plate substrate (A_{act}/A_{plate}) is equal to 1/8.

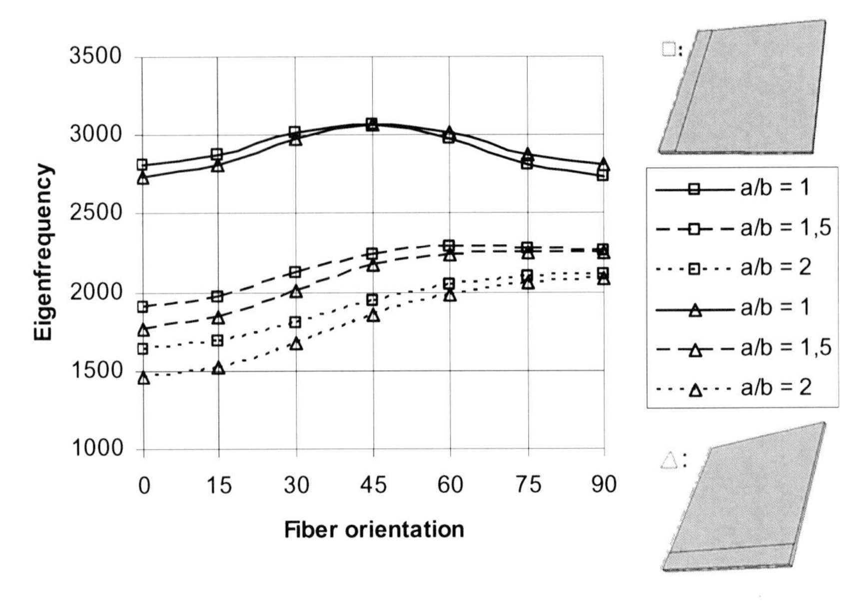

Fig. 3 Variations of the eigenfrequencies with fibre orientations for composite plate partially covered by the rectangular actuator patch, A_{act}/A_{plate}=0.125, E_z = 0.

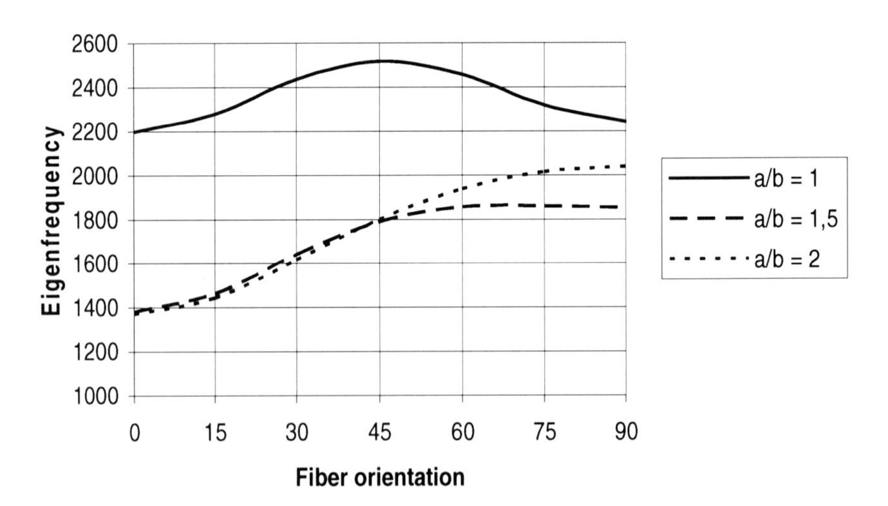

Fig. 4 Variations of the eigenfrequencies with fibre orientations for composite plate partially covered by the elliptical actuator patch, A_{act}/A_{plate}=0.125 , E_z = 0.

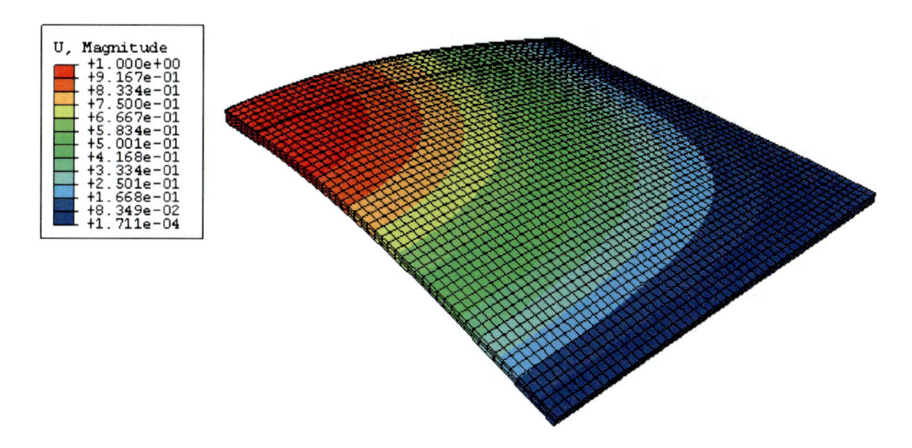

Fig. 5 Vibration mode for square composite plate partially covered by the rectangular actuator patch, A_{act}/A_{plate}=0.125 , $E_z = 0$, m=n=1.

Then, the top and bottom surfaces of the plate are subjected to the electric potential; however, their influence on the values of eigenfrequencies is negligibly small. Such a result demonstrates evidently that the maximal benefits in the optimal design of eigenfrequencies may be achieved searching for the optimal fibre orientations of the host plate, and not varying the shape of the local piezoelectric patches. However, it is worth to mention that the better electric actuation may be reached introducing two or more local, disconnected patches on the host plate, instead of one patch located at the plate center.

Acknowledgments. The Polish Research Foundation PB 1174/B/T02/2009/36 is gratefully acknowledged for financial support.

References

[1] Durr, J.K., Honke, R., Alberti, M., et al.: Development and manufacture of an adaptive light-weight mirror for space application. Smart Mater. Struct. 12, 1005–1016 (2003)

[2] Irschik, H.: A review on static and dynamic shape control of structures by piezoelectric actuation. Eng. Struct. 24, 5–11 (2002)

[3] Robbins, D.H., Reddy, J.N.: Analysis of piezoelectrically actuated beams using a layer-wise displacement theory. Comput. Struct. 41, 265–279 (1991)

[4] Chandrashekhara, K., Agarwal, A.N.: Active vibration control of laminated composite plates using piezoelectric devices: a finite element approach. J. Intell. Mater. Syst. Struct. 4, 496–508 (1993)

[5] Chattopadhyay, A., Seeley, C.E.: A higher order theory for modeling composite laminates with induced strain actuators. Composites B28, 243–252 (1997)

[6] Chee, C.Y.K., Tong, I.Y., Steven, G.P.: A mixed model for adaptive composite plates with piezoelectric for anisotropic actuation. Comput. Struct. 77, 253–268 (2000)

[7] Koconis, D.B., Kollar, L.P., Springer, G.S.: Shape control of composite plates and shells with embedded actuators. 2. Desired shape specified. J. Compos. Mater. 28, 459–482 (1994)
[8] Hsu, C.Y., Lin, C.C., Gaul, L.: Shape control of composite plates by bonded actuators with high performance configuration. J. Reinf. Plast. Compos. 16, 1692–1710 (1997)
[9] Onoda, J., Hanawa, Y.: Actuator placement optimization by genetic and improved simulated annealing algorithms. AIAA J. 31, 1167–1169 (1993)
[10] Agrawal, B.N., Treanor, K.E.: Shape control of a beam using piezoelectric actuators. Smart Mater. Sruct. 8, 729–739 (1999)
[11] Chee, C.Y.K., Tong, L.Y., Steven, G.P.: Static shape control of composite plates using a slope-displacement-based algorithm. AIAA J. 40, 1611–1618 (2002)
[12] Chee, C., Tong, L., Steven, G.P.: Piezoelectric actuator orientation optimization for static shape control of composite plates. Comput. Sruct. 55, 169–184 (2002)
[13] Jin, A., Yang, Y., Soh, C.K.: Application of fuzzy GA for optimal vibration control of smart cylindrical shells. Smart Mater. Struct. 14, 1250–1264 (2005)
[14] Agarwal, B.N., Treanor, K.E.: Shape control of a beam using piezoelectric actuators. Smart Mater. Struct. 8, 729–739 (1999)
[15] Chee, C.Y.K., Tong, L.Y., Steven, G.P.: Static shape control of composite plates using a slope–displacement based algorithm. AIAA J. 40, 1611–1618 (2002)
[16] Ajit, A., Ang, K.K., Wang, C.M.: Shape control coupled nonlinear piezoelectric beams. Smart Mater. Struct. 10, 914–924 (2001)
[17] Sun, D.C., Tong, L.Y., Wang, D.J.: An incremental algorithm for static shape control of smart structures with nonlinear piezoelectric actuators. Int. J. Solids Struct. 41, 2277–2292 (2004)
[18] Muc, A., Kędziora, P.: Variational approach in optimal design of piezoelectric sensors & actuators. In: Advanced Materials Research, vol. 47-50, pp. 1258–1261. Trans. Tech Publications, Switzerland (2008)
[19] Muc, A., Kedziora, P.: Optimal design of smart laminated composite structures. Materials and Manufacturing Processes 25, 1–9 (2010)
[20] Muc, A., Gurba, W.: Genetic algorithms and finite element analysis in optimization of composite structures. Compos. Struct. 54, 275–281 (2001)
[21] Tiersten, H.F.: Linear piezoelectric plate vibrations. Plenum Press, New York (1969)
[22] Wang, Q., Quek, S.T., Liu, X.: Analysis of piezoelectric coupled circular plate. Smart Mater. Struct. 10, 229–239 (2001)

Vibration Control of Defects in Carbon Nanotubes

A. Muc and M. Chwał

Abstract. Experimental studies of CNT mechanical properties demonstrate discrepancies with analytical predictions. Since the atomic structure of carbon nanotubes demonstrates evidently anisotropic mechanical properties an analytical molecular structural mechanics model is introduced in order to derive longitudinal and circumferential moduli of nanotubes. The identification is based on the eigenfrequencies analysis of the proposed computational model. It is combined with the FE analysis and the interatomic potentials (Tersoff-Brenner and Morse). For simply supported cylindrical shells made of a specially orthotropic material the eigenfrequencies can be easily derived in the analytical way using the Rayleigh-Ritz method as the roots of the third order algebraic equations. Detailed derivations are presented and the predicted results are shown and discussed with a few computational examples.

1 Introduction

Carbon nanotubes (CNTs) hold considerable promise as ultra-stiff high-strength fibers for use in cabling and nanocomposites. The outstanding mechanical characteristics hold for nearly perfect CNTs. Several theoretical studies have reported CNT failure strains in the range of 20–30% and failure stresses usually in excess of 100 GPa. By contrast, the few direct mechanical measurements that have been reported indicate much lower values. Most attempts to resolve the theoretical–experimental discrepancies have concentrated on the possible role of defects in limiting peak strengths.

Different approaches have been used to explore the role of vacancy defects in the fracture of CNTs under axial tension. In general, they can be divided into two groups: single- or two-atom vacancy defects Mielke et al. [1-5] or axisymmetric fracture patterns (defects) [6, 7]. However, the majority of existing works deals with the analysis of the Stone-Wales transformation that results in ductile fracture for nanotubes [8-11].

A. Muc · M. Chwał
Cracow University of Technology, Institute of Machine Design, ul. Warszawska 24, 31-155 Krakow, Poland
e-mail: olekmuc@mech.pk.edu.pl, mchwal@pk.edu.pl

G. Stépán et al. (Eds.): Dynamics Modeling & Interaction Cont., IUTAM BOOK SERIES 30, pp. 239–246.
springerlink.com

Significant challenges exist in both the micromechanical characterization of nanotubes and their composites and the modeling of the elastic and fracture behavior at the nanoscale. In general they include (a) complete lack of micromechanical characterization techniques for direct property measurement, (b) tremendous limitations on specimen size, (c) uncertainty in data obtained from indirect measurements, and (d) inadequacy in test specimen preparation techniques and lack of control in nanotube alignment and distribution. The above-mentioned problems and the description of nanocomposites fracture modeling are discussed in Refs [12-15].

In the homogenization procedures we do not have information enough about the stiffness properties, but we have information about the structural response, e.g. in terms of strain/stress distributions for the prescribed boundary conditions (force, displacements etc.), damage or eigenfrequencies. For simply supported cylindrical shells made of a specially orthotropic material the eigenfrequencies can be easily derived in the analytical way using the Rayleigh-Ritz method as the roots of the third order algebraic equations. Detailed derivations are presented and the predicted results are shown and discussed with a few computational examples. The presented method is an extension of the model proposed for perfect carbon nanotubes – see Muc [16,17].

2 Defects in Carbon Nanotubes

If CNTs have defects in the atomic network, one can expect that due to their quasi-one-dimensional atomic structure even a small number of defects will result in some degradation of their characteristics. The defects, especially vacancies, give rise to a deleterious effect – deterioration of axial mechanical properties of nanotubes. However, this deterioration in the mechanical characteristics is partly alleviated by the ability of nanotubes to heal vacancies in the atomic network by saturating dangling bonds. The defects can appear at the stage of CNT growth and purification, or later on during device or composite production. Moreover, defects in CNTs can deliberately be created by chemical treatment or by irradiation to achieve the desired functionality. Therefore, possible defects in CNTs can be classified in the following manner: 1) point defects such as vacancies, 2) topological defects caused by forming pentagons and heptagons e.g. 5-7-7-5 defect – so-called Stone-Wales defects and 3) hybridization defects caused due to functionalisation. It is possible to consider SWNTs with a single vacancy (one atom removed), with a double vacancy (two adjacent atoms knocked out) and with a triple vacancy (three adjacent atoms missing), as depicted in Fig. 1. In what follows, these configurations will be referred to as non-reconstructed defects. In each tube the non-reconstructed double vacancy defects have two axially distinguishable orientations separated by 120 degrees (only one configuration is shown in Fig. 1).

These atomic configurations are metastable but can survive for macroscopic times at low temperatures or when the atoms with dangling bonds are bonded to a surrounding medium, e.g., a polymer matrix. The other type of defects studied here are the vacancy related defects, i.e., vacancies in SWNTs relaxed to the global minimum energy configuration. In order to find these low energy

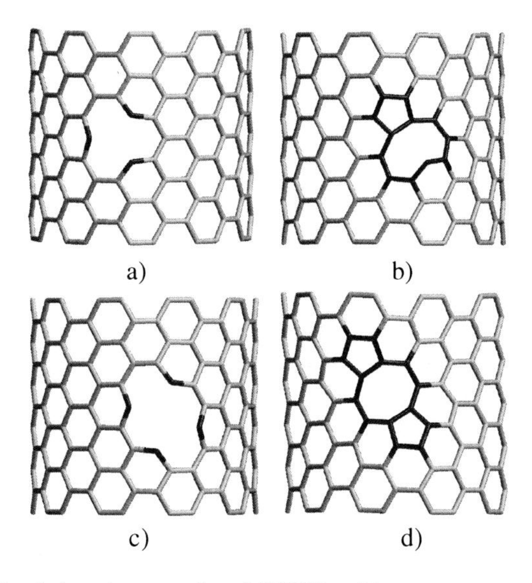

Fig. 1 Atomic networks of SWNTs with non-reconstructed (a, b) and reconstructed (c, d) single (a,c), double (b,d). Only the front wall of each tube is shown. The configurations correspond to a (10,10) armchair SWNT.

configurations, the non-reconstructed vacancies were thermally annealed at a temperature of 3000 K for 40 ps before the system was slowly cooled down at an average rate of 5 K/ps using the Berendsen thermostat. These configurations are presented in Fig. 1 (c,d) for (10,10) armchair tubes. The vacancy configurations in the other studied SWNTs are similar to those shown in Fig. 1.

3 The Interatomic Model

Let us consider the carbon nanostructure as the space-frame structure where each of the C-C bonds is represented as a beam. The stiffness of the C-C bond may be constant (this variant is further called as the model A) or variable (the model B). In the latter case the stiffness is evaluated incrementally at each step of deformations with the aid of the Tersoff-Brenner potential [18]. It is assumed that in the carbon nanostructure each carbon atom may react with the neighbourhood atoms only. At the equilibrium state the distance between carbon atoms is constant and equal to 0.142 nm. As the atom moves from the equilibrium state the non-zero reaction force is computed as the first derivative of the potential. We restrict the motion of the two atoms to one dimension, along the line connecting them, so that the atoms can only move directly towards or away from one another.

The numerical space-frame model of carbon nanotubes is presented in Fig.2. One of the ends of the tube is simply supported, whereas at the second the symmetry conditions are imposed.

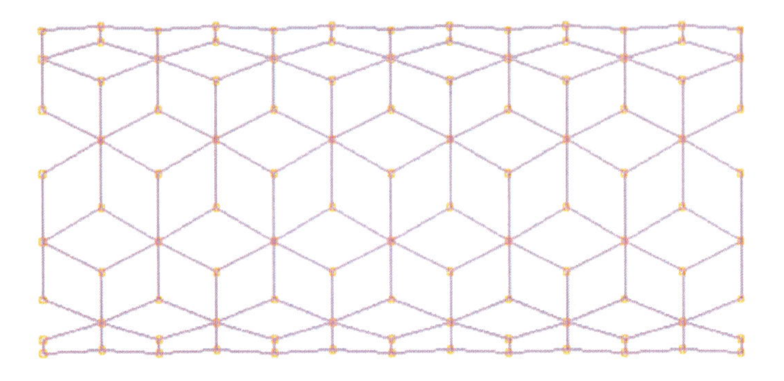

Fig. 2 Numerical model of the (5,5) armchair carbon nanotube

4 Identification Method

In the homogenization procedures we do not have information enough about the stiffness properties, but we have information about the structural response, e.g. in terms of strain/stress distributions for the prescribed boundary conditions (force, displacements etc.), damage or eigenfrequencies. For simply supported cylindrical shells made of a specially orthotropic material the eigenfrequencies can be easily derived in the analytical way using the Rayleigh-Ritz method as the roots of the following equation:

$$\psi^3 + b_0\psi^2 + c_0\psi - d_0 = 0 \tag{1}$$

where:

$$\psi = \rho R^2 h\omega^2/A_{11}, b_0 = -a_{11} - a_{22} - a_{33}, c_0 = a_{11}a_{22} + a_{11}a_{33} + a_{22}a_{33} - a_{13}^2, d_0 = a_{11}a_{22}a_{33} - a_{22}a_{13}^2,$$

$$a_{11} = \lambda_m^2,\ a_{22} = \frac{A_{66}}{A_{11}}\lambda_m^2,\ a_{33} = \frac{A_{22}}{A_{11}} + \frac{h^2}{12R^2}\lambda_m^4,\ a_{13} = -\frac{A_{12}}{A_{11}}\lambda_m,\ \lambda_m = \frac{m\pi R}{L} \tag{2}$$

ρ, R, h and L denote the nanotube density, radius, equivalent thickness and length, respectively, and m, n are wavenumbers in the longitudinal and circumferential directions. A_{ij} are the membrane stiffness matrix coefficients for specially orthotropic bodies.

In our analysis it is assumed that the nanotube can deform in the longitudinal directions only (i.e. n=0) in order to avoid local deformations of individual C-C bonds.

Among three real roots of the Eq. (1) the first one corresponds to so-called shear mode since dimensionless eigenfrequencies ψ are equal to λ^2_m A_{66}/A_{11}, whereas the next two are longitudinal ones and are dependent on the orthotropy ratio $A_{22}/A_{11} = E_{circumf}/E_{long}$ – see Fig.3. For the low orthotropy ratio the natural vibrations are dominant.

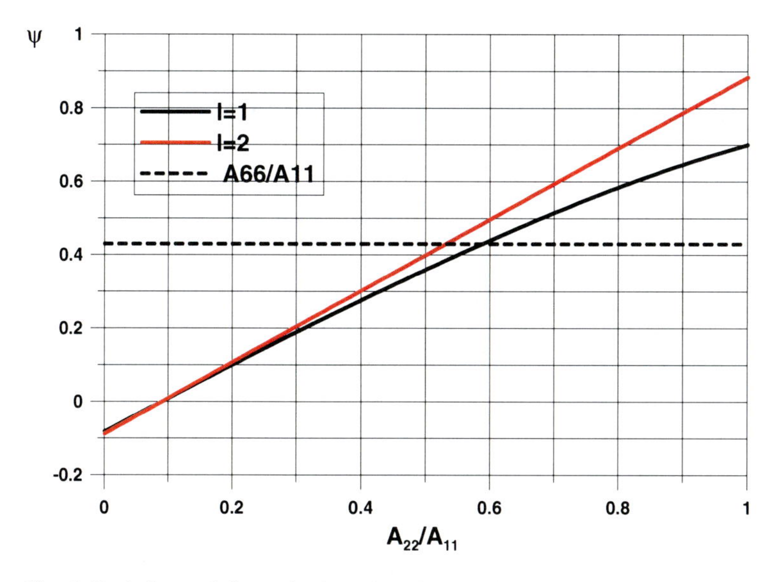

Fig. 3 Variations of dimensionless eigenfrequencies with the orthotropy ratio A_{22}/A_{11} and the wavenumber *m*

5 Numerical Results and Discussion

Let us assume that the carbon nanotube (the armchair (5,5)) has the following properties: E_{long} = 1.0 [TPa], ρ=600 [kg/m^3], L=29.5 [nm] and R =3.39 [nm]. Thus, for ψ=1 the square root of the ratio $A_{11}/(\rho h R^2)$ is equal to 12 [THz] and is the multiplier of natural frequencies – Eq. (1). As it may be seen the magnitude of natural frequencies (THz) is in the range mentioned in the literature.

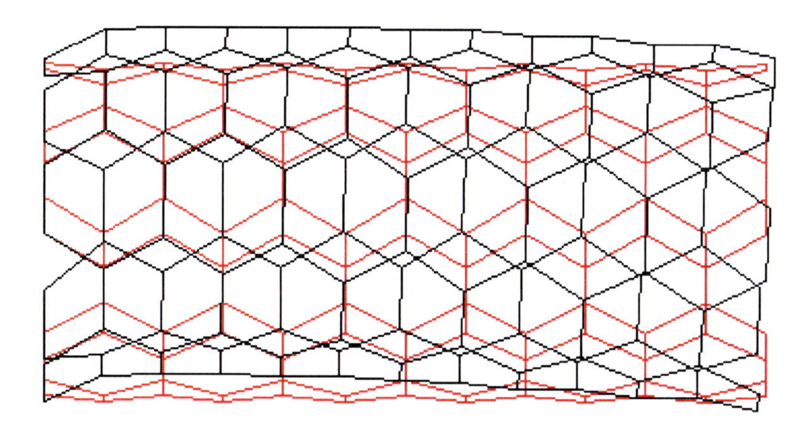

Fig. 4 Mode I – frequency = 3.472 THz (deformations not to scale)

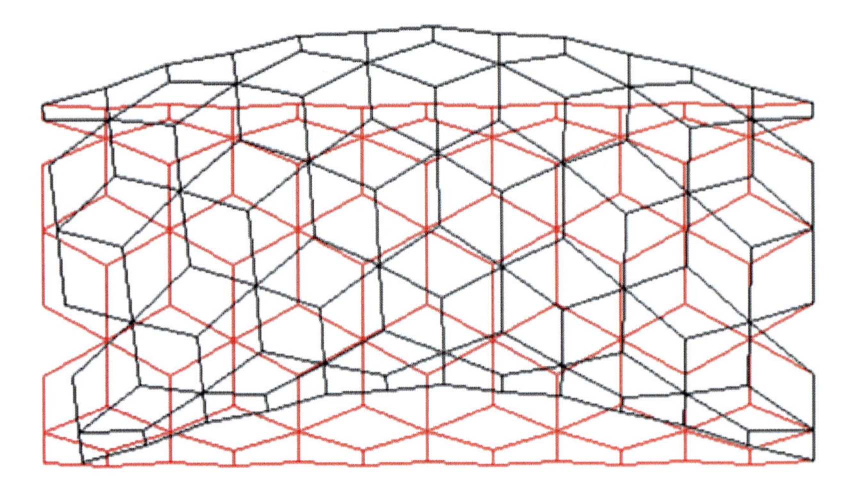

Fig. 5 Mode II – frequency = 5.719 THz (deformations not to scale)

The carbon nanotubes remain cylindrical until the critical eigenfrequency is reached at which point they deform in the longitudinal direction. Figs 4 and 5 are plots of some typical modes of deformations for two first natural frequencies. The half of nanotubes is modeled only due to symmetric boundary conditions. The natural frequencies have been obtained with the use of the NISA FE package.

Table 1 The Young moduli for armchair (5,5) carbon nanotubes

-	Model A	Model B
E_{long} [GPa]	740.90	724.71
$E_{circumf}$ [GPa]	94.36	85.32
The orthotropy ratio	0.127	0.118

Table 1 shows the Young moduli and the orthotropy ratio for the analyzed carbon nanotube. The reported by Wong [19] and Treacy [20] experimental values of the longitudinal Young modulus vary in the range 1.29 ± 0.59 [TPa] and agree very well with those identified by the eigenfrequency analysis. Of course, the obtained values are very low comparing with those computed with the use of others theoretical models. However, it should be pointed out that reduction of the value is the result of the consistent considerations that take into account orthotropic properties of carbon nanotubes. The computed value of the orthotropy ratio is quite high and is even higher than that for carbon fibre reinforced plastics so that it cannot be neglected in the analysis. The nonlinear description of carbon atoms interactions decreases values of Young's moduli.

Using the proposed model it is possible to explain the influence of the carbon nanotubes geometry and structure on Young's moduli. It is worth to note again that the proposed description of carbon nanotubes as orthotropic bodies is completely consistent with homogenization theory of skeletal or space-frame shell structures constituting (or being) in fact the structure of nanotubes.

Acknowledgments. The Polish Research Foundation PB 1174/B/T02/2009/36 is gratefully acknowledged for financial support.

References

[1] Troya, D., Mielke, S.L., Schatz, G.C.: Carbon nanotube fracture – differences between quantum mechanical mechanisms and those of empirical potentials. Chem. Phys. Lett. 382, 133–141 (2003)
[2] Mielke, S.L., et al.: The role of vacancy defects and holes in the fracture of carbon nanotubes. Chem. Phys. Lett. 390, 413–420 (2004)
[3] Liew, K.M., He, X.Q., Wong, C.H.: On the study of elastic and plastic properties of multi-walled carbon nanotubes under axial tension using molecular dynamics simulation. Acta Mater. 52, 2521–2527 (2004)
[4] Mielke, S.L., et al.: The effects of extensive pitting on the mechanical properties of carbon nanotubes. Chem. Phys. Lett. 446, 128–132 (2007)
[5] Zhang, S., et al.: Mechanics of defects in carbon nanotubes: Atomistic and multiscale simulations. Phys. Rev. 71, 115403 (2005)
[6] Duan, W.H., et al.: Molecular mechanics modeling of carbon nanotube fracture. Carbon 45, 1769–1776 (2007)
[7] Tserpes, K.I., Papanikos, P., Tsirkas, S.A.: A progressive fracture model for carbon nanotubes. Compos. Part B 37, 662–669 (2006)
[8] Nardelli, M.B., Yakobson, B.I., Bernholc, J.: Brittle and ductile behavior in carbon nanotubes. Phys. Rev. Lett. 81, 4656–4659 (1998)
[9] Lu, J., Zhang, L.: Analysis of localized failure of single-wall carbon nanotubes. Comput. Mater. Sci. 35, 432–441 (2006)
[10] Song, J., et al.: Stone–Wales transformation: Precursor of fracture in carbon nanotubes. Int. J. Mech. Sci. 48, 1464–1470 (2006)
[11] Zhang, P., et al.: An atomistic-based continuum theory for carbon nanotubes: analysis of fracture nucleation. J. Mech. Phys. Sol. 52, 977–998 (2004)
[12] Wagner, H.D., Lourie, O., Feldman, Y., Tenne, R.: Stress-induced fragmentation of multiwall carbon nanotubes in a polymer matrix. Appl. Phys. Lett. 72, 188–190 (1998)
[13] Lourie, O., Wagner, H.D.: Transmission electron microscopy observations of fracture of single-wall carbon nanotubes under axial tension. Appl. Phys. Lett. 73, 3527–3529 (1998)
[14] Shi, D.-L., et al.: Multiscale analysis of fracture of carbon nanotubes embedded in composites. Int. J. Fract. 134, 369–386 (2005)
[15] Li, C., Chou, T.-W.: Multiscale modeling of compressive behavior of carbon nanotube/polymer composites. Compos. Sci. Technol. 66, 2409–2414 (2006)
[16] Muc, A.: Design and identification methods of effective mechanical properties for carbon nanotubes. Mat. Des. 31, 1671–1675 (2010)

[17] Muc, A.: Modeling of CNTs/nanocomposites deformations and tensile fracture. In: Proc. 17th International Conference on Composite Materials (ICCM-17) Edinburgh, UK (2009)
[18] Brenner, D.W.: Empirical potential for hydrocarbons for use in simulating the chemical vapor deposition of diamond films. Phys. Rev. B 42, 9458–9471 (1990)
[19] Wong, E.W., et al.: Nanobeam mechanics: elasticity, strength, and toughness of nanorods and nanotubes. Science 277, 1971–1975 (1997)
[20] Treacy, M.M.J., et al.: High Young's modulus observed for individual carbon nanotubes. Nature 381, 680–687 (1996)

Influence of Noise on Discrete Breathers in Nonlinearly Coupled Micro-oscillator Arrays

Subramanian Ramakrishnan and Balakumar Balachandran

Abstract. Discrete Breathers, which are dynamical localizations of energy observed in the response of periodic, discrete Hamiltonian lattices to external forcing, have been recently observed in the response of micro-scale cantilever arrays. In this article, the authors invoke an analytical formalism introduced in their previous work to study the influence of white noise excitation on the formation and destruction of discrete breathers in a micro-scale cantilever array with strong intersite nonlinearities. The Fokker-Planck equation for a typical element of the array is derived. Numerical solutions to a system of approximate moment evolution equations obtained from the Fokker-Planck equation elucidate the influence of noise of varying intensity on the emergence and sustenance of discrete breathers in the array. The reported results can form the basis for developing a fundamental understanding of the influence of noise on breathers in coupled arrays of nonlinear oscillators with strong intersite nonlinearities including arrays of microelectromechanical systems.

1 Introduction

In this article, the authors study the dynamics of an array of nonlinear micromechanical oscillators subject to a combination of deterministic and white noise excitations. The key feature of the system under consideration is that each oscillator in the array is also nonlinearly coupled to its neighbors. Throughout the rest of the article, the intrinsic nonlinearity of a typical element of the array is referred to as the *onsite* nonlinearity while the nonlinearity arising from the coupling to the neighbours is referred to as the *intersite* nonlinearity. The objective of the paper is to utilize the

Subramanian Ramakrishnan
School of Dynamic Systems, University of Cincinnati
e-mail: `s.ramakrishnan@uc.edu`

Balakumar Balachandran
Department of Mechanical Engineering, University of Maryland-College Park
e-mail: `balab@umd.edu`

G. Stépán et al. (Eds.): Dynamics Modeling & Interaction Cont., IUTAM BOOK SERIES 30, pp. 247–254.
springerlink.com

analytical framework introduced by the authors in their recent study of white noise effects on the dynamics of linearly coupled arrays of micro-scale oscillators with onsite nonlinearities to the case of an array coupled by strong intersite nonlinearities in order to study the influence of noise on the formation and destruction of discrete breathers which are also known as intrinsic localized modes.

Discrete breathers (DBs) are localizations of energy observed in the response of perfectly periodic lattices of strong intrinsically nonlinear oscillators subjected to an external forcing [1, 2]. While energy localization continues to be well studied in the physics literature, the relatively recent observation of the phenomenon in micro-scale oscillator arrays impacts modern technology as well [1, 3, 4, 5]. In recent work the authors studied the dynamics of a linearly coupled micro-scale oscillator array with onsite nonlinearities with a focus on the phenomenon of energy localization [6, 7]. Using an exclusively numerical approach (which extended the work in [8]), it was concluded that white noise excitation, by itself, was unable to produce localizations in the considered array. However, in the case of an array subjected to a combined excitation with deterministic and white noise components, the numerical results indicated the existence of a threshold noise strength beyond which the DB at one location was attenuated whilst it was strengthened at another location. In addition, numerical solutions to a closed system of moment evolution equations obtained from the Fokker-Planck equation for the response of the array indicated that once a localization event occurs in the array, white noise excitation above a certain threshold of noise intensity contributes to the sustenance of the event. Moreover, it was observed that an excitation with a higher noise strength resulted in enhanced response amplitudes for oscillators in the center of the array.

While the analysis of the linear coupling case yielded interesting results, it is known that intersite nonlinearities lead to much stronger localizations in micro-scale oscillator arrays when compared with the former case and this is the prime motivation for the present work [5]. Moreover, novel results that underscore the influence of stochastic effects on energy localization have recently been reported in the literature. For instance, in their study of a hard ϕ^4 lattice with linear intersite coupling, Cubero et al. noted a threshold excitation frequency below which breather formation fails to occur in the case of purely deterministic excitation [9]. However, it was shown numerically that the concerted effect of stochastic and deterministic excitation could support breather formation below the threshold frequency. In the case of excitations above the threshold frequency, they found that while noise supported the spontaneous formation of localizations, remarkably, the same random force that induced breather formation was also able to destroy them later in time.

The rest of this article has been organized as follows. In Section 2, the system of Klein-Gordon equations representing a typical cell i of the array is introduced. In Section 3, the Fokker-Planck equation derived from the stochastic (Ito) version of the Klein-Gordon equations that account for both white noise and deterministic excitations, is presented first. Next, the system of moment evolution equations obtained from the Fokker-Planck equation is presented and the moment closure approximations are discussed. In Section 4, preliminary results obtained from numerical solutions of the approximate system of closed moment equations that elucidate the influence of white noise excitation on the formation and destruction of DBs in the array are presented. Concluding remarks are collected together and presented along with a discussion of the results in Section 5.

2 Model of the Array

Consider an array of micro-cantilever beams coupled together by the overhang region between them [3]. Each cell consists of two beams of different lengths, different spring constants, and different masses. Ignoring the mass of the overhang and assuming that the coupling between the cantilevers comprises both linear and nonlinear terms [5], the cell dynamics is governed by the nonlinear Klein-Gordon equations which, for a typical cell i are given by:

$$\begin{aligned} m_a\ddot{x}_{ai} + \frac{m_a}{\tau}\dot{x}_{ai} + k_{2a}x_{ai} + k_{4a}{x_{ai}}^3 + k_l\left(2x_{ai} - x_{bi} - x_{bi-1}\right) \\ +k_{4l}\left[\left(x_{ai}-x_{bi}\right)^3 + \left(x_{ai}-x_{bi-1}\right)^3\right] = m_a\alpha. \\ m_b\ddot{x}_{bi} + \frac{m_b}{\tau}\dot{x}_{bi} + k_{2b}x_{ai} + k_{4b}{x_{bi}}^3 + k_l\left(2x_{bi} - x_{ai+1} - x_{ai}\right) \\ +k_{4l}\left[\left(x_{bi}-x_{ai}\right)^3 + \left(x_{bi}-x_{ai+1}\right)^3\right] = m_b\alpha. \end{aligned} \tag{1}$$

In equation 1 , the subscripts a and b correspond to the different cantilever lengths, m_a and m_b are their respective masses, τ is a time constant, k_{2a} and k_{2b} are the associated harmonic spring constants, k_{4a} and k_{4b} are the associated nonlinear spring constants, k_l is the linear coupling constant, and k_{4l} is the nonlinear coupling constant. The entire array is subjected to a uniform acceleration α which is achieved using piezoelectric actuation.

3 Analytical Framework

The details of the Fokker-Planck formalism invoked in this article may be found in the standard works on stochastic analysis (see, for instance, [10]). While in principle the Fokker-Planck equation may be solved to obtain the probability density function that characterizes the system dynamics, the challenges encountered in solving the equation for nonlinear systems often require the use of methods such as the one of moment approximations. Here the averaged dynamics is represented by a system of moment evolution equations derived from the Fokker-Planck equation. Typically, nonlinear systems give rise to an infinite hierarchy of moment equations, closure of which is achieved through a set of moment approximations. The closed system of approximate moment equations thus obtained is then numerically solved.

3.1 Fokker-Planck Equation

Denoting the noise strength by σ, the deterministic forcing function by $f(t)$, the spring constants by $k_{2a}/m_a = K_{2a}$ and so forth and defining a constant $c = 1/\tau$, the Fokker-Planck equation corresponding to the addition of stochastic forcing terms to Eqn.1 is derived in the Ito sense as:

$$
\begin{aligned}
\frac{\partial P}{\partial t} = &- \left[y_a \frac{\partial P}{\partial x_a} + y_b \frac{\partial P}{\partial x_b} \right] + \left[c y_a + K_{2a} x_a + K_{4a} {x_a}^3 + K_{Ia} \left\{ 2x_a - x_b - x_{bi-1} \right\} \right] \frac{\partial P}{\partial y_a} \\
&+ \left[K_{4Ia} \left\{ (x_a - x_b)^3 + (x_a - x_{bi-1})^3 \right\} - f(t) \right] \frac{\partial P}{\partial y_a} + 2cP \\
&+ \left[c y_b + K_{2b} x_b + K_{4b} {x_b}^3 + K_{Ib} \left\{ 2x_b - x_a - x_{ai+1} \right\} - f(t) \right] \frac{\partial P}{\partial y_b} \\
&+ \frac{\sigma^2}{2} \left[\left(\frac{\partial}{\partial y_a} \right)^2 + \left(\frac{\partial}{\partial y_b} \right)^2 \right] P + \left[K_{4Ib} \left\{ (x_b - x_a)^3 + (x_b - x_{ai+1})^3 \right\} \right] \frac{\partial P}{\partial y_b} .
\end{aligned}
\tag{2}
$$

3.2 Moment Evolution Equations

Denoting the moments by angled brackets, the moment evolution equations for the a type cantilever at cell i are derived from the Fokker-Planck equation 2 as:

$$
\begin{aligned}
\frac{d}{dt} \left\langle {x_a}^n {y_a}^m \right\rangle = &\, n \left\langle {x_a}^{n-1} {y_a}^{m+1} \right\rangle - cm \left\langle {x_a}^n {y_a}^m \right\rangle - mK_{2a} \left\langle {x_a}^{n+1} {y_a}^{m-1} \right\rangle \\
&- m (K_{4a} + 2K_{4Ia}) \left\langle {x_a}^{n+3} {y_a}^{m-1} \right\rangle - 2mK_{Ia} \left\langle {x_a}^{n+1} {y_a}^{m-1} \right\rangle \\
&+ mK_{Ia} \left\langle {x_a}^n {y_a}^{m-1} x_b \right\rangle + mK_{Ia} \left\langle {x_a}^n {y_a}^{m-1} x_{bi-1} \right\rangle + mK_{4Ia} \left\langle {x_a}^n {y_a}^{m-1} {x_b}^3 \right\rangle \\
&+ 3mK_{4Ia} \left\langle {x_a}^{n+2} {y_a}^{m-1} x_b \right\rangle - 3mK_{4Ia} \left\langle {x_a}^{n+1} {y_a}^{m-1} {x_b}^2 \right\rangle + mK_{4Ia} \left\langle {x_a}^n {y_a}^{m-1} {x_{bi-1}}^3 \right\rangle \\
&+ 3mK_{4Ia} \left\langle {x_a}^{n+2} {y_a}^{m-1} x_{bi-1} \right\rangle - 3mK_{4Ia} \left\langle {x_a}^{n+1} {y_a}^{m-1} {x_{bi-1}}^2 \right\rangle + mf(t) \left\langle {x_a}^n {y_a}^{m-1} \right\rangle \\
&+ \frac{m(m-1)}{2} \sigma^2 \left\langle {x_a}^n {y_a}^{m-2} \right\rangle .
\end{aligned}
\tag{3}
$$

By symmetry, a similar set of moment evolution equations are obtained for the b type cantilever as well. A closed system of moment evolution equations are then obtained by employing the following approximations: i) moments of order three and higher are neglected except for the third order moment of displacement and ii) moments of products involving variables in neighboring cells as well as different cantilever types are approximated as the product of moments of the respective variables. These approximations result in a system of eight moment evolution equations for each cantilever in the array that are numerically solved.

4 Numerical Solutions of Moment Evolution Equations

The following set of parameter values are assumed for the micro-cantilever array: $m_a = 5.46E - 13$ kg; $m_b = 4.96E - 13$ kg; $\tau = 8.75E - 3$ s; $k_{2a} = 0.303$ N/m; $k_{2b} = 0.353$ N/m; $k_4 = k_{4a} = k_{4b} = 5E8N/m^3; k_I = 0.0241$ N/m; $k_{4I} = k_{4Ia} = k_{4Ib} = \beta k_4$, where β is a factor that determines the strength of the nonlinear intersite coupling and $\alpha_0 = 10E4m/s^2$. The dynamics of a representative array of 50 cells (100 cantilevers) are simulated with random initial displacements of the order of magnitude $10E - 7$ m. The excitation profile (identical to the one employed in [6, 7]) is designed to promote localizations in the array and consists of three distinct phases. The forcing frequency of the deterministic excitation is linearly increased in the first phase, held constant in the second phase where the array is subjected to a harmonic excitation, and then switched off in the final phase. The results are presented in Figures 1 and 2 with the color code indicating mean coupling energy in units of Nm.

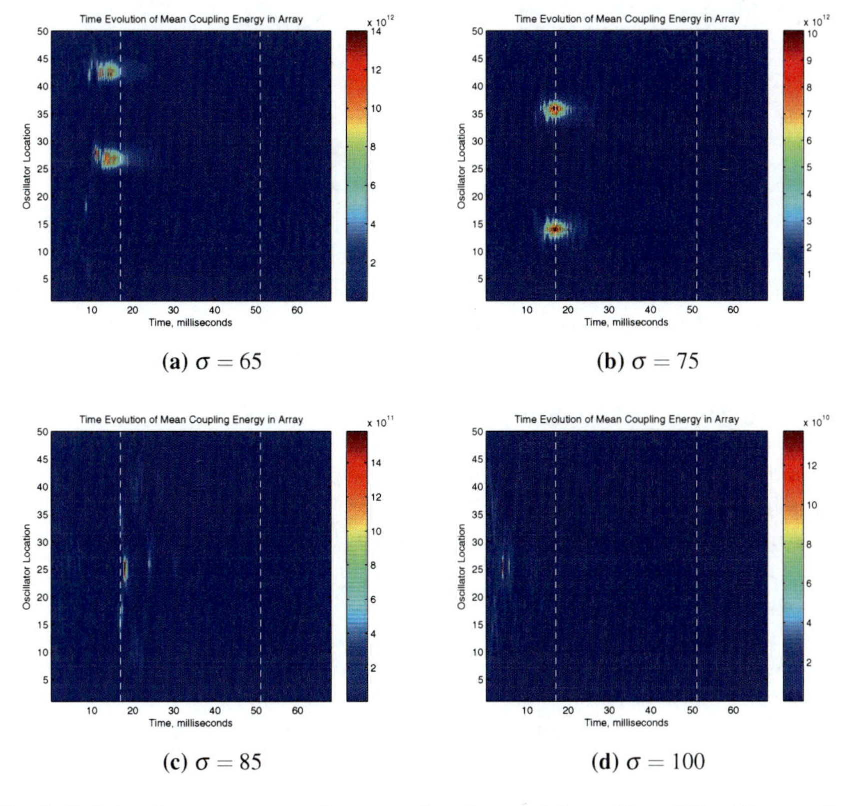

(a) $\sigma = 65$ **(b)** $\sigma = 75$

(c) $\sigma = 85$ **(d)** $\sigma = 100$

Fig. 1 Variation in response as noise strength σ is varied from 65 to 100 while coupling strength factor is held constant at $\beta = 0.25$.

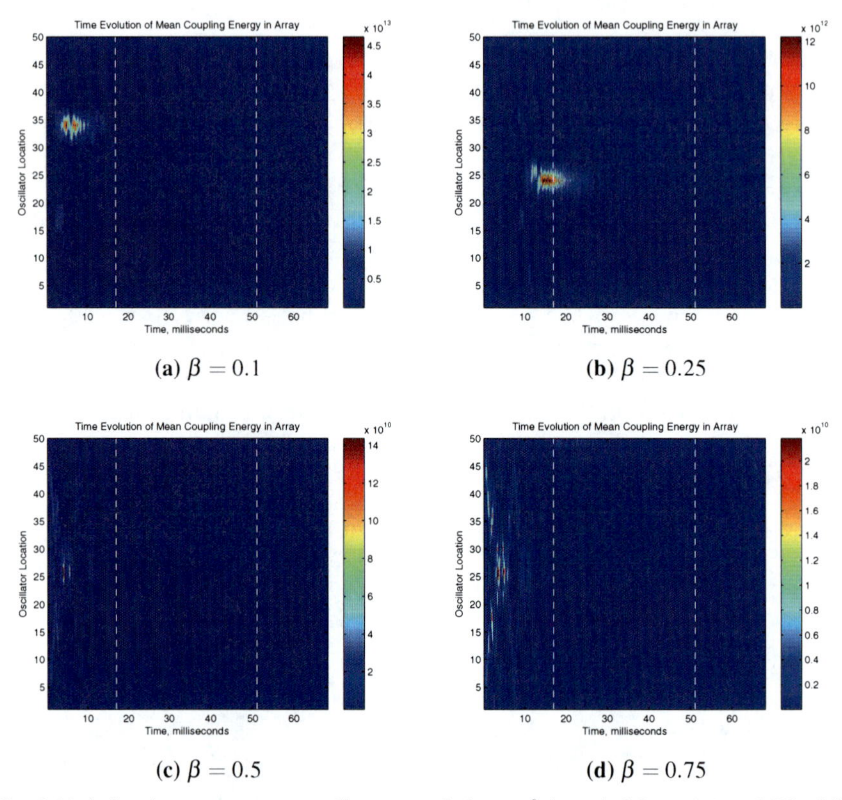

(a) $\beta = 0.1$ (b) $\beta = 0.25$

(c) $\beta = 0.5$ (d) $\beta = 0.75$

Fig. 2 Variation in response as coupling strength factor β is varied from 0.1 to 0.75 while noise strength is held constant at $\sigma = 70$.

5 Discussion and Conclusions

The numerical results presented in Figures 1 and 2 provide insights into the effects of noise as well as the strength of the nonlinear intersite coupling on the formation and destruction of DBs in the micro-cantilever array. The first set of results correspond to the cases of noise intensities $\sigma = 65, 75, 85$ and 100 with the coupling strength factor held constant at $\beta = 0.25$ in all the cases. This set of results, presented in Figures 1a to 1d, indicates the existence of high energy DBs in the presence of intersite nonlinearities. A comparison of the cases $\sigma = 65$ (Fig.1a) and $\sigma = 75$ (Fig. 1b) indicates that a variation in the intensity of excitation noise can result in the vanishing of breathers at two locations (in the neighborhoods of oscillators 27 and 43) accompanied by the emergence of breathers at new locations (in the neighborhoods of oscillators 36 and 14 in the present case). The localizations appears at around t=13 ms and survive until around t=20 ms. This shift in location as a result of a relatively small variation in excitation noise intensity as well as the relatively high energy of the formed breathers is in sharp contrast to the case of a linearly coupled array wherein weak localizations were observed only in oscillators around the center of the array [6, 7]. When the noise intensity is further increased to $\sigma = 85$ a short lived breather appears in the neighborhood of oscillator 25 (see Fig. 1c). The mean energy of this breather is lower by an order of magnitude. Increasing the intensity of noise further to $\sigma = 100$ results in the appearance of very short lived breathers in the neighborhood of oscillator 25 at around t=4 ms. Importantly, the mean energy of these weak breathers is lower by two orders of magnitude when compared to the cases of $\sigma = 65$ and $\sigma = 75$, despite the increase in noise intensity. Taken together, this set of results suggests the novel potential of noise as a mechanism for manipulating the emergence of DBs at particular locations in a micro-cantilever array. Furthermore, the results lead to the interesting conclusion that increasing the noise intensity does not imply a concomitant, monotonic, increase in the breather energy. This is consistent with the conclusion of Cubero et al. [9] that high intensity noise which promotes localization is also capable of destroying breathers.

The second set of results corresponds to the cases of coupling strength factors $\beta = 0.1, 0.25, 0.5$ and 0.75 with the noise intensity held constant at $\sigma = 70$ in all the cases. This set of results is presented in Figures 2a to 2d. Comparison of the results for cases of $\beta = 0.1$ and $\beta = 0.25$ (Fig.2a and Fig.2b) indicates a change of location of the breathers from the neighborhood of oscillator 34 to the neighborhood of oscillator 24 as well as the later emergence of the localization in the case of the stronger coupling. Moreover, in the case of $\beta = 0.25$ the breathers persist for a short duration even after the excitation frequency is held constant (the forcing frequency is held constant in the time window bracketed by the broken lines in Fig.2b). As seen from Figures 2c and 2d, a further increase in the coupling strength discourages the formation of breathers in the array. While there is a weak tendency for energy to localize in these cases (particularly at the center of the array around oscillator 25) the breathers hardly persist and their energy is lower by three orders of magnitude. Hence this set of results suggests that stronger coupling tends to attenuate the formation of breathers.

In conclusion, in this article the authors addressed the effects of noise and coupling strength on the formation and destruction of DBs in micro-oscillator arrays characterized by both onsite and intersite nonlinearities. Numerical simulations of approximate moment evolution equations obtained from the Fokker-Planck equation indicated that, white noise of higher intensity, in combination with deterministic excitation, can promote the formation of high energy discrete breathers. The results also showed that variation of noise intensity can result in the emergence of breathers at new locations within the array thereby suggesting the possibility of using noise as a mechanism to move energy localizations across the array. In addition, a further increase in noise intensity led to a decrease in the breather energy indicating that noise can also destroy localizations. Furthermore, a second set of results indicated that variations in the coupling strength of intersite nonlinearities can also result in the emergence of breathers at new locations within the array. Moreover, the results suggest that stronger intersite nonlinear coupling attenuates the formation of breathers. A set of related questions such as the emergence of breathers under sub-threshold and super-threshold excitation frequencies is currently under investigation and the results will be reported in future publications.

Acknowledgements. Partial support received for this work through NSF Grant No. 0826173 is gratefully acknowledged.

References

1. Campbell, D.K., Flach, S., Kivshar, V.S.: Localizing energy through nonlinearity and discreteness. Phys. Today 57, 43–49 (2004)
2. Sievers, A.J., Takeno, S.: Intrinsic localized modes in anharmonic crystals. Phys. Rev. Lett. 61, 970–973 (1988)
3. Sato, M., Hubbard, B.E., Sievers, A.J., Ilic, B., Czaplewski, D.A., Craighead, H.G.: Observation of locked intrinsic localized vibrational modes in a micromechanical oscillator array. Phys. Rev. Lett. 90(044102), 1–4 (2003)
4. Sato, M., Hubbard, B.E., English, L.Q., Sievers, A.J., Ilic, B., Czaplewski, D.A., Craighead, H.G.: Study of intrinsic localized vibrational modes in micromechanical oscillator arrays. Chaos 13, 702–715 (2003)
5. Sato, M., Hubbard, B.E., Sievers, A.J.: Nonlinear energy localization and its manipulation in micromechanical oscillator arrays. Rev. Mod. Phys. 78, 137–157 (2006)
6. Ramakrishnan, S., Balachandran, B.: Intrinsic localized modes in micro-scale oscillator arrays subjected to deterministic excitation and white noise. In: Proc. of IUTAM Symposium on Multifunctional Material Structures and Systems. IUTAM Book Series, vol. 19, pp. 325–334. Springer, Heidelberg (2010)
7. Ramakrishnan, S., Balachandran, B.: Energy localization and white noise induced enhancement of response in a micro-scale oscillator array. Nonlinear Dynamics (2010), doi:10.1007/s11071-010-9694-6
8. Dick, A.J., Balachandran, B., Mote Jr., C.D.: Intrinsic localized modes in microresonator arrays and their relationship to nonlinear vibration modes. Nonlinear Dynamics 54, 13 (2008)
9. Cubero, D., Cuevas, J., Kevrekidis, P.G.: Nucleation of breathers via stochastic resonance in nonlinear lattices. Phys. Rev. Lett. 102, 205–505 (2009)
10. Gardiner, C.W.: Handbook of Stochastic Methods. Springer, Heidelberg (1983)

Modeling Dry Friction

The papers in this section explore some models of dry friction and investigate the dynamics of systems with considerable frictional effects. A. V. Karapetyan have published several works on friction modeling, and his present paper introduces a new model of friction taking into account all types of friction (sliding, pivoting and rolling). This time, the dynamics of a homogeneous sphere on a rough plane with all types of friction is in the focus. The dynamics of a doublespherical tippe-top considering different friction models are analyzed in the paper by A. Zobova in detail. By considering Coulomb friction, G. Kono *et al.* investigate the intricate dynamic behavior of cleaning blades in laser printers subjected to frictional vibrations.

New Models of Friction and Their Applications in Rigid Body Dynamics*

A.V. Karapetyan

Abstract. The new model of friction taking into account all types of friction (sliding, pivoting and rolling) is given. The dynamics of a homogeneous sphere on the horizontal plane with all types of friction is investigated.

1 Calculation of Friction Force and Moment

The friction model proposed in this paper will be illustrated for the problem of the motion of a homogeneous sphere of radius a along a stationary horizontal plane. Following Contensou [1, 2], we will replace the point of contact of the sphere with the plane with a contact patch, assuming, unlike Contensou, that it has the shape of a spherical segment [3] rather than of a circle. In other words, we will assume that, when the sphere comes into contact with the plane, both the sphere and the bearing surface undergo elastic deformations. These deformations determine the two parameters of the contact patch: $\delta = a/R \in [0,1]$ and $\varepsilon = r/a \in [0,1]$, where R is the radius of the sphere, a segment of which determines the contact patch, and r is the radius of this segment (see Fig. 1, where a side view of the contact patch is shown on the left, and a top view on the right). Note that such a spherical segment approximates a segment $x^2 + y^2 \leq r^2$ of a paraboloid $z = (x^2 + y^2)/(2R)$ up to terms $\varepsilon^4\delta^2$. The case $\delta = 0$ $(R = \infty)$ corresponds to the Contensou-Zhuravlev model, in which the bearing surface is assumed to be non-deformable, and the case $\varepsilon = 0$ $(r = 0)$ corresponds to the Coulomb model, in which the sphere is assumed to be non-deformable. The case $\delta = 1$ $(R = a)$ also corresponds to the assumption of sphere non-deformability but requires a separate analysis because it is not described by the Coulomb model.

A.V. Karapetyan
Lomonosov Moscow State University, Moscow, Russia
e-mail: avkarapetyan@yandex.ru

* This research is supported by the Russian Foundation for Basic Research (09-08-00925, 10-01-00292).

G. Stépán et al. (Eds.): Dynamics Modeling & Interaction Cont., IUTAM BOOK SERIES 30, pp. 257–264.
springerlink.com

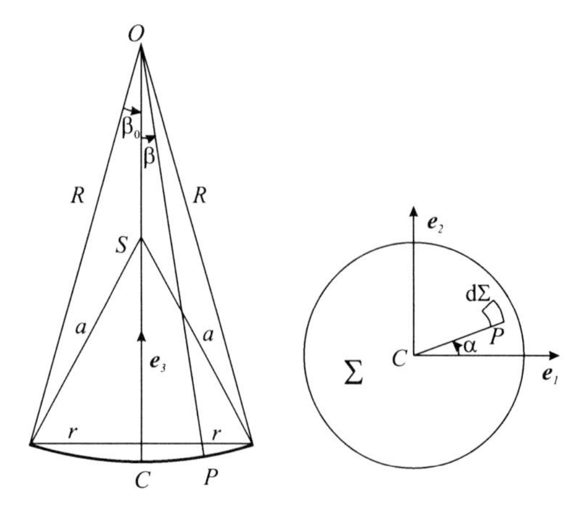

Fig. 1 Contact patch: a side view and a top view

Let P be an arbitrary point of the contact patch (a spherical segment Σ), and let its position be defined by the angles $\alpha \in [0, 2\pi]$ and $\beta \in [0, \beta_0]$,where $\beta_0 = \arcsin \mu$, $\mu = r/R = \varepsilon\delta$ (see Fig.1, in which the point is the centre of the sphere, a segment of which is Σ, and S and are the centre of the sphere and its projection onto *Sigma*).

We will introduce an orthonormalized reference frame $\mathbf{e_1}, \mathbf{e_2}, \mathbf{e_3}$ such that the unit vector $\mathbf{e_1}$ is directed along the sliding velocity $\mathbf{u} = u\mathbf{e_1}$ of the sphere (the velocity of point of the sphere), the unit vector $\mathbf{e_2}$ is orthogonal to the sliding velocity $\mathbf{u}$ and, like $\mathbf{e_1}$, lies in the horizontal plane and the unit vector $\mathbf{e_3}$ is directed along a rising vertical. We will find the velocity of the point P from Euler's formula

$$\mathbf{v}_P = \mathbf{v}_C + [\omega, \rho]; \mathbf{v}_C = \mathbf{u} = u\mathbf{e_1}, \omega = \omega_1\mathbf{e_1} + \omega_2\mathbf{e_2} + \omega_3\mathbf{e_3}$$

where ω is the angular velocity of the sphere and ρ is the radius vector of point P relative to the centre of the contact patch:

$$\rho = \rho_1\mathbf{e_1} + \rho_2\mathbf{e_2} + \rho_3\mathbf{e_3}; \rho_1 = R\sin\beta\cos\alpha, \rho_2 = R\sin\beta\sin\alpha, \rho_3 = R(1-\cos\beta)$$

Thus

$$\begin{aligned}\mathbf{v}_P &= v_1\mathbf{e_1} + v_2\mathbf{e_2} + v_3\mathbf{e_3}\\ v_1 &= u + R\omega_2(1-\cos\beta) - R\omega_3\sin\beta\sin\alpha\\ v_2 &= -R\omega_1(1-\cos\beta) + R\omega_3\sin\beta\cos\alpha\\ v_3 &= R\sin\beta(\omega_1\sin\alpha - \omega_2\cos\alpha)\end{aligned}$$

We will find the sliding velocity $\mathbf{u_P}$ of the point P, bearing in mind that this point is sliding over the spherical segment Σ (rather than over a horizontal plane, as in the model ([1, 2]). For this, we will decompose the velocity of the point P into two components – the tangential component and the normal component to Σ at point P:

$$\mathbf{v}_P = \mathbf{u}_P + (\mathbf{v}_P, \mathbf{n}_P)\mathbf{n}_P; \mathbf{n}_P = -\sin\beta\cos\alpha\mathbf{e_1} - \sin\beta\sin\alpha\mathbf{e_2} + \cos\beta\mathbf{e_3}$$

where $\mathbf{n}_P$ is the unit vector of the normal to the segment Σ at the point P, directed towards the concavity of this segment.

Thus

$$\mathbf{u}_P = u_1\mathbf{e_1} + u_2\mathbf{e_2} + u_3\mathbf{e_3}$$

$$\begin{aligned} u_1 &= u(1 - sin^2\beta\cos^2\alpha) + R\omega_1\sin^2\beta\sin\alpha\cos\alpha + \\ &+ R\omega_2(1-\cos\beta-\sin^2\beta\cos^2\alpha) - R\omega_3\sin\beta\sin\alpha \\ u_2 &= -usin^2\beta\sin\alpha\cos\alpha - R\omega_1(1-\cos\beta-\sin^2\beta\sin^2\alpha) - \\ &-R\omega_2\sin^2\beta\sin\alpha\cos\alpha + R\omega_3\sin\beta\cos\alpha \\ u_3 &= u\sin\beta\cos\beta\cos\alpha + R\sin\beta(1-\cos\beta)(\omega_1\sin\alpha - \omega_2\cos\alpha) \\ u_p &= [u^2(1-\sin^2\beta\cos^2\alpha) + R^2(\omega_1^2+\omega_2^2)(1-\cos\beta)^2 + R^2\omega_3^2\sin^2\beta + \\ &+2uR\omega_1\sin^2\beta\sin\alpha\cos\alpha + 2uR\omega_2(1-\cos\beta-\sin^2\beta\cos^2\alpha) - \\ &-2uR\omega_3\sin\beta\sin\alpha - 2R^2(\omega_1\cos\alpha+\omega_2\sin\alpha)\omega_3(1-\cos\beta)\sin\beta]^{1/2} \end{aligned} \tag{1}$$

According to Hertz's theory, the normal pressure density of the sphere at the point P is given by the formula

$$\nu(P) = \frac{3N}{2\sigma}\sqrt{1-\frac{\rho^2}{{\rho_0}^2}}, \sigma = 2\pi R^2(1-\cos\beta_0), \rho = 2R\sin\frac{\beta}{2}, \rho_0 = 2R\sin\frac{\beta_0}{2}$$

where N is the normal pressure, σ is the contact patch area, ρ is the distance from the point C to the point P and ρ_0 is the distance from the point C to the contact patch boundary. The pressure of the sphere on an elementary area $d\Sigma$ of the contact patch, constructed at the point P, is given by the formula $N(P) = \nu(P)d\sigma$, where $d\sigma = R^2\sin\beta d\beta d\alpha$ is the area $d\Sigma$. Thus

$$N(P) = \frac{3N(\cos\beta-\cos\beta_0)^{1/2}}{4\pi(1-\cos\beta_0)^{3/2}}\sin\beta d\beta d\alpha$$

Assuming that the sliding friction force $d\mathbf{F}(P)$ acting on the elementary area $d\Sigma$ and applied at the point P satisfies Coulomb's law, we have $d\mathbf{F}(P) = -kN(P)\mathbf{u}_P/u_P$ where $k > 0$ is the friction coefficient. The resulting friction force acting on the sphere and applied at the centre of the contact patch (at the point C) is given by the formula

$$\mathbf{F} = \int_\Sigma d\mathbf{F}(\mathbf{P}) = F_1\mathbf{e_1} + F_2\mathbf{e_2} \tag{2}$$

$$F_i = -k\frac{3N}{4\pi(1-\cos\beta_0)^{3/2}}\int_0^{\beta_0}(\cos\beta-\cos\beta_0)^{1/2}\sin\beta d\beta\int_0^{2\pi}\frac{u_i}{u_P}d\alpha, i = 1,2.$$

Similarly, the principal moment of the friction forces about the point is given by the formula

$$\mathbf{M}_\mathrm{C} = \int_{d\Sigma}[\rho, \mathbf{dF}(\mathbf{P})] = M_1\mathbf{e_1} + M_2\mathbf{e_2} + M_3\mathbf{e_3} \tag{3}$$

$$M_j = -k\frac{3N}{4\pi(1-\cos\beta_0)^{3/2}}\int_0^{\beta_0}(\cos\beta-\cos\beta_0)^{1/2}\sin\beta d\beta\int_0^{2\pi}\frac{w_j}{u_P}d\alpha, j=1,2,3$$

$$w_1 = R[u\sin^2\beta\sin\alpha\cos\alpha + R\omega_1(1-\cos\beta)^2 - R\omega_3(1-\cos\beta)\sin\beta\cos\alpha]$$

$$w_2 = R[u(1-\cos\beta-\sin^2\beta\cos^2\alpha) + R\omega_2(1-\cos\beta)^2 - R\omega_3(1-\cos\beta)\sin\beta\sin\alpha]$$

$$w_3 = R\sin\beta[-u\sin\alpha - R(1-\cos\beta)(\omega_1\cos\alpha+\omega_2\sin\alpha) + R\omega_3\sin\beta]$$

Bearing in mind that $R = a/\delta$ and $\sin\beta_0 = \varepsilon\delta$, and taking relations (1)-(3) into account, we conclude that the resulting friction force $\mathbf{F}$ and the principal moment $\mathbf{M}_C$ of the friction forces about the point depend on the parameters of the contact patch:

$$\mathbf{F} = \mathbf{F}(\delta,\varepsilon),\ \mathbf{M_C} = \mathbf{M_C}(\delta,\varepsilon) \tag{4}$$

Here, the moment $\mathbf{M}_C$ depends continuously on these parameters, while the force $\mathbf{F}$ depends continuously on the parameter δ and discontinuously [2] on the parameter ε (in the vicinity of the zero value of ε).

Furthermore, from relations (2) and (3) it follows that

$$\mathbf{F}(\delta,0) = -kN\mathbf{e_1},\ \mathbf{M_C}(\delta,0) = 0 \tag{5}$$

$$\mathbf{F}(0,\varepsilon) = F_1^{(0)}\mathbf{e_1},\ \mathbf{M_C}(0,\varepsilon) = M_3^{(0)}\mathbf{e_3} \tag{6}$$

Here [1, 2]

$$\begin{aligned}
F_1^{(0)} &= -k\frac{3N}{2\pi}\int_0^1 s\sqrt{1-s^2}ds\int_0^{2\pi}\frac{u-r\omega_3\sin\alpha}{u_0}d\alpha \\
M_3^{(0)} &= -k\frac{3Nr}{2\pi}\int_0^1 s^2\sqrt{1-s^2}ds\int_0^{2\pi}\frac{r\omega_3 s-u\sin\alpha}{u_0}d\alpha \\
r &= a\varepsilon, u_0 = (u^2-2ur\omega_3 s\sin\alpha + r^2\omega_3^2 s^2)^{1/2}
\end{aligned} \tag{7}$$

where u_0 is the value of the modulus of the sliding velocity of the point P when $\delta = 0$.

Thus, in the case of a point contact of the sphere with the bearing surface, the model proposed reduces to the Coulomb model (5) (we recall that the limit $\mathbf{F}(\delta,\varepsilon)$ does not exist when $\varepsilon \to +0$), and in the case of a plane contact patch it is converted into the model (6) [1, 2].

Note also that the proposed model of friction, unlike the model [1, 2], possesses complete dissipation:

$$(\mathbf{F},\mathbf{u}) + (\mathbf{M_C},\omega) < 0$$

for any $\mathbf{u}$ and ω not equal to zero simultaneously.

It is curious that the parameter δ occurs in expressions (4) only in a product with the parameter ε, i.e., relations (4) can be represented in the form ($\mu = \varepsilon\delta$)

$$\mathbf{F} = \mathbf{F}(\mu,\varepsilon),\ \mathbf{M_C} = \mathbf{M_C}(\mu,\varepsilon) \tag{8}$$

where the force $\mathbf{F}$ depends continuously on the on parameter μ. Here, the parameter ε occurs in expressions (8) only in a product with the radius of the sphere a, i.e., relations (8) can be represented in the form ($r = a\varepsilon$ is the radius of the contact patch)

$$\mathbf{F} = \mathbf{F}(\mu, r), \; \mathbf{M}_\mathrm{C} = \mathbf{M}_\mathrm{C}(\mu, r) \tag{9}$$

Obviously, the moment $\mathbf{M}_C$ depends continuously on μ and r, while the force $\mathbf{F}$ depends continuously on μ and discontinuously on r (in the vicinity of a zero value of r).

Furthermore, the radius of the contact patch r occurs in the expression for the friction force $\mathbf{F}$ only in a product with the angular velocity ω of the sphere, i.e., $\mathbf{F}$ and $\mathbf{M}_C$ depend on the sliding velocity $\mathbf{u}$, the vector $\mathbf{v} = r\omega$ and the parameter μ:

$$\mathbf{F} = \mathbf{F}(\mathbf{u}, \mathbf{v}, \mu), \quad \mathbf{M}_\mathrm{C} = r\mathbf{M}(\mathbf{u}, \mathbf{v}, \mu) \tag{10}$$

Note also that F_2 and M_1 vanish if $\omega_1 = 0$, and M_3 vanishes if $\omega_1 = \omega_3 = 0$.

Thus, in the general case, the resulting friction force has a component perpendicular to the sliding velocity of the sphere ($F_2 \neq 0$), while the principal moment of the friction forces about the center of the contact patch has horizontal components both along the sliding velocity $M_1 \neq 0$) and perpendicular to it ($M_2 \neq 0$).

2 Investigation of a Homogeneous Sphere Dynamics on a Horizontal Plane with Friction

We write the equations of motion of the sphere on the horizontal plane using the general theorems of dynamics

$$m\dot{\mathbf{u}}_s = (N - mg)\mathbf{e}_3 + \mathbf{F} \tag{11}$$

$$\frac{2}{5}ma^2\dot{\omega} = [-a\mathbf{e}_3, \mathbf{F}] + \mathbf{M}_\mathrm{C} \tag{12}$$

Here $\mathbf{u}_s = \mathbf{u} + [\omega, a\mathbf{e}_3]$ is the velocity of the sphere center and g is the gravity acceleration.

We suppose that the sphere can not leave the supporting plane (i.e. $(\mathbf{u}_\mathrm{s}, \mathbf{e}_3) \equiv 0$), so $N = mg$ (see equation (11). Taking into account that the reference frame $\mathbf{e}_1$, $\mathbf{e}_2$, $\mathbf{e}_3$ can rotate about the vertical and denotiny its angular velocity as $\Omega = \Omega\mathbf{e}_3$ we write down equations (11), (12) in this reference frame

$$\dot{u} + a(\dot{\omega}_2 + \omega_1\Omega) = f_1, \; u\Omega - a(\dot{\omega}_1 - \omega_2\Omega) = f_2 \tag{13}$$
$$a(\dot{\omega}_1 - \omega_2\Omega) = \frac{5f_2}{2} + \frac{5m_1}{2a}, \; a(\dot{\omega}_2 + \omega_1\Omega) = -\frac{5f_1}{2} + \frac{5m_2}{2a}, \; a\dot{\omega}_3 = \frac{5m_3}{2a}$$

Here $f_i = F_i/m$, $m_j = M_j/m$ and F_i M_j are given by relations (2), (3), where N has to be replaced by mg ($i = 1, 2; j = 1, 2, 3$).

Let us introduce variables $\Omega_{1,3} = a\omega_{1,3}$, $\Omega_2 = a\omega_2 + 5u/7$ and write down the system (13) in the form

$$\dot{u} = \frac{7f_1}{2} - \frac{5m_2}{2a}, \ \dot{\Omega}_3 = \frac{5m_3}{2a} \tag{14}$$

$$u\Omega = \frac{7f_2}{2} + \frac{5m_1}{2a}, \ \dot{\Omega}_1 = \Omega\Omega_2 + \frac{5m_1}{7a}, \ \dot{\Omega}_2 = -\Omega\Omega_1 + \frac{5m_2}{7a} \tag{15}$$

the system (14), (15) admits the ungrowing function

$$H = \frac{1}{2}\left(\frac{2}{7}u^2 + \frac{7}{5}\Omega_1^2 + \frac{7}{5}\Omega_2^2 + \frac{2}{5}\Omega_3^2\right) \leq h \tag{16}$$

where h is the initial value of H.

Let us prove that the sphere will stop in a finite time. Really, the total derivaty by time of function H with respect to system (14), (15) can be presented in the form $\dot{H} = -\sqrt{(\mathbf{Ap},\mathbf{p})}$, where $\mathbf{A}$ is the 4×4 matrix of the positive quadratic form and $\mathbf{p}$ is the 4-vector with components $u,\Omega_1,\Omega_2,\Omega_3$. Let $\sqrt{2H}$ be the norm of the vector $\mathbf{p}$, then $p\dot{p} = -\sqrt{(\mathbf{Ap},\mathbf{p})} \leq -cp$, where c is a positive constant. So $p \leq \sqrt{2h} - ct$ and the time t_R of motion of the sphere is limited ($t_R \leq \sqrt{2h}/c$).

Let us prove now that sliding velocity and angular velocity of the sphere vanish in one and the same moment of time. Eliminating Ω from equations (14), (15) we can represented these equations in the form

$$\begin{aligned}
\frac{d\Omega_1}{du} &= \frac{(7af_2+5m_1)\Omega_2/u+10m_1/7}{7af_1-5m_2} = Y_1\\
\frac{d\Omega_2}{du} &= \frac{-(7af_2+5m_1)\Omega_1/u+10m_2/7}{7af_1-5m_2} = Y_2\\
\frac{d\Omega_3}{du} &= \frac{5m_3}{7af_1-5m_2} = Y_3
\end{aligned} \tag{17}$$

Evidently, the right sides of the system (17) depend on z_1,z_2,z_3, where $z_j = \Omega_j/u$ $(j-1,2,3)$. Introducing variables $\tau = -\ln(u/u(0))$ $(u(0)\neq 0)$, z_j instead variables u, Ω_j we represent the system (17) in the form ($z' = dz/dt$)

$$z_j' = z_j - Y_j = Z_j(z_1,z_2,z_3), \ j = 1,2,3 \tag{18}$$

System (18) admits invariant set $z_1 = z_3 = 0$ $(\tau \in [0,+\infty))$ because $Z_1(0,z_2,0) = Z_3(0,z_2,0) \equiv 0$ (see (2), (3)). In original variables this invariant set corresponds to solutions

$$\omega_1 \equiv 0, \ \omega_3 \equiv 0, \ \Omega \equiv 0, \ u = u^0(t), \ \omega_2 = \omega_2^0(t) \tag{19}$$

of system (13). Functions $u^0(t)$ and $\omega_2^0(t)$ satisfy to the system

$$\dot{u} + a\dot{\omega}_2 = f_1^0, \ a\dot{\omega}_2 = -5f_1^0/2 + 5m_2^0/2a \tag{20}$$

Here

$$f_1^0 = -k\gamma\int_0^{\beta_0} b(\beta)d\beta\int_0^{2\pi}\frac{u_1^0}{u_p^0}d\alpha, \quad m_2^0 = -k\gamma\int_0^{\beta_0} b(\beta)d\beta\int_0^{2\pi}\frac{w_2^0}{u_p^0}d\alpha \tag{21}$$

$$\begin{aligned}
\gamma = (3g/4\pi)(1-\cos\beta_0)^{-3/2}, \quad b(\beta) = (\cos\beta - \cos\beta_0)^{1/2}\sin\beta \\
u_1^0 = u(1-\sin^2\beta\cos^2\alpha) + a\omega_2(1-\cos\beta-\sin^2\beta\cos^2\alpha)/\delta \\
w_2^0 = a[u(1-\cos\beta-\sin^2\beta\cos^2\alpha)/\delta + a\omega_2(1-\cos\beta)^2/\delta^2] \\
u_p^0 = \left[u^2(1-\sin^2\beta\cos^2\alpha) + 2u(a\omega_2)(1-\cos\beta-\sin^2\beta\cos^2\alpha)/\delta + \right. \\
\left. +(a\omega_2)^2(1-\cos\beta)^2/\delta^2\right]^{1/2}
\end{aligned} \tag{22}$$

Taking into account relations (21) and (22) we conclude that system (20) can not have solutions, satisfying to conditions

$$u^0(t)\omega_2^0(t) = 0, \ (u^0(t))^2 + (a\omega_2^0(t))^2 \neq 0$$

Really, if there exists $\theta \in (0, t_R)$ such that $u^0(t) \equiv 0, \ \omega_2^0(t) \neq 0, \ (t \in [t_R - \theta, t_R))$ then

$$f_1^0 = -\frac{kg(1-\cos\beta_0)}{5}, \ m_2^0 = -\frac{2kga(1-\cos\beta_0)}{5\delta}$$

and equations (20) take the form

$$a\dot{\omega}_2 = -\frac{kg(1-\cos\beta_0)}{5}, \ a\dot{\omega}_2 = -\frac{kg(1-\cos\beta_0)}{2\delta}(2-\delta)$$

These equations contradict one to another, because $\delta < 1$.

Similarly we can consider the case $u^0(t) \neq 0, \omega_2^0(t) \equiv 0 \ (t \in [t_R - \theta, t_R))$.

Let us consider now the general case

$$z_1^2(0) + z_3^2(0) = z_0^2 \neq 0, \ z_0^2 = a^2(\omega_1^2(0) + \omega_3^2(0))/u^2(0)$$

Introducing variables z and φ by formulae $z_1 = z\cos\varphi$, $z_3 = \cos\varphi$ we represent the system (18) in the form

$$\begin{aligned}
z' = Z(z,\varphi,z_2), \ \varphi' = \Phi(z,\varphi,z_2), \ z_2' = Z_2(z\cos\varphi, z_2, z\sin\varphi) \\
Z = Z_1(z\cos\varphi, z_2, z\sin\varphi)\cos\varphi + Z_3(z\cos\varphi, z_2, z\sin\varphi)\sin\varphi \\
\Phi = [Z_3(z\cos\varphi, z_2, z\sin\varphi)\sin\varphi - Z_1(z\cos\varphi, z_2, z\sin\varphi)\cos\varphi]/z
\end{aligned} \tag{23}$$

The right sides of system (23) are smooth for sufficiently small values of parameter μ and $Z(0,\varphi,z_2) \equiv 0$. If $z(\tau)$ is a bounded function on $\tau \in [0,+\infty)$ or $z(\tau)$ tends to infinity more slowly than e^τ, then $\lim_{u\to 0}(\omega_1^2 + \omega_3^2) = 0$, i.e. ω_1, ω_3, u and therefore ω_2 vanish the same time. If $z(\tau)$ tends to infinity more quickly than e^τ then u vanishes earlier than $\omega_1^2 + \omega_3^2$. In this case one can show [4] that system (13) has now solutions satisfying to conditions

$$u(t) \equiv 0, \ \omega_1^2(t) + \omega_3^2(t) \neq 0, \ \omega_2(t) \neq 0 \ (t \in [t_R - \theta, t_R))$$

Thus sliding and rotations of the sphere vanish the same finite time.

The given above analysis can be illustrated by numerical calculations (see Fig.2). These calculations [4] demonstrate that the time of full stop of the sphere can be

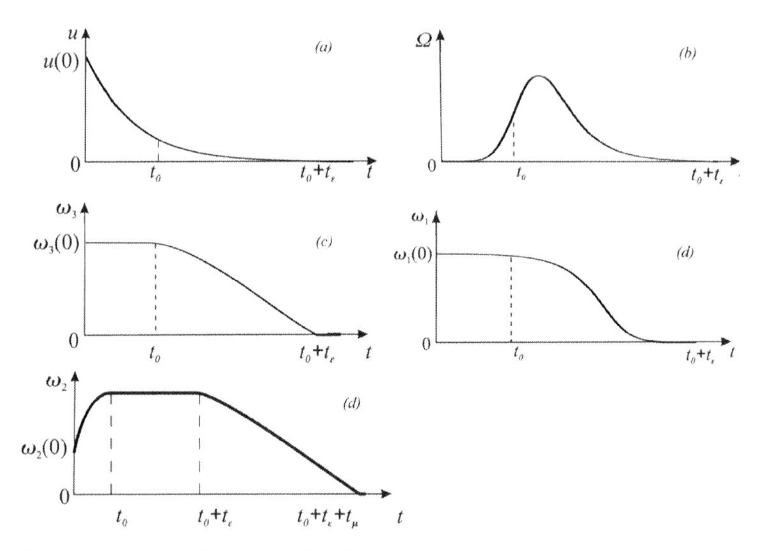

Fig. 2 Time dependencies of the phase variables

devided into three intervals $t_R = t_0 + t_\varepsilon + t_\mu$, where

$$t_0 \sim \frac{2u(0)}{7kg}, \; t_\varepsilon \sim \frac{32a\omega_3(0)}{15\pi kg}\frac{1}{\varepsilon}, \; t_\mu \sim \frac{a\omega_2(0)}{kg}\frac{1}{\varepsilon\mu}$$

During the first interval the sliding velocity (u) decreases "quickly" to values of the ε order, the horizontal component of the angular velocity, orthogonal to the sliding velocity (ω_2) increases, and the other components of the angular velocity (ω_1, ω_3) are almost constant. During the second interval u "slowly" and ω_1, ω_3 "quickly" decrease to values of the μ order, and ω_2 is almost constant. During the third interval all variables "slowly" decrease to zero values.

Thus, in general case, the sliding, pivoting and rolling of a sphere vanish in one and the same finite moment of time.

References

1. Contensou, P.: Couplage entre frottement de glissement et frottement de pivotement dans la theorie de la toupee. In: Kreiselprobleme. Gyrodym. Symp., pp. 210–216 (1963)
2. Zhuravlev, V.F.: A model of dry friction in the problem of the rolling of solids. Prikl. Mat. Mekh. 62(5), 762–767 (1998)
3. Karapetyan, A.V.: A two-parameter friction model. Prikl. Mat. Mekh. 73(4), 512–519 (2009)
4. Ishkhanyan, M.V., Karapetyan, A.V.: Dynamics of a homogenious sphere on the horizontal plane with friction of sliding, pivoting and rolling. In: Mechanics of Solids, vol. (2), pp. 3–14 (2010)

Different Models of Friction in Double-Spherical Tippe-Top Dynamics

A.A. Zobova

Abstract. Dynamics of a two-spherical tippe-top on a rough horizontal plane is considered. The tippe-top is bounded by a non-convex surface that consists of two-spheres and a cylinder; the axis of cylinder coincides with the common spheres' axes of symmetry. Being fast spun around its axis of symmetry the tippe-top overturns from the bottom (the big sphere) to the leg (that is modeled by a small sphere) and some time it returns back to stable equilibrium position. Different models of friction are examined analytically and numerically. Numerical investigation is done in assumption that supporting plane is slightly deformable, that allows one to describe rolling and impacts by the same system of dynamical equations. Results of numerical investigation coincide with analytical conclusions.

1 Statement of the Problem and Dynamical Equations

Let us consider a heavy solid body (a top) on a rough horizontal plane (Fig. 1). The top consists of two spherical segments with radii r_1 and r_2 ($r_1 > r_2$), the segments are connected by a rod, that goes through their centers O_1 and O_2 (this model is taken from [5]). Let the unit vector of the dynamical and geometrical axis of symmetry $\mathbf{e} = \overrightarrow{O_1O_2}/O_1O_2$ make an angle of nutation θ with the up-going vertical $\boldsymbol{\gamma}$. The top touches the plane in point C_1 by the first spherical segment if $\theta \in [0, \pi - \alpha)$, and by the second segment in point C_2 if $\theta \in (\pi - \alpha, \pi]$; the top lies on two points if $\theta = \pi - \alpha$.

The center of mass lies on the symmetry axis O_1O_2 such that $O_iS = c_i$ ($i = 1, 2$) and $c_2 = (r_1 - r_2)/\cos\alpha + c_1$. The vectors, that point the contact points from the center of mass, are the following

$$\mathbf{r}_i = \overrightarrow{SC_i} = -r_i\boldsymbol{\gamma} + c_i\mathbf{e} = -r_i(\boldsymbol{\gamma} - b_i\mathbf{e}),\ i = 1, 2$$

A.A. Zobova
Lomonosov Moscow State University, Moscow, Russia
e-mail: azobova@gmail.com

G. Stépán et al. (Eds.): Dynamics Modeling & Interaction Cont., IUTAM BOOK SERIES 30, pp. 265–272.
springerlink.com

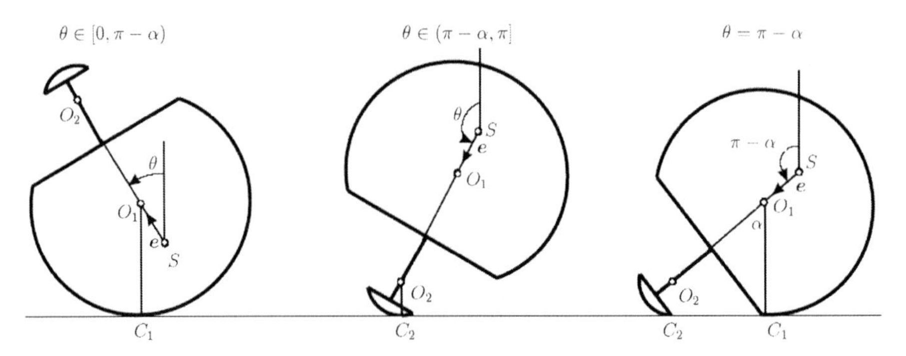

Fig. 1 Geometry of the tippe-top

Let m be the mass of the top, and $\mathbf{J} = \text{diag}(J_1, J_2, J_3)$ $(J_1 = J_2)$ be the tensor of inertia of the top with respect to the center of mass S. So dimensionless parameters of the top are the following:

$$\alpha \in (0, \pi/2),\ b_1 = c_1/r_1 \in (0,1),\ b_2 = c_2/r_2,\ a = J_1/J_3 \in [1/2, +\infty)$$

Let us suppose that, besides the gravity $-mg\boldsymbol{\gamma}$ and the reaction of the supporting plane $N\boldsymbol{\gamma}$, the dry friction forces (including sliding, rolling and spinning frictions) act on the top. Then the governing dynamical equations with respect to the principal axes of the top are the following:

$$m\dot{\mathbf{v}} + [\boldsymbol{\omega}, m\mathbf{v}] = (N - mg)\boldsymbol{\gamma} + \mathbf{F} \tag{1}$$

$$\mathbf{J}\dot{\boldsymbol{\omega}} + [\boldsymbol{\omega}, \mathbf{J}\boldsymbol{\omega}] = [\mathbf{r}, N\boldsymbol{\gamma} + \mathbf{F}] + \mathbf{M} \tag{2}$$

$$\dot{\boldsymbol{\gamma}} + [\boldsymbol{\omega}, \boldsymbol{\gamma}] = 0 \tag{3}$$

$$(\mathbf{v} + [\boldsymbol{\omega}, \mathbf{r}], \boldsymbol{\gamma}) = 0 \tag{4}$$

Here $\mathbf{F}$ is the friction force and $\mathbf{M}$ is the friction torque. Eq. (1) is the impulse equation, eq. (2) is the torque equation, eq. (3) shows that $\boldsymbol{\gamma}$ is fixed in absolute coordinate frame, and eq. (4) is a holonomic constraint (though it is expressed in differential form), that means that the top lies on the plane: the velocity of the contact point $\mathbf{u}$ is horizontal. Dot corresponds to the derivative in the coordinate frame of principal axes of the inertia tensor. The unknown functions of this system are $\mathbf{v}$, $\boldsymbol{\omega}$, $\boldsymbol{\gamma}$, N, and to enclose this system we should add the models for friction force and friction torque $\mathbf{F} = \mathbf{F}(\mathbf{v}, \boldsymbol{\omega}, \boldsymbol{\gamma}, N)$, $\mathbf{M} = \mathbf{M}(\mathbf{v}, \boldsymbol{\omega}, \boldsymbol{\gamma}, N)$.

2 Different Models of Friction

In rigid body dynamics the following models of the interaction of the rigid body and supporting surface can be used:

1. Smooth plane ($\mathbf{F},\ \mathbf{M} = 0$)
2. Absolutely rough plane (non-holonomic constraint)
3. Rough plane with friction

   a. Dry Coulomb friction
   b. Contensou-Zhuravlev friction [7]
   c. Two-parametric friction, proposed by A.V. Karapetyan [4]

Here we use and compare only one-valued models, that are simpler for numerical simulations, comparatively to set-valued model [5].

The first model of smooth plane means that friction force and torque equal zero (ideal holonomic constraint). The equations of motion admit the same first integrals as the Lagrangian top (full mechanical energy, projections of the kinetic moment on the vertical axis and on the symmetry axis), and it can be analytically shown [6] that the nutation angle θ is a periodic function of time. So the model of smooth plane does not explain the overturn motions of the top.

The model of absolutely rough plane is the following: it is supposed that the velocity of contact point C_i ($i = 1$ when the top lies on the bottom and $i = 2$ when it lies on the leg) is equal to zero. Then the tangent component of reaction force is calculated from the dynamics equation and constraint, and it depends on the $\mathbf{v}$, $\boldsymbol{\omega}$, $\boldsymbol{\gamma}$, N. If the constraint is supposed to be ideal, then the friction torque is assumed to be zero. In this case the equations admit the energy integral, and two integrals that are linear with respect to the components of $\boldsymbol{\omega}$. It also can be shown analytically [6] that in this model the nutation angle θ is a periodic function of time, and the overturn motions are also impossible.

Other models of the plane-top interaction, mentioned above, are dissipative, so that the full mechanical energy

$$H = H(\mathbf{v},\boldsymbol{\omega},\boldsymbol{\gamma}) = \frac{1}{2}m\mathbf{v}^2 + \frac{1}{2}(\mathbf{J}\boldsymbol{\omega},\boldsymbol{\omega}) - mg(\mathbf{r}_i,\boldsymbol{\gamma}) \leq h$$

is non-increasing function of time:

$$\dot{H} = (\mathbf{F},\mathbf{u}) + (\mathbf{M},\boldsymbol{\omega}) \leq 0$$

If the friction torque ($\mathbf{M} = 0$) is neglected, then another first integral exists for all models of friction, if the force is applied to the contact point of the spherical segment: that is the Jellet integral — the projection of the kinetic moment on the radius vector of the contact-point C_i:

$$K_i = -\frac{1}{r_i}(\mathbf{J}\boldsymbol{\omega},\mathbf{r}_i) \equiv k_i$$

These two properties of the dynamical equations allow to construct effective potential of the system by means of the modified Routh theory (for more details see [2]). The properties of the critical points of the effective potential determine the, stability and bifurcation of the steady-state motions, that are motions with constant nutation angle θ and zero sliding velocity.

The full analysis of all steady-state motions, including precession motions, of double-spherical tippe-top is provided in [8]. It is shown that if $b_1 > 1 - a$ and $b_2 > a - 1$, then the rotations on the bottom ($\theta \equiv 0$) are stable only with small angular velocity, and with the increasing of the angular velocity they become unstable; the rotations on the leg ($\theta \equiv -\pi$) are stable only with large angular velocity. The analysis of Smale bifurcation diagram for this region of parameters (the scheme of the investigation can be found in [3]) shows that the overturn motions exist when the friction force and friction torque fulfill the following conditions:

1. $(\mathbf{F},\mathbf{u}) < 0$ when $\mathbf{u} \neq 0$; $\mathbf{F} = 0$ when $\mathbf{u} = 0$; $(\mathbf{M},\boldsymbol{\omega}) \leq 0$.
2. $(\mathbf{M},\boldsymbol{\omega}) \ll (\mathbf{F},\mathbf{u})$ when $\mathbf{u} \neq 0$.
3. $\left|\dfrac{(\mathbf{M},\mathbf{r}_i)}{(\mathbf{J}\boldsymbol{\omega},\mathbf{r}_i)}\right| \ll \sqrt{\dfrac{g}{r_i}}$

These conditions are rather natural and are fulfilled for wide range of friction models. The first one means that the friction is dissipative (note that the vanishing of the friction force when there is no sliding is meaningful and this condition is fulfilled for the friction models that are used in problems of rigid body dynamics, the discussion of this fact can be seen in [7]), the second one – that the power of friction torque is much less then the power of friction force, and the third one means that friction torque changes the kinetic moment slow comparatively to the scale of time. Besides, if $\mathbf{M} \equiv 0$, than there exists only the first overturn from the bottom to the leg, and after that the top spins on the leg infinitely long (the dissipation is partial, and energy conserves on finite steady motions). If $\mathbf{M} \not\equiv 0$, then the second overturn occurs and the top stops in the stable equilibrium position $\theta = 0$ with the center of mass in the lowest position.

The quantitative results were obtained by numerical simulation for the models 3a-3c, the used exact formulae and the calculation method are discussed in the following section.

3 Method of Numerical Calculation

The aim of numerical calculation is to solve Cauchy problem for the equations (1-4) with the initial conditions that correspond to the fast spinning of the top on the bottom and to obtain trajectories with the overturns.

The holonomic constraint (4) means that during all the time of motion the top lies on the supporting plane. Then the normal reaction N should be calculated from the equations of motion: it depends on the phase variables $\mathbf{v},\boldsymbol{\omega},\boldsymbol{\gamma}$. But in fact this condition should be replaced with the unilateral constraint $(\mathbf{v} + [\boldsymbol{\omega},\mathbf{r}],\boldsymbol{\gamma}) \geq 0$, because the top can jump off the plane. So one should monitor the sign of the normal reaction N on each step of numerical integration, and when it becomes negative, switch on the equations of free rigid body in the gravity field ($N=0,\ \mathbf{F}=0,\ \mathbf{M}=0$). When the top falls on the plane one should use some model of the impact of the body and the plane. (Note, that for steady-state motion the normal reaction is positive and equals the weight of the top). So we have to integrate the system with switches and impacts.

To simplify this problem and to avoid the problem of the impact model choice we propose the following scheme. Let's introduce two new variables ε_1 and ε_2 that are the distances between the lowest points C_1 and C_2 of the segments and the plane. They are connected with the height of the center z_s of mass by the equation:

$$\varepsilon_i = z_S + c_i(\boldsymbol{\gamma},\mathbf{e}) - r_i, \quad i=1,2$$

Then let's introduce simple viscoelastic Kelvin-Voigt model of the normal reaction force:

$$N_i = \begin{cases} 0, \text{when} \quad \varepsilon_i \geq 0 \\ -n_1\varepsilon_i, \text{when} \quad \varepsilon_i < 0, \dot{\varepsilon}_i \geq 0 \\ -n_1\varepsilon_i - n_2\dot{\varepsilon}_i, \text{when} \quad \varepsilon_i < 0, \dot{\varepsilon}_i < 0 \end{cases} \tag{5}$$

The force N_i is applied to the point C_i, N_1 acts when $\theta \in [0, \pi-\alpha]$, N_2 acts when $\theta \in [\pi-\alpha, \pi]$. So we added the new variable z_S, and we should add the corresponding differential equation:

$$\dot{z_S} = (\mathbf{v},\boldsymbol{\gamma})$$

This suggestion allows to calculate the motions of the system by rather simple numerical algorithm without switches. Impacts are modeled by means of this algorithm and model (5).

For the friction forces we take the following formulae. The classical Coulomb friction force 3a depends only on the sliding velocity $\mathbf{u}$, and the friction torque equals zero. We take two regular variants of it, that are widely used in the problems of robotics and rigid body dynamics. The first one is signum with the gap:

$$\mathbf{F}(\mathbf{u}) = \begin{cases} 0, \text{when} \quad |\mathbf{u}| \leq \delta \\ -f_0 N \frac{\mathbf{u}}{|\mathbf{u}|}, \text{when} \quad |\mathbf{u}| > \delta \end{cases} \tag{6}$$

The second one is an arctangent function:

$$\mathbf{F}(\mathbf{u}) = -\frac{2f_0}{\pi} N \arctan(c_f|\mathbf{u}|)\frac{\mathbf{u}}{|\mathbf{u}|} \tag{7}$$

Here and further f_0 is the friction coefficient that is constant and depends only on the materials of contacting surfaces.

The Contensou-Zhuravlev friction 3b was proposed in [1] and developed in [7]. In this model it is supposed that the contact patch of the sphere segment and the plane is planar circle with radius ρ, and the Coulomb friction acts on each small area within this patch. The resulting force and torque are obtained by integrating these forces on the circular patch. So the formulae for this model are the following:

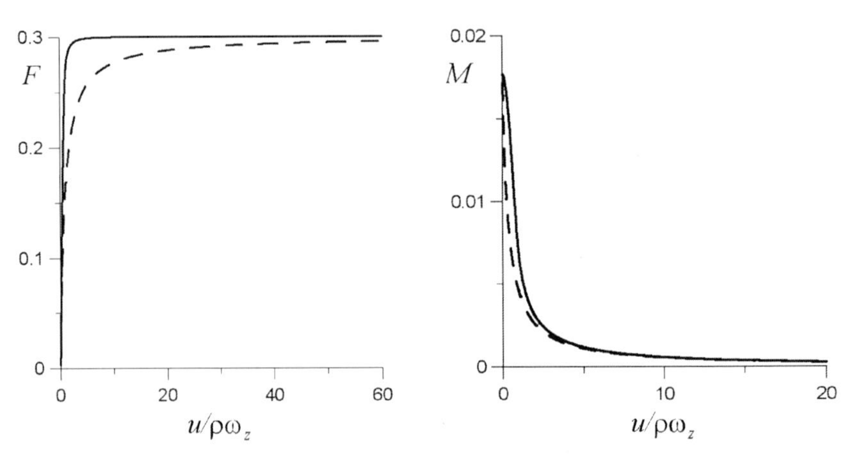

Fig. 2 Padé approximation (dashed line) and integral formula (solid line) for the Contensou-Zhuravlev friction force and torque

$$\mathbf{F} = -\frac{3f_0N}{2\pi}\left(\int_0^1\int_0^{2\pi} s\sqrt{1-s^2}\frac{u-\rho s\omega_z\sin\alpha}{\sqrt{u^2-2\rho s\omega_z u\sin\alpha+\rho^2s^2\omega_z^2}}d\alpha ds\right)\frac{\mathbf{u}}{|\mathbf{u}|}$$

$$\mathbf{M} = -\frac{3f_0\rho N}{2\pi}\left(\int_0^1\int_0^{2\pi} s\sqrt{1-s^2}\frac{\rho s\omega_z-u\sin\alpha}{\sqrt{u^2-2\rho s\omega_z u\sin\alpha+\rho^2s^2\omega_z^2}}d\alpha ds\right)\boldsymbol{\gamma} \quad (8)$$

Here $u = |\mathbf{u}|$, $\omega_z = (\boldsymbol{\omega},\boldsymbol{\gamma})$. These integrals can be taken in close algebraic form. But often ([7]) these formulae are approximated by Padé approximation (the n-th order Padé approximation is approximation by the ratio of two forms of the n-th order, that preserve the behavior of the function in the neighborhood of zero point and in the infinity). Here we use also the first-order Padé approximation of the friction force and torque (8):

$$\mathbf{F} = -\frac{f_0N\mathbf{u}}{|\mathbf{u}|+\beta\rho|\omega_z|}, \quad \mathbf{M} = M\boldsymbol{\gamma} = -\frac{3\pi\rho f_0N}{16}\frac{\rho\omega_z}{\rho|\omega_z|+\alpha|\mathbf{u}|}\boldsymbol{\gamma} \quad (9)$$

Here $\alpha = \frac{15\pi}{16}$, $\beta = \frac{8}{3\pi}$. The comparison of this approximation and the exact integral formulae is presented at the Fig. 2.

The friction model 3c is the further extension of the Contensou-Zhuravlev friction. The planar circular patch is changed by the spherical cap patch. In this model there is also component of the friction force that is perpendicular to the sliding velocity $\mathbf{u}$, and the friction torque has not only vertical component, but also two horizontal components. So this friction model has the full dissipation on every motion of the body. The exact formulae are given in [4].

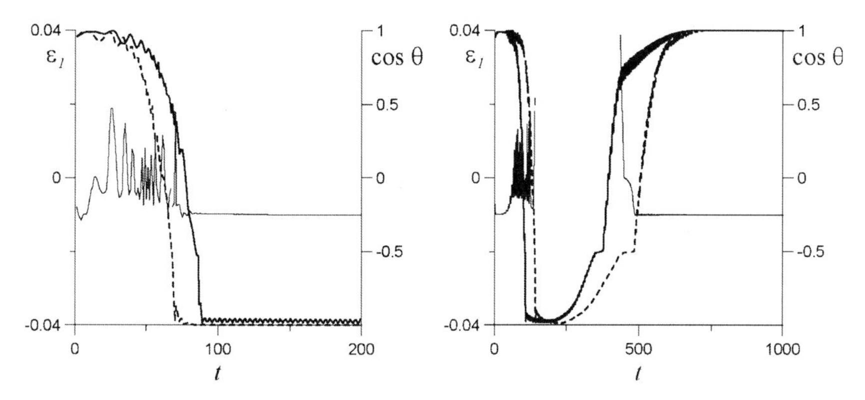

Fig. 3 Time-dependencies of the coordinates: dry Coulomb friction (left), Contensou-Zhuravlev friction (right).

4 Numerical Results and Conclusions

In order to obtain dimensionless equations let's scale mass, length and time so that $g = 1$, $r_1 = 1$, $m = 1$, the other geometrical and dynamical parameters of the top are the following: $b_1 = 0.3$, $r_2 = 0.2$, $J_1 = 0.592$, $J_3 = 0.518$, $\alpha = \frac{\pi}{3}$ For these parameters slow rotation $\theta = 0$ are stable, fast rotations $\theta = 0$ are unstable, and fast rotations $\theta = \pi$ are stable. Friction parameters are $f_0 = 0.3$, $\rho = 0.1$, $\delta = 0.05$, $c_f = 1000$, normal force parameters are $n_1 = 100$, $n_2 = 80$.

The initial conditions are the following: $\cos\theta = 0.95$, $\boldsymbol{\omega} = \omega_0 \mathbf{e}$, $\omega_0 = 15$, $\varepsilon_1 = 0$, $\mathbf{v} = 0$. The results of numerical calculations are presented in the Fig. 3. Time axis is horizontal, vertical axis on the left corresponds to ε_i, vertical axis on the right corresponds to $\cos\theta$, bold curves stand for $\cos\theta$, thin curves represent ε_1 when $\cos\theta \in [-\cos\alpha,\ 1]$ and ε_2 when $\cos\theta \in [-1,\ -\cos\alpha]$.

As it was mentioned above, the analytical investigation shows that if $\mathbf{M} \equiv 0$ then only the first overturn from the bottom to the leg occurs. On the left part of the Fig. 3 the motions of the top are presented when the friction force is given by eq.7 (solid curve) and by eq.6 (dashed curve). The overturn from the bottom to the leg takes place rather quickly, but before the overturn multiple impacts occur — ε_1 changes its sign a lot of times even while the top lies on the bottom. In natural experiments we can detect these jumps by specific sound of bouncing. (On the picture only one curve for ε_1 is presented, because they do not differ very much qualitatively for these models). The Coulomb friction with the gap (dashed curve) shows that after the overturn the top rotates on the leg in vertical position infinitely long, $\mathbf{e}$ is aligned vertically, $\mathbf{v}$ equals zero. The arctangent Coulomb friction (solid line) shows fast oscillations that damp very slowly. In both case during the rotations on the leg $\varepsilon_2 = -0.01$, and the normal reaction force equals the weight of the top.

So arctangent approximation of the Coulomb friction model, that is used very often in robotics and that is the smooth regularization of dry friction, gives fast oscillations instead of regular steady motion. If we decrease the coefficient c_f in the argument of arctangent, then the time of overturn will increase considerably, but oscillations will damp very fast. Thereby this regularization should be used very accurately in practical problems.

The motions that are obtained using the friction models (8, 9) are shown on the right part of the Fig. 3 (solid and dashed curves respectively). The first overturn occurs when $t = 100$ and $t = 150$ that is slower then for the Coulomb friction. It could be explained by the fact that for large vertical angular velocities the Contensou friction force is less then the Coulomb friction (see Fig. 2). During the first overturn multiple impacts occur, like in previous case, and the second overturn is rather smooth: ε_i are negative when the top lies on the appropriate part of its surface, and there is a period of time when the top contacts the plane by two points C_1 and C_2 (horizontal part of the curves $\cos\theta$ at the value -0.5). Qualitatively the curves have the same behavior, but quantitatively they have slight differences. The overturns motions look like one received in [5], but here one-valued simpler models of friction are used.

We show numerically that the model 3c that is proposed by A.V. Karapetyan [4] gives practically the same results as the model (8), the differences in $\cos\theta$ are less than 0.3%, and it is more significant at the second part of the motion during the second overturn when $\mathbf{u}$ vanishes.

Thus different models of friction are examined in the double spherical tippe-top dynamics, both analytically and numerically. It is shown that different friction models can give qualitatively different results in rigid body dynamics.

Acknowledgements. This work is supported by Russian Foundation for Basic Research (grant 10-01-00292, 09-08-00925), and by the President Council of Russian Federation (young researcher grant MK-698.2010.1).

References

1. Contensou, P.: Couplage entre frottement de glissement et frottement de pivotement dans la theorie de la toupee. In: Kreiselprobleme. Gyrodym. Symp., pp. 210–216 (1963)
2. Karapetyan, A.V.: Invariant Sets of Mechanical Systems. In: Modern Methods of Analytical Dynamics and their Applications. Springer, Wien-NY (1998)
3. Karapetyan, A.V.: Global qualitative analysis of tippe top dynamics. Mechanics of Solids 43(3), 342–348 (2008)
4. Karapetyan, A.V.: A two-parameter friction model. J. Appl. Math. Mech. 73(4), 367–370 (2009)
5. Leine, R.I., Gloker, C.: A set-valued force law for spatial Coulomb – Contensou friction. Europ. J. Mech. A/Solids. 22(2), 193–216 (2003)
6. Routh, E.J.: The advanced part of a treatise on the dynamics of a system of rigid bodies. MacMillan, London (1884)
7. Zhuravlev, V.F.: A model of dry friction in the problem of the rolling of solids. J. Appl. Math. Mech. 62(5), 762–767 (1998)
8. Zobova, A.A., Karapetyan, A.: Analysis of the steady motions of the tippe top. J. Appl. Math. Mech. 73(6), 623–630 (2009)

Frictional Vibration of a Cleaning Blade in Laser Printers

Go Kono, Yoshinori Inagaki, Tsuyoshi Nohara, Minoru Kasama,
Toshihiko Sugiura, and Hiroshi Yabuno

Abstract. It is known that chatter vibration of a cleaning blade in laser printers, caused by the friction between the cleaning blade and the photoreceptor, occasionally produces a squeaking noise. This research aims to analyze the dynamics of the cleaning blade, from the viewpoint of mode-coupled vibrations.

The dynamics of the cleaning blade are theoretically analyzed using an essential 2DOF link model, with emphasis placed on the contact between the blade and the photoreceptor. The cleaning blade is assumed to always be in contact at one point with a moving floor surface, which is given a displacement δ from its initial position in the vertical direction. This causes the vertical load N and the frictional force μN to continuously act on the bottom end. By solving the equations governing the motion of the analytical model, five patterns of static equilibrium states are obtained, and

Go Kono
Keio University, 3-14-1 Hiyoshi, Yokohama, Japan
e-mail: gogogoh20@hotmail.com

Yoshinori Inagaki
Keio University, 3-14-1 Hiyoshi, Yokohama, Japan
e-mail: y_inag_2001@yahoo.co.jp

Tsuyoshi Nohara
Mitubishi Heavy Industries, Ltd, 1200 Higashi Tanaka, Komaki, Japan
e-mail: tsuyoshi.nohara@mhi.co.jp

Minoru Kasama
Fuji Xerox Co., Ltd, Sakai 430, Nakai-machi, Japan
e-mail: minoru.kasama@fujixerox.co.jp

Toshihiko Sugiura
Keio University, 3-14-1 Hiyoshi, Yokohama, Japan
e-mail: sugiura@mech.keio.ac.jp

Hiroshi Yabuno
Keio University, 3-14-1 Hiyoshi, Yokohama, Japan
e-mail: yabuno@mech.keio.ac.jp

G. Stépán et al. (Eds.): Dynamics Modeling & Interaction Cont., IUTAM BOOK SERIES 30, pp. 273–281.
springerlink.com

the effect of friction on the static states is discussed. It is shown that one of five patterns corresponds to the shape of the cleaning blade, and it is clarified through linear stability analysis that this state becomes dynamically unstable, only when friction is present. This unstable vibration is a bifurcation classified as Hamiltonian-Hopf bifurcation, and confirms the occurence of mode-coupled self-excited vibration with a constant frictional coefficient.

1 Introduction

Laser printers, which have recently shown advancements in image quality and printing speed, have come into wide corporate and individual use. Because laser printers are often placed near these users, it is imperative that they operate quietly. However, it is known that chatter vibration of a cleaning blade, which is the part of a toner cartridge used to clean the photoreceptor, occasionally produces an uncomfortable noise. The cleaning blade is pressed in the counter direction of the rotation of the photoreceptor to clean its surface, and the chatter occurs due to this contact. In order to eliminate this noise, it is first necessary to determine the mechanism of this chatter.

Though there have been several past studies on the vibration of the cleaning blade, the excitation mechanism of such vibration has not been theoretically clarified yet, except for the fact that is a self-excited-vibration induced by friction. Nakamura[7]showed the mechanism where the frictional force between the photoreceptor and the cleaning blade excites a twisting vibration of the photoreceptor, causing the photoreceptor to emit the noise. Kawamoto[5]focused on a squeaking noise occurring just after the start of rotation, or just before the rotation stops. The mechanism was explained as a nonlinear phenomenon consisting of both a self-excited vibration due to negative damping of dry friction, and a forced vibration generated due to the vibration of the photoreceptor induced by the alternative electrostatic force of the charger roller. Both studies showed that frictional vibration is generated by negative damping of dry friction.

On the other hand, Kasama et al.[4] clarified that this vibration is caused by coupled-mode flutter through finite element analysis. It was shown that the coupling between a shrinking mode and a bending mode induced by friction is the cause of the self-excited vibration. The validity of this study was confirmed by comparing the frequency of the unstable mode with the frequency of the vibration observed in the actual machine. The mode coupling principle, which has also been presented as the cause of chatter in machining operations[2][6], is the focus of this study.

In this study, a Two-Degree-Of-Freedom(2DOF) model has been introduced, as an essential model of the cleaning blade. Formulation of such model allows for the qualitative examination of the excitation mechanism through parameteric study. First, the governing equations of the 2DOF system are derived. Next, the static equilibrium states are calculated, and their stability are analyzed in order to confirm the occurence of mode-coupling chatter. This method where a system with reduced degrees of freedom is considered in studies of nonconservative problems[1] associated with elastic beams, has been used in studies of the vibrations of pipes conveying fluid[9], as well as wiper blades[3].

2 Analytical Model and Its Governing Equations

A cleaning blade can be regarded as an elastic beam, but a more simplified and essential analytical model is required to investigate the excitation mechanism of the cleaning blade. In this study, a 2DOF model is introduced, while keeping the contact between the blade and the photoreceptor in emphasis.

2.1 Analytical Model

The 2DOF analytical model is shown in Fig. 1. The system is composed of two linked rigid bodies, a mass M, a vertical spring k_1, and the two rotational springs k_2 and k_3. The length of the two rigid bodies is l, and their mass per unit length is m. k_1 corresponds to the tensile rigidity, and two rotational springs correspond to the flexural rigidity. This allows for the bending and shrinking of the blade to be considered, which were the coupled modes of the excitation mechanism presented by Kasama et al. [4].

At first, the shape of the system is Fig. 1 (a), where the springs are not stretched or pressed. After giving a vertical displacement δ from the initial setting position, the shape of the system becomes Fig. 1 (b). The vertical displacement of mass M is x in x-y plane and the rotational angles of the two rigid bodies about their centerline are θ and ϕ, respectively. The bottom of the link is assumed to always be in contact with the floor surface, which corresponds to the photoreceptor,

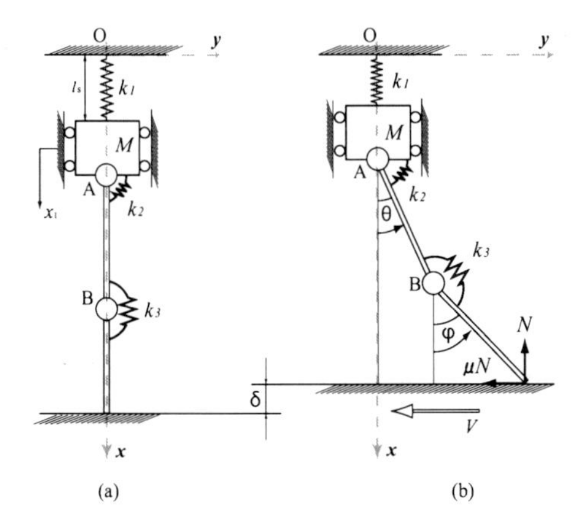

Fig. 1 Two Degree-of-Freedom Analytical Model

moving at constant velocity V. This causes the vertical load N and the frictional force μN to continuously act on the rigid body. The friction coefficient μ will be placed as constant, independent of the floor velocity. In such case with a constant frictional coefficient, Kasama et.al [4] clarified by finite element method that the self-excited vibration is generated by the mode-coupling flutter, not by the negative damping.

2.2 Equations of Motion

Considering the properties of the presented model, the governing equations have been derived using Newton's law as follows:

$$(M+2ml)\ddot{x}+k_1x-\frac{3}{2}ml^2\ddot{\theta}\sin\theta-\frac{3}{2}ml^2\dot{\theta}^2\cos\theta-\frac{1}{2}ml^2\ddot{\phi}\sin\phi$$
$$-\frac{1}{2}ml^2\dot{\phi}^2\cos\phi=-N \tag{1}$$

$$\frac{4}{3}ml^3\ddot{\theta}+k_2\theta-k_3(\phi-\theta)+\frac{1}{2}ml^3\ddot{\phi}\cos(\phi-\theta)-\frac{1}{2}ml^3\dot{\phi}^2\sin(\phi-\theta)$$
$$-\frac{3}{2}ml^2\ddot{x}\sin\theta=Nl\sin\theta-\mu Nl\cos\theta \tag{2}$$

$$\frac{1}{3}ml^3\ddot{\phi}+k_3(\phi-\theta)+\frac{1}{2}ml^3\ddot{\theta}\cos(\phi-\theta)+\frac{1}{2}ml^3\dot{\theta}^2\sin(\phi-\theta)$$
$$-\frac{1}{2}ml^2\ddot{x}\sin\phi=Nl\sin\phi-\mu Nl\cos\phi, \tag{3}$$

where $(\dot{\cdot})$ denotes the derivative with respect to time t. Furthermore, the restraint condition, where the blade is always in contact with the floor, is expressed as follows:

$$x=2l-l\cos\theta-l\cos\phi-\delta. \tag{4}$$

Next, these governing equations have been nondimensionalized. Using the dimensionless variables, which carry an asterisk, as defined below,

$$N=\frac{k_3}{l}\eta^*,\quad x=lx^*,\quad t=\sqrt{\frac{ml^3}{3k_3}}t^* \tag{5}$$

the dimensionless governing equations in the x-y plane can be expressed as follows:

$$\beta \ddot{x}^* + \alpha x^* - \frac{9}{2}\ddot{\theta}\sin\theta - \frac{9}{2}\dot{\theta}^2\cos\theta - \frac{3}{2}\ddot{\phi}\sin\phi - \frac{3}{2}\dot{\phi}^2\cos\phi = -\eta^* \quad (6)$$

$$4\ddot{\theta} + (K+1)\theta + \frac{3}{2}\ddot{\phi}\cos(\phi-\theta) - \phi - \frac{3}{2}\dot{\phi}^2\sin(\phi-\theta) - \frac{9}{2}\ddot{x}^*\sin\theta$$
$$= \eta^*\sin\theta - \mu\eta^*\cos\theta \quad (7)$$

$$\ddot{\phi} + \phi + \frac{3}{2}\ddot{\theta}\cos(\phi-\theta) - \theta + \frac{3}{2}\dot{\theta}^2\sin(\phi-\theta) - \frac{3}{2}\ddot{x}^*\sin\phi$$
$$= \eta^*\sin\phi - \mu\eta^*\cos\phi \quad (8)$$

$$x^* = 2 - \cos\theta - \cos\phi - \delta^*. \quad (9)$$

In these equations, K,α,β and δ^* are non-dimensional parameters expressed as follows:

$$K = \frac{k_2}{k_3},\ \alpha = \frac{k_1 l^2}{k_3},\ \beta = 3\left(\frac{M}{ml} + 2\right),\ \delta^* = \frac{\delta}{l}. \quad (10)$$

3 Stability Analysis

In this section, the occurence of mode-coupling instability in friction-induced vibrations of the blade is examined. First, the static equilibrium states, and their variance with increasing displacement δ are determined. The effect of the frictional force on the bifurcation of the static states is shown. Furthermore, by considering the linear terms of the disturbances from the static states, the stability of these states are investigated.

3.1 Static Equilibrium States

By removing the terms that depend on time from the governing equations, and expressing the variables as $\theta \equiv \theta_s$, $\phi \equiv \phi_s$, $x \equiv x_s$ and $\eta \equiv \eta_s$, the equations of static equilibrium are expressed as follows:

$$(K+1)\theta_s - \phi_s - \eta_s\sin\theta_s + \mu\eta_s\cos\theta_s = 0 \quad (11)$$

$$\phi_s - \theta_s - \eta_s\sin\phi_s + \mu\eta_s\cos\phi_s = 0 \quad (12)$$

$$\alpha x_s + \eta_s = 0 \quad (13)$$

$$x_s = 2 - \cos\theta_s - \cos\phi_s - \delta. \quad (14)$$

The asterisks denoting the dimensionless variables are omitted in these equations and henceforward. The unknown variables, θ_s, ϕ_s, x_s and η_s, have been calculated for both cases when $\mu = 0$ and $\mu = 0.3$. The values of the other dimensionless parameters are $K = 0.8$ and $\alpha = 5.0$, and the parameter δ, or the initial displacement, has been increased continuously from 0 to 1.2. The results of each variable with respect to increasing δ, and the shapes of the corresponding static states are shown in Fig.2 and Fig.3.

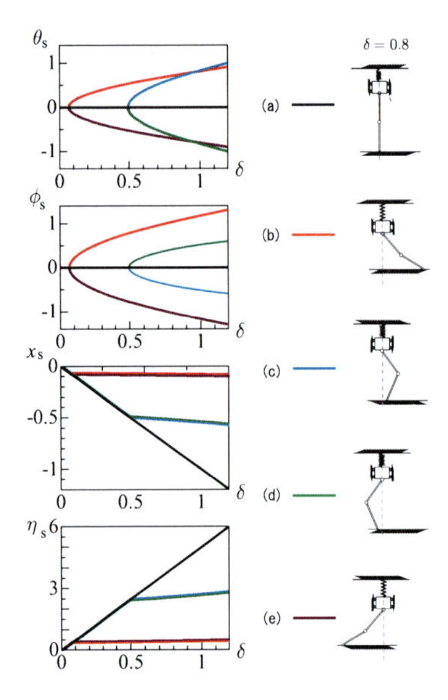

Fig. 2 Static Equilibrium States ($\mu = 0$)

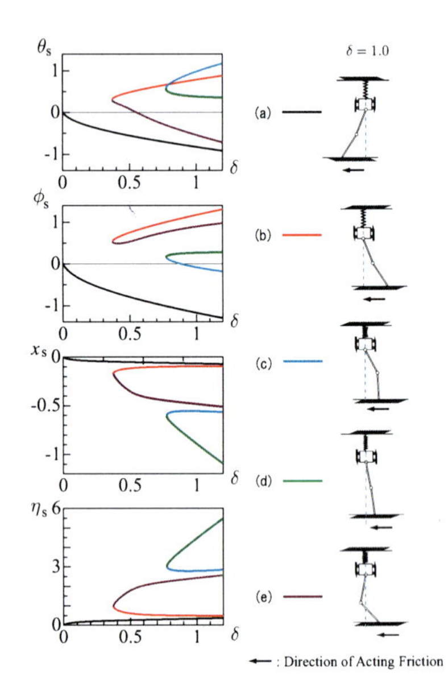

Fig. 3 Static Equilibrium States ($\mu = 0.3$)

The results clarified that there are five static equilibrium states for both cases. When friction is not present, this analysis corresponds to the buckling phenomenon due to compression. It can be seen that the rotational angles of state (b) and (e), as well as (c) and (d), are symmetrical with respect to the x axis, while the inclusion of frictional force leads to a change in the bifurcation. Furthermore, one state (b) corresponds to the shape of the cleaning blade.

3.2 Linear Stability Analysis

To determine the stability of the static equilibrium states, a time-dependent disturbance is added to the static state. After eliminating x and η from the governing equations, θ and ϕ are represented as $\theta \equiv \theta_s + \theta_d$ and $\phi \equiv \phi_s + \phi_d$. By subtracting the static equilibrium state, and neglecting the nonlinear terms with respect to θ_d and ϕ_d in the equations, the linear differential equations about θ_d and ϕ_d can be expressed in the vector and the matrix form as follows:

$$A\ddot{\mathbf{v}} + B\mathbf{v} = 0 \tag{15}$$

where

$$\mathbf{v} = \begin{bmatrix} \theta_d \\ \phi_d \end{bmatrix}, \quad A = \begin{bmatrix} a_{11} & a_{12} \\ a_{21} & a_{22} \end{bmatrix}, \quad B = \begin{bmatrix} b_{11} & b_{12} \\ b_{21} & b_{22} \end{bmatrix} \tag{16}$$

and

$$a_{11} = 4 + (\beta - 9)\sin^2\theta_s - \frac{1}{4}(2\beta - 9)\mu \sin 2\theta_s \tag{17}$$

$$a_{12} = \frac{1}{2}(2\beta - 9)\sin\theta_s \sin\phi_s + \frac{1}{2}\cos\theta_s\{3\cos\phi_s + (3-2\beta)\mu\sin\phi_s\} \tag{18}$$

$$a_{21} = \frac{3}{2}\cos\phi_s\cos\theta_s + \frac{1}{2}(2\beta - 9)\sin\theta_s(\sin\phi_s - \mu\cos\phi_s) \tag{19}$$

$$a_{22} = 1 + (\beta - 3)\sin^2\phi_s + \frac{1}{4}\mu(3-2\beta)\sin 2\phi_s \tag{20}$$

$$b_{11} = (K+1) + \alpha\{(\sin\theta_s - \mu\cos\theta_s)\sin\theta_s + (2 - \cos\theta_s - \cos\phi_s - \delta)(\cos\theta_s + \mu\sin\theta_s)\} \tag{21}$$

$$b_{12} = -1 + \alpha(\sin\theta_s - \mu\cos\theta_s)\sin\phi_s \tag{22}$$

$$b_{21} = -1 + \alpha(\sin\phi_s - \mu\cos\phi_s)\sin\theta_s \tag{23}$$

$$b_{22} = 1 + \alpha\sin\phi_s(\sin\phi_s - \mu\cos\phi_s) + \alpha(\cos\phi_s + \mu\sin\phi_s)(2 - \cos\theta_s - \cos\phi_s - \delta) \tag{24}$$

Letting $\mathbf{v} = (\theta_{amp}, \phi_{amp})^t \exp(\lambda t)$ and substituting them into eq(15), this can be cast into an eigenvalue problem. The eigenvalues λ are the roots of the following characteristic equation:

$$\lambda^4 + \frac{a_{22}b_{11} - a_{21}b_{12} - a_{12}b_{21} + a_{11}b_{22}}{-a_{12}a_{21} + a_{11}a_{22}}\lambda^2 + \frac{b_{12}b_{21} - b_{11}b_{22}}{a_{12}a_{21} - a_{11}a_{22}} = 0 \tag{25}$$

The eigenvalue λ is divided into the damping coefficient part λ_r and the natural frequency part λ_i where $\lambda = \lambda_r + i\lambda_i$. When λ_r is a real positive number and $\lambda_i \neq 0$, the system has a state of dynamical instability, where unstable vibration is generated in the system.

Using the displacement from initial setting position δ as a system parameter, the values of both λ_r and λ_i were calculated. The values of the other dimensionless parameters are $K = 0.8$, $\alpha = 5.0$ and $\beta = 7.0$. From this analysis, it was found that the static states when $\mu = 0$ do not show dynamical instability, meaning that self-excited vibration does not occur, while some states showed static instability, which can be explained as correspondent to the buckling phenomenon. When $\mu = 0.3$, the state of dynamical instability appears in state (b) as shown in Fig. 3, which corresponds to the shape of the cleaning blade. The Argand diagrams of the complex eigenvalues

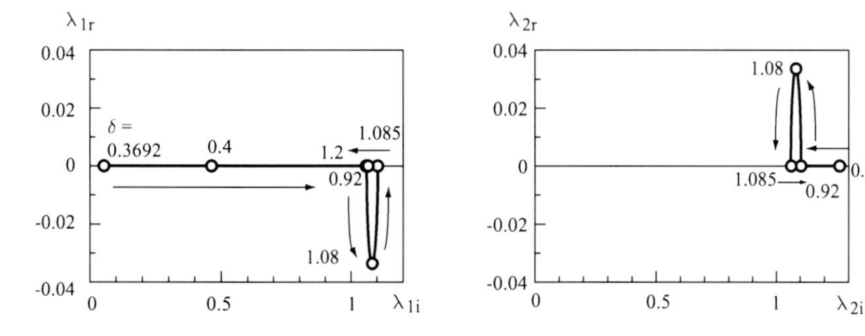

Fig. 4 Argand Diagrams of state (b) when $\mu = 0.3$

of state (b) are shown in Fig. 4. The two eigenvalues with positive imaginary parts shown, as well as the conjugates of these eigenvalues, are the roots of eq(25).The real part of λ_2 is positive, confirming the occurence of mode-coupling self-excited vibration in this system, where the frictional coefficient is constant. Furthermore, this condition was found to occur only when δ is between 0.926 and 1.085, which leads to a bifurcation classified as Hamiltonian-Hopf bifurcation [8].

4 Conclusions

The mechanism of the vibration of a cleaning blade in laser printers, in particular the occurence of mode-coupling instability, has been investigated using a simple analytical model. A 2DOF model, which incorporates the bending and shrinking of the blade as well as the friction due to the contact with the photoreceptor, was presented and analyzed. By comparing both cases when friction is nonexistent and present, the effect of friction on the motion of the blade was clarified.

By deriving the static states, it was found that there are five states of static equilibrium on this system, while one state corresponds to the shape of the cleaning blade. It was shown that the presence of friction greatly affects the bifurcation of the variables.

Furthermore, from linear stability analysis, it was found that when friction is not present, there are no states of dynamical instability, only static instability. This phenomenon can be correspondent to simple buckling, where the blade becomes unstable at certain displacements but does not vibrate. However, the inclusion of friction leads to the static state corresponding to the shape of the cleaning blade becoming dynamically unstable, confirming the occurence of mode-coupling vibration due to friction. Furthermore, the bifurcation of the argand diagram of this unstable state is one classified as Hamiltonian-Hopf bifurcation.

Acknowledgements. The authors would like to gratefully acknowledge Dr. M.Yoshizawa, a retired professor of Keio University, for valuable discussions, assistance, and advice.

References

1. Bolotin, V.V.: Nonconservative Ploblems of the Theory of Elastic Stability. Pergamon, London (1963)
2. Gasparetto, A.: Eigenvalue Analysis of Mode-Coupling Chatter for Machine-Tool Stabilization. Journal of Vibration and Control 7, 181–197 (2001)
3. Grenouillat, R., Leblanc, C.: Simulation of Chatter Vibrations for Wiper Systems. In: SAE Papers, 2002-01-1239 (2002)
4. Kasama, M., Yoshizawa, M., Yu, Y., Itoh, T.: Coupled-Mode Flutter of a Cleaning Blade System in a Laser Printer. Journal of System Design and Dynamics 2(3), 849–860 (2008)
5. Kawamoto, H.: Chatter Vibration of a Cleaner Blade in Electrophotography. Journal of Imaging Science and Technology 40-1, 8–13 (1996)
6. Koenigsberger, F., Tlusty, J.: Machine Tool Structures. Elsevier, Amsterdam (1970)

7. Nakamura, K.: Generation Mechanism of Unusual Noise in the Laser-Beam Printer. Transactions of the Japan Society of Mechanical Engineers, Series C 62(601), 3428–3433 (1996)
8. Païdousis, M.P.: Fluid-Structure Interactions Slender Structure and Axial, vol. 1, pp. 68–69. Academic Press, San Diego (1998)
9. Rousselet, J., Herrmann, G.: Flutter of Articulated Pipes at Finite Amplitude. Journal of Applied Mechanics 44(1), 154–158 (1977)

Thermoelasticity Aspects

Some refinements and improvements of classical models and methods in the fields of elastic and thermoplastic problems are discussed in the papers of this section. The paper by J.P. Meijard deals with the refinement of classical beam theory for beams with a large aspect ratio of their cross-sections. Dynamic contact problems for shells with moderately large deflections are analyzed by I. Bock in detail. Finally, the paper by S. Po-jen *et al.* presents new solutions for scattering in thermoelastic half-plane.

Refinements of Classical Beam Theory for Beams with a Large Aspect Ratio of Their Cross-Sections

J.P. Meijaard

Abstract. In order to obtain a required stiffness ratio between compliant and stiff directions, beams with a large ratio of width to depth are needed. For the simulation of multibody dynamic systems, mostly the Bernoulli–Euler assumptions are made. Some known but mostly neglected effects for beams with a rectangular cross-section, leaf springs, are considered here. The effect of shear stiffness, as in Timoshenko's beam theory, is important for beams that are short in comparison to their width. For beams under torsion, the effect of constrained warping can become important. The constrained anticlastic bending for large deflections is a non-linear effect. Clamping is never perfect and the support stiffness or restrained transverse contraction needs to be considered. Residual stresses have a marked influence if they exceed the buckling load, which can easily occur if the aspect ratio is large.

1 Introduction

In multibody dynamics, the Euler–Bernoulli beam theory with Saint-Venant's torsion and shear theory is most often used for modelling flexible beams. Here, some effects that are usually neglected are studied. In particular, beams with a narrow rectangular cross-section where the length need not be much larger than the width are considered. Boundary conditions at clamped ends result in the suppression of transverse contraction and the warping of the cross-section. For large deflections in the flexible direction, the bending stiffness becomes a non-linear function of the curvature. These effects become important in an accurate analysis of the behaviour and especially in the buckling analysis of compliant mechanisms [1]. The purpose of this investigation is to increase the accuracy of beam theory without adding to the complexity or increasing the number of degrees of freedom.

J.P. Meijaard
Laboratory of Mechanical Automation and Mechatronics,
Faculty of Engineering Technology, University of Twente, Enschede, The Netherlands
e-mail: J.P.Meijaard@utwente.nl

G. Stépán et al. (Eds.): Dynamics Modeling & Interaction Cont., IUTAM BOOK SERIES 30, pp. 285–292.
springerlink.com

2 Beam Refinements

This section starts with some considerations about the equations of a plate loaded in its plane and perpendicular to its plane [2, 3, 4]. Then these are applied to formulate some effects observed in the behaviour of beams.

2.1 Equations and Energy Functionals

A beam with a narrow rectangular cross-section is considered, see Fig. 1. The axis of the beam is along the local x-axis, the local y-axis is in the direction of the width and the local z-axis in the direction of the thickness. The origin is at an end of the beam. The width b is assumed to be at least an order of magnitude larger than the thickness t, whereas the length l need not be much larger than the width. In many respects, the beam may be seen as a plate that is loaded in its plane and perpendicular to its plane.

The in-plane equilibrium equations for the stresses are satisfied by introducing the Airy stress function $\Phi(x,y)$,

$$\sigma_x = \Phi_{,yy}\,, \quad \sigma_y = \Phi_{,xx}\,, \quad \tau_{xy} = -\Phi_{,xy}\,, \tag{1}$$

where a subscript comma followed by variables denotes partial derivatives with respect to these variables. The compatibility condition for the case of plane stress yields the biharmonic equation

$$\Delta\Delta\Phi = \Phi_{,xxxx} + 2\Phi_{,xxyy} + \Phi_{,yyyy} = 0\,. \tag{2}$$

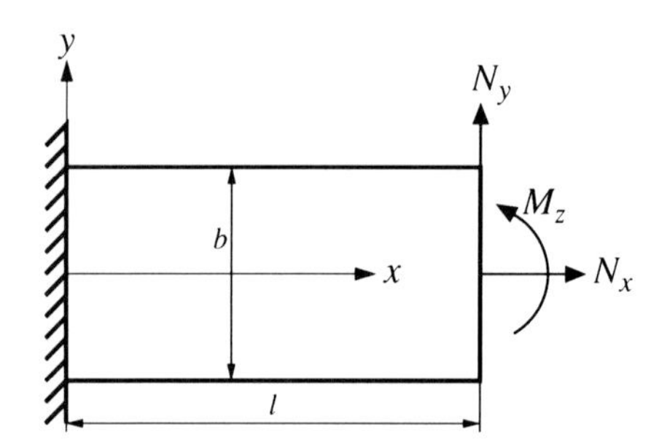

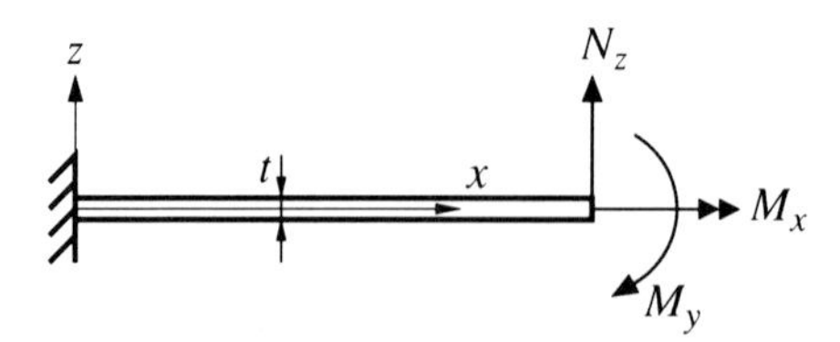

Fig. 1 Beam with a narrow rectangular cross-section clamped at one end and with a general loading at the other end

The in-plane strains follow from the constitutive equations,

$$\varepsilon_x = (\sigma_x - \nu\sigma_y)/E\,, \quad \varepsilon_y = (-\nu\sigma_x + \sigma_y)/E\,, \quad \gamma_{xy} = 2(1+\nu)\tau_{xy}/E\,, \tag{3}$$

where E is Young's modulus and ν Poisson's ratio. The displacements u in the x-direction and v in the y-direction are found from integrating these strains. Boundary conditions are either prescribed stresses or prescribed displacements.

For a plate loaded perpendicular to its middle plane with no distributed load, the displacement w in the z-direction satisfies the biharmonic equation

$$\Delta\Delta w = w_{,xxxx} + 2w_{,xxyy} + w_{,yyyy} = 0\,. \tag{4}$$

The internal moments per unit of length are given by the constitutive equations

$$(M_x, M_y, M_{xy}) = -\frac{Et^3}{12(1-\nu^2)}(w_{,xx} + \nu w_{,yy},\, \nu w_{,xx} + w_{,yy},\, (1-\nu)w_{,xy})\,, \tag{5}$$

and the transverse shear forces per unit of length are

$$Q_x = M_{x,x} + M_{xy,y}\,, \quad Q_y = M_{xy,x} + M_{y,y}\,. \tag{6}$$

At the boundary we have a normal unit vector with components n_x, n_y and a tangential unit vector with components $t_x = -n_y$, $t_y = n_x$. With the normal couple, $M_n = M_x n_x^2 + 2M_{xy}n_x n_y + M_y n_y^2$, and the twist couple, $M_t = M_x n_x t_x + M_{xy}(n_x t_y + n_y t_x) + M_y n_y t_y$, per unit of length, the dynamic boundary conditions are a prescribed value of M_n and a prescribed value of the equivalent shear force per unit of length, $Q_x n_x + Q_y n_y + M_{t,s}$, where s is the arc length along the boundary. At corners, concentrated loads equal to the negative jump of M_t can occur.

For the in-plane problem, the potential energy is

$$\frac{Et}{2(1-\nu^2)}\iint \left[\varepsilon_x^2 + \varepsilon_y^2 + 2\nu\varepsilon_x\varepsilon_y + \frac{1-\nu}{2}\gamma_{xy}^2\right]\mathrm{d}x\mathrm{d}y - \int (f_x u + f_y v)\mathrm{d}s\,, \tag{7}$$

where f_x and f_y are the prescribed surface tractions. For the out-of plane problem, we have the potential energy

$$\frac{Et^3}{24(1-\nu^2)}\iint \left[w_{,xx}^2 + w_{,yy}^2 + 2\nu w_{,xx}w_{,yy} + 2(1-\nu)w_{,xy}^2\right]\mathrm{d}x\mathrm{d}y$$
$$+\int \left[M_n w_{,n} - (f_z + T_{t,s})w\right]\mathrm{d}s\,, \tag{8}$$

where f_z is the force per unit of length and T_t is the twisting couple per unit of length.

2.2 Shear

For an in-plane loading by a moment M_z, a normal force N_x and a transverse load N_y at $x = l$, the stresses are found as

$$\sigma_x = -\frac{M_z + N_y(l-x)}{I_z} y + \frac{N_x}{A}, \quad \sigma_y = 0, \quad \tau_{xy} = \frac{N_y}{I_z}\left(\frac{1}{8}b^2 - \frac{1}{2}y^2\right). \tag{9}$$

If the average displacements and rotation at $x = 0$ are zero, the corresponding displacements are

$$u = \frac{N_x}{EA}x - \frac{M_z}{EI_z}xy - \frac{N_y}{EI_z}\left(lxy - \frac{1}{2}x^2 y\right) + \frac{N_y}{EA}\frac{(6+3\nu)}{10}\left(y - \frac{20}{3}\frac{y^3}{b^2}\right), \tag{10}$$

$$v = -\frac{N_x}{EA}\nu y + \frac{M_z}{EI_z}\left[\frac{1}{2}x^2 + \frac{\nu}{2}\left(y^2 - \frac{b^2}{12}\right)\right]$$
$$+ \frac{N_y}{EI_z}\left[\frac{1}{2}lx^2 - \frac{1}{6}x^3 + \frac{\nu}{2}(l-x)\left(y^2 - \frac{b^2}{12}\right)\right] + \frac{N_y}{EA}\frac{(12+11\nu)x}{5}. \tag{11}$$

From this we see that the shear coefficient agrees with that given by Cowper [5]: $k_y = 10(1+\nu)/(12+11\nu)$.

These solutions have to be modified for a perfect clamping near $x = 0$. The stresses are affected over a boundary zone with a length of the order of the width of the beam. From numerical calculations for the case of tension or compression and $\nu = 0.3$, it follows that the elongation is modified as

$$\bar{u}(l) = (l - 0.01223\,b)N_x/(EA). \tag{12}$$

The numerical factor is about proportional to ν^2. For bending by a moment and shear force, the numerically obtained modified relations for the average end deflection and rotation at $x = l$ and $\nu = 0.3$ are

$$\bar{v}(l) = \left[\frac{l^3}{3} - 0.00544\,l^2 b + \left(\frac{12+11\nu}{60} + 0.00105\right)lb^2 - 0.0040\,b^3\right]\frac{N_y}{EI_z}$$
$$+ \left(\frac{l^2}{2} - 0.00544\,lb + 0.00295\,b^2\right)\frac{M_z}{EI_z}, \tag{13}$$

$$\bar{\psi}(l) = \left(\frac{l^2}{2} - 0.00544\,lb - 0.00205\,b^2\right)\frac{N_y}{EI_z} + (l - 0.00544\,b)\frac{M_z}{EI_z}. \tag{14}$$

The shear coefficient appears to be decreased by the clamping, and the angle of rotation is diminished. The failing of the reciprocity relation is due to the inconsistency of the definition of the average deflection and the application of the shear force according to Eq. (9): the deflection depends on the way in which the force N_y is applied at the free end. The additional displacement is here $\nu b^2 M_z/(60EI_z)$, which agrees with the numerical results. The term quadratic in the length in the displacement will dominate the effect of shear for slender beams if $l > 50b$. The corrections in the term with the moment and the corresponding corrections in the terms with the shear force are approximately proportional to ν^2, whereas the other corrections have a weak dependence on ν.

If the clamping is considered as a connection to an elastic half-space with the same material properties as those of the beam, the correction terms are of the order of $t\ln(b/t)$, which has to be compared with the order b or $b\nu^2$ of the correction terms due to the hindering of the free warping and the transverse contraction, which are, in principle, of a larger order. This elastic clamping compensates a part of the increase of stiffness due to the other effects.

2.3 Constrained Warping

In the torsion problem, the loading is given by a couple of forces M_x/b at the two free corner points. The solution is

$$w = \frac{6(1+\nu)M_x}{Ebt^3}xy\,, \tag{15}$$

with a specific twist of $6(1+\nu)M_x/(Ebt^3)$. In order to investigate the influence of the clamping, an assumed displacement field satisfying the boundary conditions at $x = l$,

$$w = \frac{6(1+\nu)M_x}{Ebt^3}y\big(x+\frac{1}{\lambda}(e^{-\lambda x}-1)\big)\,, \tag{16}$$

is substituted in the potential energy expression (8) and the parameter λ is varied to obtain a minimum; this yields $\lambda = \sqrt{24(1-\nu)}/b$. If, on the other hand, the additional incompatible assumption $w_{,yy} = -\nu w_{,xx}$, instead of $w_{,yy} = 0$, is made, we obtain $\lambda = \sqrt{24/(1+\nu)}/b$. These results were obtained in [6]. The true value is between these two. From numerical calculations for $\nu = 0.3$, we obtain for the twist angle

$$\bar{\varphi}(l) = \frac{6(1+\nu)M_x}{Ebt^3}(l-0.2419\,b)\,, \tag{17}$$

whereas $1/\sqrt{24(1-\nu)} = 0.2449$ and $\sqrt{(1+\nu)/24} = 0.2327$.

For out-of-plane bending by a moment M_y and a shear force F_z, the effect of transverse shear can be neglected, as in the standard plate bending theory. The displacement distribution is

$$w = \frac{M_y}{EI_y}\Big[-\frac{1}{2}x^2+\frac{\nu}{2}\big(y^2-\frac{b^2}{12}\big)\Big]+\frac{N_z}{EI_y}\Big[\frac{1}{2}lx^2-\frac{1}{6}x^3-\frac{\nu}{2}(l-x)\big(y^2-\frac{b^2}{12}\big)\Big]\,, \tag{18}$$

with the average displacement and rotation at the loaded end

$$\bar{w}(l) = -\frac{M_yl^2}{2EI_y}+\frac{N_zl^3}{3EI_y}\,, \quad \bar{\chi}(l) = \frac{M_yl}{EI_y}-\frac{N_zl^2}{2EI_y}\,. \tag{19}$$

With a fully clamped end, where the anticlastic deformation of the cross-section is prevented, we numerically find for the end deflection and rotation, with $\nu = 0.3$,

$$\bar{w}(l) = \frac{M_y}{EI_y}\left(\frac{-l^2}{2} + 0.0368\,lb - 0.0144\,b^2\right)$$

$$+\frac{N_z}{EI_y}\left(\frac{l^3}{3} - 0.0368\,l^2 b + 0.0144\,lb^2 + 0.0093\,b^3\right), \tag{20}$$

$$\bar{\chi}(l) = \frac{M_y}{EI_y}(l - 0.0368\,b) + \frac{N_z}{EI_y}\left(\frac{-l^2}{2} + 0.0368\,lb - 0.0144\,b^2\right). \tag{21}$$

If the beam is connected to an elastic half-space, the results change by an amount of the order t against an order b for the warping and transverse contraction corrections, so this effect is even smaller than for the in-plane case.

2.4 Large Deflections

Ashwell [7] has given an analysis of the bending moment needed to bend a plate strip into a cylindrical shell. The result is that the bending moment needed is $M_y = f_e Ebt^3/(12R)$, where R is the radius of the cylindrical surface and f_e is a numerical stiffness factor which is equal to one for large values of R, i.e. $R \gg b^2/t$, and equal to $1/(1-\nu^2)$ for small values of R. This secant stiffness factor is given by

$$f_e = \frac{2 + \nu^2 f_1(\mu) - 4\nu^2 f_2(\mu)}{2(1-\nu^2)}, \tag{22}$$

with

$$\mu = \sqrt[4]{3(1-\nu^2)\frac{b^4}{R^2 t^2}}, \tag{23}$$

$$f_1(\mu) = \frac{\sinh 2\mu - \sin 2\mu - 4\mu \sinh\mu \sin\mu + 2\cosh\mu \sin\mu - 2\sinh\mu \cos\mu}{2\mu(\sinh\mu + \sin\mu)^2}, \tag{24}$$

$$f_2(\mu) = \frac{\cosh\mu - \cos\mu}{\mu(\sinh\mu + \sin\mu)}. \tag{25}$$

Note that $f_1(0) = f_1(\infty) = f_2(\infty) = 0$, $f_2(0) = 1/2$. The tangent stiffness is given by

$$\frac{\mathrm{d}M_y}{\mathrm{d}(1/R)} = \left(f_e + \frac{\mu}{2}\frac{\mathrm{d}f_e}{\mathrm{d}\mu}\right)\frac{Ebt^3}{12} \tag{26}$$

Also for slowly varying curvatures, the above relations can be used. The secant and tangent stiffness factors are shown in Fig 2.

Another non-linear effect may result from the non-uniform torsion at large deflections of a beam. An approximation as Eq. (16) can be used,

$$w = y\bar{\varphi}(x), \tag{27}$$

where $\bar{\varphi}(x)$ is the twist angle. This results in a modified relation between the twisting moment and the twist angle,

$$M_x = \frac{Ebt^3}{6(1+\nu)}\left(\frac{\mathrm{d}\bar{\varphi}}{\mathrm{d}x} - \frac{b^2}{24(1-\nu)}\frac{\mathrm{d}^3\bar{\varphi}}{\mathrm{d}x^3}\right). \tag{28}$$

This relation for non-uniform torsion can also be used for torsional oscillations.

3 Residual Stresses

Classical results can be severely influenced by the presence of largely unknown residual stresses. In thin rolled metal sheets, these can easily exceed the buckling load [8]. Their influence was also noticed in the experiments reported in [1]. A longitudinal residual stress σ_x can only influence the bending stiffness about the y-axis if the buckling load is exceeded, because the integral of this stress over the cross-section is zero and so gives no contribution to the pre-buckling geometric stiffness. The influence on the torsional rigidity, however, is also felt below the buckling load. A longitudinal tension at the edges and a compression at the centre increase the stiffness, and vice versa, a longitudinal compression at the edges and a tension at the centre decrease the stiffness.

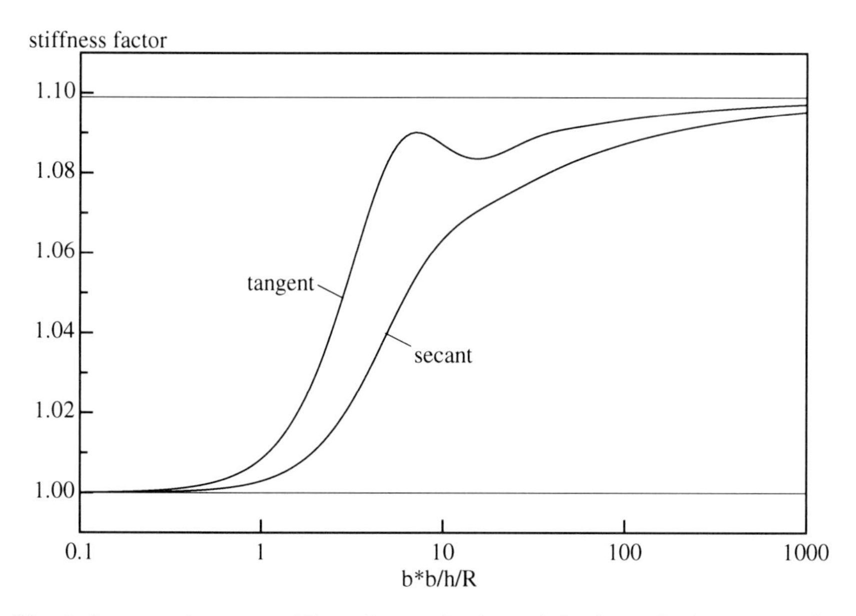

Fig. 2 Secant and tangent stiffness factors for large deflections of a beam according to Eqs. (22) and (26)

4 Conclusions

For an accurate dynamic analysis of compliant mechanisms, a proper description of the stiffness properties of its deforming elements is more difficult and more important than a precise description of the mass properties. The end conditions of leaf springs have to be considered carefully. The constraining of the warping and the transverse contraction can be of equal or more importance than the support stiffness. For slender beams, the corrections due to end effects can be more important than the effect of shear flexibility, although both are relatively small in this case. Residual stresses can have a marked influence on the stiffness of leaf springs with a high aspect ratio of their cross-sections, especially if the buckling load is exceeded. This is inconvenient, because these stresses are largely unknown.

The effects of the end conditions can be included in an analysis by modifying the stiffness properties of the beam elements at the ends of a leaf spring.

References

1. Meijaard, J.P., Brouwer, D.M., Jonker, J.B.: Analytical and experimental investigation of a parallel leaf spring guidance. Multibody Syst. Dyn. 23, 77–97 (2010)
2. Love, A.E.H.: A treatise on the mathematical theory of elasticity, 4th edn. Cambridge University Press, Cambridge (1927)
3. Biezeno, C.B., Grammel, R.: Technische Dynamik (2. Aufl.). Springer, Berlin (1953)
4. Timoshenko, S., Woinowsky-Krieger, S.: Theory of plates and shells, 2nd edn. McGraw-Hill, New York (1959)
5. Cowper, G.R.: The shear coefficient in Timoshenko's beam theory. ASME J. Appl. Mech. 33, 335–340 (1966)
6. Timoschenko, S.: On the torsion of a prism, one of the cross-sections of which remains plane. Proc. Lond. Math. Soc. 20(2), 389–397 (1922)
7. Ashwell, D.G.: The anticlastic curvature of rectangular beams and plates. J. Roy. Aer. Soc. 54, 708–715 (1950)
8. Fischer, F.D., Rammerstorfer, F.G., Friedl, N.: Residual stress-induced center wave buckling of rolled strip metal. ASME J. Appl. Mech. 70, 84–90 (2003)

Dynamic Contact Problems for Shells with Moderately Large Deflections

Igor Bock

Abstract. We deal with an initial-boundary value problem describing the perpendicular vibrations of Kármán-Donnell shells with a rigid inner obstacle. The elastic as well the viscoelastic materials are considered. A weak formulation of the problems are in the form of the hyperbolic variational inequalities. We solve the problem using the penalization method.

1 Introduction and Notation

Contact problems represent an important but complex topic of applied mathematics. Its complexity profounds if the dynamic character of the problem is respected. For elastic problems there is only a very limited amount of results available (cf. [5] and there cited literature). The presented results in the elastic case extend the research made in [4], where the problem for an elastic von Kármán plate in a dynamic contact with a rigid obstacle was considered. Viscosity makes possible to prove the existence of solutions for a broader set of problems for membranes, bodies as well as for linear models of plates. The von Kármán plate made of a short memory material in a dynamic contact was studied in [3]. The aim of the present paper is to extend these results to the nonlinear von Kármán-Donnell shells. Our results also extend the research made for the quasistatic contact problems for viscoelastic shells (cf. [2]).

Let $\Omega \subset R^2$ be a bounded convex polygonal or $C^{3,1}$ domain with a boundary Γ and $I \equiv (0,T)$ be a bounded time interval. The unit outer normal vector is denoted by $\mathbf{n} = (n_1, n_2)$, $\tau = (-n_2, n_1)$ is the unit tangent vector. The displacement is denoted by $\mathbf{u} \equiv (u_i)$. A strain tensor is defined as

$\varepsilon_{ij}(\mathbf{u}) = \frac{1}{2}(\partial_i u_j + \partial_j u_i + \partial_i u_3 \partial_j u_3) - k_{ij} u_3 - x_3 \partial_{ij} u_3,\ i,j = 1,2$ with $k_{12} = 0$ and the curvatures $k_{ii} > 0,\ i = 1,2.$

Igor Bock
Dept. of Mathematics, Faculty of Electrical Engineering and Information Technology, Slovak University of Technology, 81219 Bratislava, Slovak Republic
e-mail: igor.bock@stuba.sk

G. Stépán et al. (Eds.): Dynamics Modeling & Interaction Cont., IUTAM BOOK SERIES 30, pp. 293–300.
springerlink.com

Further, we set $[u,v] \equiv \partial_{11}u\partial_{22}v + \partial_{22}u\partial_{11}v - 2\partial_{12}u\partial_{12}v$. In the sequel, we denote by $W_p^k(M)$, $k \geq 0$, $p \in [1,\infty]$ the Sobolev spaces defined on a domain or an appropriate manifold M. By $\mathring{W}_p^k(M)$ the spaces with zero traces are denoted. If $p = 2$ we use the notation $H^k(M)$, $\mathring{H}^k(M)$. The duals to $\mathring{H}^k(M)$ are denoted by $H^{-k}(M)$. By $\mathscr{H}$, $\mathring{\mathscr{H}}$ we denote the spaces $L_\infty(I;H^2(\Omega))$, $L_\infty(I;\mathring{H}^2(\Omega))$, respectively.

2 Contact of an Elastic Shell

Emploing the Einstein summation convention, the constitutional law has the form $\sigma_{ij}(\mathbf{u}) = E(1-\mu^2)^{-1}\big((1-\mu)\varepsilon_{ij}(\mathbf{u}) + \mu\delta_{ij}\varepsilon_{kk}(\mathbf{u})\big)$. The constants $E > 0$ and $\mu \in (0,\frac{1}{2})$ are the Young modulus of elasticity and the Poisson ratio, respectively. We shall use the abbreviation $b = h^2/(12\rho(1-\mu^2))$, where $h > 0$ is the shell thickness and ρ is the density of the material.

The compact imbedding $H^2(\Omega) \hookrightarrow\hookrightarrow C(\overline{\Omega})$ plays the crucial role. Neglecting the rotary inertia of the plate enables us to achieve the dual estimate of the acceleration term after the penalization.

Let the function $f : Q \mapsto R$ represent the perpendicular load acting on the shell. We obtain for the bending function u, the unknown contact force g and the Airy stress function v the classical formulation composed of the system the form in a similar way as in [6], or [7], where the dynamic problems for von Kármán-Donnell shells without contacts were considered. Assuming a shell with a middle surface Ω, free on its boundary we have the initial-boundary value problem

$$\left.\begin{array}{l} \ddot{u} + bE\Delta^2 u - [u,v] - k_{11}\partial_{22}v - k_{22}\partial_{11}v = f + g, \\ u \geq 0,\ g \geq 0,\ ug = 0, \\ \Delta^2 v + E\left(\frac{1}{2}[u,u] + k_{11}\partial_{22}u + k_{22}\partial_{11}u\right) = 0 \end{array}\right\} \text{ on } Q, \tag{1}$$

$$\left.\begin{array}{l} u \geq 0,\ \Sigma(u) \geq 0,\ u\Sigma(u) = 0, \\ \mathscr{M}(u) = 0,\ v = 0 \text{ and } \partial_n v = 0 \end{array}\right\} \text{ on } S, \tag{2}$$

$$u(0,\cdot) = u_0 \geq 0,\ \dot{u}(0,\cdot) = u_1 \text{ on } \Omega. \tag{3}$$

For $u, y \in L_2(I;H^2(\Omega))$ we define the bilinear form A and a cone $\mathscr{C}$ by

$$A(u,y) := b\big(\partial_{kk}u\partial_{kk}y + \mu(\partial_{11}u\partial_{22}y + \partial_{22}u\partial_{11}y) + 2(1-\mu)\partial_{12}u\partial_{12}y\big)$$

$$\mathscr{C} := \{y \in \mathscr{H};\ y \geq 0\}. \tag{4}$$

Then the variational formulation of the problem (1)-(3) has the following form:

Find $\{u,v\} \in C \times L_2(I;\mathring{H}^2(\Omega))$ *such that*

$$\int_Q (EA(u,y_1-u)+\ddot{u}(y_1-u)-([u,v]+k_{11}\partial_{22}v+k_{22}\partial_{11}v)(y_1-u))\,dx\,dt$$
$$\geq \int_Q f(y_1-u)\,dx\,dt\;\forall y_1\in\mathscr{C}, \tag{5}$$
$$\int_\Omega \big(\Delta v\Delta y_2+(\frac{1}{2}E[u,u]+k_{11}\partial_{22}u+k_{22}\partial_{11}u)y_2\big)\,dx=0\;\forall y_2\in\mathring{H}^2(\Omega). \tag{6}$$

We define the bilinear operator $\Phi: H^2(\Omega)^2 \to \mathring{H}^2(\Omega)$ and the linear operators Δ_k : $H^2(\Omega)\mapsto L_2(\Omega)$, $L: H^2(\Omega)\to\mathring{H}^2(\Omega)$ by means of

$$\int_\Omega \Delta\Phi(u,v)\Delta\varphi\,dx=\int_\Omega [u,v]\varphi\,dx\;\forall\varphi\in\mathring{H}^2(\Omega), \tag{7}$$
$$\Delta_k v=k_{11}\partial_{22}v+k_{22}\partial_{11}v\;\forall v\in H^2(\Omega), \tag{8}$$
$$\int_\Omega \Delta Lu\Delta\varphi\,dx=\int_\Omega \Delta_k u\varphi\,dx\;\forall\varphi\in\mathring{H}^2(\Omega). \tag{9}$$

The equation (7) has a unique solution, because $[u,v]\in L_1(\Omega)\hookrightarrow H^2(\Omega)^*$. The well-defined operator Φ is compact and symmetric. Moreover $\Phi: H^2(\Omega)^2\to W_p^2(\Omega)$, $2<p<\infty$ and

$$\|\Phi(u,v)\|_{W_p^2(\Omega)}\leq c\|u\|_{H^2(\Omega)}\|v\|_{W_p^1(\Omega)}\;\forall u\in H^2(\Omega),\,v\in W_p^1(\Omega). \tag{10}$$

The right-hand side of the equation (9) represents the linear bounded functional over $\mathring{H}^2(\Omega)$ and hence the operator $L: H^2(\Omega)\mapsto\mathring{H}^2(\Omega)$ is uniquely defined. Moreover it is compact due to the compact imbedding $H^1(\Omega)\hookrightarrow\hookrightarrow H^2(\Omega)$. Further it fulfils $L: H^2(\Omega)\mapsto W_p^2(\Omega)$, $2<p<\infty$ and

$$\|Lu\|_{W_p^2(\Omega)}\leq c\|u\|_{H^2(\Omega)}\;\forall u\in H^2(\Omega). \tag{11}$$

We reformulate the problem (5),(6) into

Problem $\mathscr{P}$. *Find* $u\in\mathscr{C}$ *such that* $\ddot{u}\in\mathscr{H}^*$, *the initial conditions (3) are satisfied in a certain generalized sense, and the inequality*

$$\langle\ddot{u},y-u\rangle_0+$$
$$\int_Q E\big(A(u,y-u)+\big([u,\frac{1}{2}\Phi(u,u)+Lu]+\Delta_k(\frac{1}{2}\Phi(u,u)+Lu)\big)(y-u)\big)\,dx\,dt$$
$$\geq\int_Q f(y-u)\,dx\,dt\;\forall\, y\in\mathscr{C}. \tag{12}$$

Here $\langle\cdot,\cdot\rangle_0$ denotes the duality pairing between $\mathscr{H}$ and its dual $\mathscr{H}^*$ as a natural extension of the scalar product in $L_2(Q)$. We denote further by $\langle\cdot,\cdot\rangle_*$ duality pairing between $H^2(\Omega)$ and its dual $H^2(\Omega)^*$.

For any $\eta>0$ we define the *penalized problem*:

Problem $\mathscr{P}_\eta$. *Find* $u \in L_\infty(I, H^2(\Omega))$ *such that* $\ddot{u} \in L_2(I;(H^2(\Omega))^*)$, *the equation*

$$\begin{aligned}&\int_0^T \langle \ddot{u}, y\rangle_* + \int_Q E\big(A(u,z) + ([u, \tfrac{1}{2}\Phi(u,u) + Lu] + \Delta_k(\tfrac{1}{2}\Phi(u,u) + Lu))z\big)\,dx\,dt \\ &= \int_Q (f + \eta^{-1}u^-)z\,dx\,dt\end{aligned} \tag{13}$$

holds for any $z \in L_2(I;(H^2(\Omega))$ *and the initial conditions (3) remain valid.*

Lemma. *Let* $f \in L_2(Q)$, $u_0 \in H^2(\Omega)$, *and* $u_1 \in L_2(\Omega)$. *Then there exists a solution* u *of the problem* $\mathscr{P}_\eta$.

Proof. Let us denote by $\{w_i \in H^2(\Omega); i = 1,2,...\}$ a basis of $H^2(\Omega)$ orthonormal in $L_2(\Omega)$. We construct the Galerkin approximation u_m of a solution in a form

$$u_m(t) = \sum_{i=1}^{m} \alpha_i(t) w_i,\ \alpha_i(t) \in R,\ i = 1,...,m,\ m \in N,$$

$$\begin{aligned}&\int_\Omega (\ddot{u}_m(t)w_i + EA(u_m(t), w_i) + E[u_m(t), w_i](\frac{1}{2}\Phi(u_m,u_m) + Lu_m)(t)\,)dx + \\ &\int_\Omega \Delta(\frac{1}{2}\Phi(u_m,u_m) + Lu_m)\Delta L w_i\,dx = \int_\Omega (f(t) + \eta^{-1}u_m(t)^-)w_i\,dx,\end{aligned} \tag{14}$$

$$u_m(0) = u_{0m},\ \dot{u}_m(0) = u_{1m},\ u_{0m} \to u_0 \text{ in } H^2(\Omega),\ u_{1m} \to u_1 \text{ in } L_2(\Omega). \tag{15}$$

After multiplying the equation (14) by $\dot{\alpha}_i(t)$, summing up with respect to i , taking in mind the definitions of the operators Φ, L and integrating we obtain the *a priori* estimates not depending on m:

$$\begin{aligned}&\|\dot{u}_m\|^2_{L_\infty(I;L_2(\Omega))} + \|u_m\|^2_{L_\infty(I;H^2(\Omega))} + \|\Phi(u_m,u_m)\|^2_{L_\infty(I;H^2(\Omega))} \\ &+ \|Lu_m\|^2_{L_\infty(I;H^2(\Omega))} + \eta^{-1}\|u_m^-\|^2_{L_\infty(I;L_2(\Omega))} \le c \equiv c(f,u_0,u_1).\end{aligned} \tag{16}$$

Moreover, the estimate (10), (11) and the definitions of the operators Δ_k and L imply

$$\|\Phi(u_m,u_m)\|_{L_\infty(I;W_p^2(\Omega))} + \|Lu_m\|_{L_\infty(I;W_p^2(\Omega))} \le c_p \equiv c_p(f,u_0,u_1)\forall p > 2, \tag{17}$$

$$\Big\|[u_m, \frac{1}{2}\Phi(u_m,u_m) + Lu_m]\Big\|_{L_2(I;L_r(\Omega))} \le c_r \equiv c_r(f,u_0,u_1). \tag{18}$$

From the equation (14) we obtain straightforwardly the estimate

$$\|\ddot{u}_m\|^2_{L_2(I;V_m^*)} \le c_\eta,\ m \in N, \tag{19}$$

where $V_m \subset H^2(\Omega)$ is the linear hull of $\{w_i\}_{i=1}^m$. Applying the estimates (16)-(19), the compact imbedding theorem and the interpolation, we obtain for any $p \in [1,\infty)$, a subsequence of $\{u_m\}$ (denoted again by $\{u_m\}$), and a function u the convergence

$$
\begin{aligned}
&u_m \rightharpoonup^* u \text{ in } \mathscr{H},\ \dot{u}_m \rightharpoonup^* \dot{u} \text{ in } L_\infty(I;L_2(\Omega)),\ \ddot{u}_m \rightharpoonup \ddot{u} \text{ in } \left(L_2(I;H^2(\Omega))\right)^*,\\
&u_m \to u \text{ in } C(I;H^{1-\varepsilon}(\Omega)) \cap L_\infty(I;H^{2-\varepsilon}(\Omega)),\ \varepsilon > 0,\\
&\tfrac{1}{2}\Phi(u_m,u_m) + Lu_m \to \tfrac{1}{2}\Phi(u,u) + Lu \text{ in } L_2(I;H^2(\Omega)),\\
&\tfrac{1}{2}\Phi(u_m,u_m) + Lu_m \rightharpoonup^* \tfrac{1}{2}\Phi(u,u) + Lu \text{ in } L_\infty(I;W_p^2(\Omega)),\\
&\left[u_m, \tfrac{1}{2}\Phi(u_m,u_m) + Lu_m\right] \rightharpoonup \left[u, \tfrac{1}{2}\Phi(u,u) + Lu\right] \text{ in } L_2(I;L_r(\Omega))
\end{aligned}
\tag{20}
$$

implying that a function u fulfils the identity (13). The initial conditions (3) follow due to (15) and the proof of the existence of a solution is complete.

The estimates (16), (17) imply

$$
\begin{aligned}
&\|\dot{u}_\eta\|^2_{L_\infty(I;L_2(\Omega))} + \|u_\eta\|^2_{L_\infty(I;H^2(\Omega))} + \|\Phi(u_\eta,u_\eta)\|^2_{L_\infty(I;W_p^2(\Omega))} + \|Lu_\eta\|^2_{L_\infty(I;W_p^2(\Omega))}\\
&+\eta^{-1}\|u_\eta^-\|^2_{L_\infty(I;L_2(\Omega))} \le c \equiv c(f,u_0,u_1)
\end{aligned}
\tag{21}
$$

with u_η a solution of the penalized problem.

In order to start the limit process to the original problem we rewrite the penalized problem (13) into the operator form

$$
\ddot{u}_\eta + B(u_\eta) - \eta^{-1}u_\eta^- = f \tag{22}
$$

with the operator $B : H^2(\Omega) \to H^2(\Omega)^*$ defined by

$$
\langle B(v), w\rangle = E\int_\Omega \left(A(v,w) + [\frac{1}{2}\Phi(v,v) + Lv, w]v + (\frac{1}{2}\Phi(v,v) + Lv)\Delta_k w\right)dx
$$

Let us multiply the equation (22) by $z = 1$. We get

$$
0 \le \int_Q \eta^{-1}u_\eta^-\,dx\,dt = \int_\Omega \ddot{u}_\eta(T,\cdot)\,dx - \int_\Omega u_1\,dx - \int_Q f\,dx\,dt \le C,
$$

where C is independent of η (cf. (21)). Since $B(u_\eta)$ takes its estimate in (21) and $L_1(\Omega) \hookrightarrow L_\infty(\Omega)^* \hookrightarrow H^2(\Omega)^*$ we get the crucial estimate of the acceleration term $\|\ddot{u}_\eta\|_{\mathscr{H}^*} \le C$. Hence there is a sequence $\eta_k \searrow 0$ such that for $u_k \equiv u_{\eta_k}$ hold

$$
\left.
\begin{aligned}
&u_k \rightharpoonup^* u \text{ in } \mathscr{H},\ \dot{u}_k \rightharpoonup^* \dot{u} \text{ in } L_\infty(I;L_2(\Omega)),\ \ddot{u}_k \rightharpoonup^* \ddot{u} \text{ in } \mathscr{H}^*,\\
&u_k \to u \ \text{ in } C(I;H^{1-\varepsilon}(\Omega)) \cap L_\infty(I;H^{2-\varepsilon}(\Omega)),\ \varepsilon > 0,\\
&\tfrac{1}{2}\Phi(u_k,u_k) + Lu_k \to \tfrac{1}{2}\Phi(u,u) + Lu \text{ in } L_2(I,H^2(\Omega)),\\
&\tfrac{1}{2}\Phi(u_k,u_k) + Lu_k \rightharpoonup^* \tfrac{1}{2}\Phi(u,u) + Lu \text{ in } L_\infty(I;W_p^2(\Omega)),\\
&\eta_k^{-1}u_k^- \rightharpoonup^* g \ \text{ in } L_\infty(Q)^* \hookrightarrow \mathscr{H}^*,
\end{aligned}
\right\}
\tag{23}
$$

where $g \ge 0$ is the corresponding contact force.

The performed convergence implies that the limit u satisfies the variational inequality (12). The initial condition for u is satisfied in the sense of a weak limit in $H^2(\Omega)$ while that for $\dot{u}$ is satisfied in the sense of the integration by parts. We have proved the following

Theorem 1. *Let $u_0 \in H^2(\Omega)$, $u_1 \in L_2(\Omega)$ and $f \in L_2(Q)$. Then there exists a solution of the contact Problem $\mathscr{P}$.*

3 Contact of a Viscoelastic Sell

The constitutional law has the form

$\sigma_{ij}(\mathbf{u})=\frac{\mathbf{E_1}}{\mathbf{1}-\mu^2}\partial_t\big((\mathbf{1}-\mu)\varepsilon_{\mathbf{ij}}(\mathbf{u})+\mu\delta_{\mathbf{ij}}\varepsilon_{\mathbf{kk}}(\mathbf{u})\big)+\frac{\mathbf{E_0}}{\mathbf{1}-\mu^2}\big((\mathbf{1}-\mu)\varepsilon_{\mathbf{ij}}(\mathbf{u})+\mu\delta_{\mathbf{ij}}\varepsilon_{\mathbf{kk}}(\mathbf{u})\big).$

The constants $E_0,\ E_1>0$ are the Young modulus of elasticity and the modulus of viscosity, respectively. In contrast to the elastic shell we involve the rotation inertia expressed by the term $a\Delta\ddot{u}$ in the first equation of the considered system with $a=\frac{h^2}{12}$. It will play the crucial role in the deriving a strong convergence of the sequence of velocities $\{\dot{u}_m\}$ in the appropriate space. We concentrate again on the case of a free plate. The classical formulation is the initial-value problem

$$\left.\begin{aligned}&\ddot{u}+a\Delta\ddot{u}+b(E_1\Delta^2\dot{u}+E_0\Delta^2u)-[u,v]-\Delta_k v=f+g,\\&u\ge 0,\ g\ge 0,\ ug=0,\\&\Delta^2 v+E_1\partial_t(\tfrac{1}{2}[u,u]+k_{11}\partial_{22}u+k_{22}\partial_{11}u)\\&+E_0(\tfrac{1}{2}[u,u]+k_{11}\partial_{22}u+k_{22}\partial_{11}u)=0\end{aligned}\right\}\ \text{on } Q, \tag{24}$$

the boundary conditions

$$\begin{aligned}&u\ge 0,\ \Sigma_1(u)\ge 0,\ u\Sigma_1(u)=0,\\&\mathscr{M}_1(u)=0,\ v=0 \text{ and } \partial_n v=0 \text{ on } S\end{aligned} \tag{25}$$

and the initial conditions

$$u(0,\cdot)=u_0\ge 0,\ \dot{u}(0,\cdot)=u_1 \text{ on } \Omega. \tag{26}$$

We introduce a cone $\mathscr{C}_1$ as

$$\mathscr{C}_1:=\{y\in H^{1,2}(Q);\ \dot{y}\in L_2(I;H^1(\Omega)),\ y\ge 0\}. \tag{27}$$

Then the variational formulation of the problem (24–26) has the following form:

Find $\{u,v\}\in C_1\times L_2(I;\mathring{H}^2(\Omega))$ *such that* $\dot{u}\in L_2(I;H^2(\Omega))$ *and the system*

$$\begin{aligned}&\int_Q (E_1A(\dot{u},y_1-u)+E_0A(u,y_1-u)-([u,v]+k_{11}\partial_{22}v+k_{22}\partial_{11}v)(y_1-u))\,dx\,dt-\\&\int_Q (a\nabla\dot{u}\cdot\nabla(\dot{y}_1-\dot{u})+\dot{u}(\dot{y}_1-\dot{u}))\,dx\,dt+\int_\Omega (a\nabla\dot{u}\cdot\nabla(y_1-u)+\dot{u}(y_1-u))(T,\cdot)\,dx\\\ge&\int_\Omega (a\nabla u_1\cdot\nabla(y_1(0,\cdot)-u_0)+u_1(y_1(0,\cdot)-u_0))\,dx+\int_Q f(y_1-u)\,dx\,dt,\end{aligned} \tag{28}$$

$$\begin{aligned}&\int_\Omega \Delta v\Delta y_2\,dx=\\&-\int_\Omega\Big(E_1\partial_t(\frac{1}{2}[u,u]+k_{11}\partial_{22}u+k_{22}\partial_{11}u)+E_0(\frac{1}{2}[u,u]+k_{11}\partial_{22}u+k_{22}\partial_{11}u)\Big)\,y_2\,dx\end{aligned} \tag{29}$$

is satisfied for all $(y_1,y_2)\in\mathscr{C}_1\times\mathring{H}^2(\Omega)$.

The Airy stress function is expressed in a form

$$v = -E_1\partial_t(\frac{1}{2}\Phi(u,u)+Lu)-E_0(\frac{1}{2}\Phi(u,u)+Lu)$$

and we reformulate the system (28) ,(29) into the following variational inequality:

Problem $\mathscr{P}_1$. *Find* $u\in\mathscr{C}_1$ *such that* $\dot u\in L_2(I;H^2(\Omega))$ *and*

$$\begin{aligned}
&\int_Q (E_1A(\dot u,y-u)+E_0A(u,y-u))\,dx\,dt\\
&+\int_Q[u,E_1\partial_t(\tfrac{1}{2}\Phi(u,u)+Lu)+E_0(\tfrac{1}{2}\Phi(u,u)+Lu)](y-u)\,dx\,dt\\
&+\int_Q\Delta_k\left(E_1\partial_t(\tfrac{1}{2}\Phi(u,u)+Lu)+E_0(\tfrac{1}{2}\Phi(u,u)+Lu)\right)(y-u)\,dx\,dt\\
&-\int_Q(a\nabla\dot u\cdot\nabla(\dot y-\dot u)+\dot u(\dot y-\dot u))\,dx\,dt+\int_\Omega(a\nabla\dot u\cdot\nabla(y-u)+\dot u(y-u))(T,\cdot)\,dx\\
&\geq\int_\Omega(a\nabla u_1\cdot\nabla(y(0,\cdot)-u_0)+u_1(y(0,\cdot)-u_0))\,dx+\int_Q f(y_1-u)\,dx\,dt\;\forall y\in\mathscr{C}_1.
\end{aligned}\tag{30}$$

For any $\eta>0$ we define the *penalized problem*
Problem $\mathscr{P}_{1,\eta}$. *Find* $u\in H^{1,2}(Q)$ *such that* $\dot u\in L_2(I;H^2(\Omega))$, $\ddot u\in L_2(I;H^1(\Omega))$,

$$\begin{aligned}
&\int_Q\left(\ddot u z+a\nabla\ddot u\cdot\nabla z+E_1A(\dot u,z)+E_0A(u,z)\right)dx\,dt\\
&+\int_Q[u,E_1\partial_t(\tfrac{1}{2}\Phi(u,u)+Lu)+E_0(\tfrac{1}{2}\Phi(u,u)+Lu)]z\,dx\,dt\\
&+\int_Q\Delta_k\left(E_1\partial_t(\tfrac{1}{2}\Phi(u,u)+Lu)+E_0(\tfrac{1}{2}\Phi(u,u)+Lu)\right)z\,dx\,dt\\
&=\int_Q(f+\eta^{-1}u^-)z\,dx\,dt\;\forall z\in L_2(I;H^2(\Omega))
\end{aligned}\tag{31}$$

and the conditions (26) remain valid.

Again, after applying the Galerkin method, we obtain the existence and uniqueness of a solution to the penalized problem with the *a priori* estimates

$$\begin{aligned}
&\|\dot u\|^2_{L_2(I;H^2(\Omega))}+\|\dot u\|^2_{L_\infty(I;H^1(\Omega))}+\|u\|^2_{L_\infty(I;H^2(\Omega))}\\
&+\|\partial_t\Phi(u,u)\|^2_{L_2(I;H^2(\Omega))}+\|\partial_t Lu\|^2_{L_2(I;H^2(\Omega))}\leq c\equiv c(f,u_0,u_1).
\end{aligned}\tag{32}$$

Moreover the estimates (10), (11) imply

$$\|\partial_t\Phi(u,u)\|_{L_2(I;W^2_p(\Omega))}+\|\partial_t Lu\|_{L_2(I;W^2_p(\Omega))}\leq c_p\equiv c_p(f,u_0,u_1)\;\forall p>2.\tag{33}$$

The estimates are η independent. Since for a fixed $\eta>0$ the penalty term $\eta^{-1}u^-$ belongs to $H^1(Q)$, this together with (32) yields an estimate of $\ddot u-a\Delta\ddot u$ in $L_2(I;H^2(\Omega)^*)$. Applying the *a priori* estimates of solutions to the penalized problem we obtain

Theorem 2. *Let* $f\in L_2(Q)$, $u_i\in H^2(\Omega)$, $i=0,\ 1$. *Then there exists a solution* $u\in H^{1,2}(Q)$ *of the contact Problem* $\mathscr{P}_1$.

Proof. We perform the limit process $\eta\searrow 0$ and write u_η for the solution of the problem $\mathscr{P}_{1,\eta}$. To get the crucial estimate for the penalty, we put $z=1$ in (31) and obtain the estimate

$$\|\eta^{-1}u_\eta^-\|_{L_1(Q)}\leq c(f,u_0,u_1).\tag{34}$$

The imbedding $H^2(\Omega)\hookrightarrow L_1(\Omega)$ and the *a priori* estimates (21) and (33) imply the relative compactness in $L_2(I;H^1(\Omega)^*)$ of the system $\{\varphi_\eta;\eta>0\}$ for the functionals $\varphi_\eta: w\mapsto\int_Q a\nabla\ddot u_\eta\nabla w+\ddot u_\eta w\,dx\,dt$.

The *a priori* estimates (32), (33), the last relative compactness and the standard theory of linear elliptic equations yield the existence of a sequence $\eta_k \searrow 0$ such that for $u_k \equiv u_{\eta_k}$ the following convergence hold for any real $p \geq 1$:

$$\begin{aligned} &\ddot{u}_k \rightharpoonup \ddot{u} \text{ in } L_2(I;H^2(\Omega)),\ \dot{u}_k \to \dot{u} \text{ in } L_2(I;W_p^1(\Omega)),\ u_k \to u \text{ in } C(I;W_p^1(\Omega)),\\ &\tfrac{1}{2}\partial_t \Phi(u_k,u_k) + \partial_t L u_k \rightharpoonup \tfrac{1}{2}\partial_t \Phi(u,u) + \partial_t L u \text{ in } L_2(I;W_p^2(\Omega)). \end{aligned} \tag{35}$$

Inserting the test function $z = y - u_k$ in (31) for $y \in \mathscr{C}_1$, performing the integration by parts in the terms containing $\ddot{u}$, applying the convergence (35) and the weak lower semicontinuity verifies that the limit u is a solution of the original problem $\mathscr{P}_1$.

4 Conclusion

The method of proofs of the existence of solutions to dynamic contact problems of shells presented here seems to be the most powerful available. It can lead to similar results for a large class of beam, plate and shell models.

In order to obtain the numerical scheme for solving the above problems we can them transform to first-order with respect to time systems and then apply the method of time discretization together with the linearization in a similar way as in [1].

Acknowledgements. The work presented here was supported by the Ministry of Education of Slovak Republic under VEGA grants 1/0021/10 and 1/0093/10 .

References

1. Bock, I.: On the semidiscretization and linearization of a pseudoparabolic von Kármán system for viscoelastic plates. Math. Meth. Appl. Sci. 29, 557–573 (2006)
2. Bock, I.: On a pseudoparabolic system for a viscoelastic shallow shell. PAMM Proc. Appl. Math. Mech. 6, 621–622 (2006)
3. Bock, I., Jarušek, J.: Unilateral dynamic contact of viscoelastic von Kármán plates. Advances in Math. Sci. and Appl. 16, 175–187 (2006)
4. Bock, I., Jarušek, J.: Solvability of dynamic contact problems for elastic von Kármán plates. SIAM J. Math. Anal. 41, 37–45 (2009)
5. Eck, C., Jarušek, J., Krbec, M.: Unilateral Contact Problems in Mechanics. Variational Methods and Existence Theorems. In: Monographs & Textbooks in Pure & Appl. Math., vol. 270. Taylor & Francis Group, Boca Raton (2005)
6. Voľmir, A.G.: Gibkije plastinky i oboločky, Gosizdat, Moskva (1956) (in Russian)
7. Vorovič, I.I., Lebedev, I.P.: Existence of solutions in nonlinear theory of shallow shells. Applied Mathematics and Mechanics 36(4), 691–704 (1972)

Application of Steepest Descent Path Method to Lamb's Solutions for Scattering in Thermo-elastic Half-Plane

Po-Jen Shih, Sheng-Ping Peng, Chau-Shioung Yeh, Tsung-Jen Teng, and Wen-Shinn Shyu

Abstract. When an incidence impinges an alluvial valley in half-plane, wave interactions of three inhomogeneities are considered on thermoelastic coupling effects, and the stress concentration along continuous interface is demonstrated. Because of the inhomogeneities, the scattering waves can be deduced by three part, the incidence sources in the half-plane, reflection waves simulated by the image sources in the mirror image half-plane, and the refraction wave inside the alluvial valley. For in-plane problem, two coupled longitudinal waves, of which one is predominantly elastic and the other is predominantly thermal, and a transversal wave are adopted to analyze scattering. This work uses a Rayleigh series of Lamb's formal integral solutions as a simple basis set. The corresponding integrations of the basis set are calculated numerically by applying a modified steepest descent path integral method, which provides strongly convergence in numerical integrations. Moreover, Betti's third identity and orthogonal conditions are applied to obtain a transition matrix for solving the scattering. The results at the surface of a semicircular alluvial valley embedded in half-plane are demonstrated to show the displacement fields and the temperature gradient fields. They also indicate that softer alluvial valley is associated with a substantially greater amplification at the interface of the alluvial valley.

1 Introduction

Investigations of scattering in a half-space or in a half-plane almost focus on elastic materials, and most neglect thermal effects. Many studies those involve the

Po-Jen Shih
National University of Kaohsiung, Kaohsiung, Taiwan

Sheng-Ping Peng · Chau-Shioung Yeh
National Taiwan University, Taipei, Taiwan

Tsung-Jen Teng
National Center for Research on Earthquake Engineering, Taipei, Taiwan

Wen-Shinn Shyu
National Pingtung University of Science and Technology, Pingtung, Taiwan

G. Stépán et al. (Eds.): Dynamics Modeling & Interaction Cont., IUTAM BOOK SERIES 30, pp. 301–308.
springerlink.com

coupled thermoelastic theory have emphasized general wave propagation in a half-space and scattering in infinite-space, but few investigations have addressed the scattering of a buried inclusion. The Green-Lindsay theory [1] and the Lord-Shulman theory [2] are two major parts for thermoelastic theory [3-5]. Half-space studies [6-10] typically discuss the properties of waves generated by various sources in planes. However, few studies have discussed scattering by an alluvial valley. The reason maybe a high frequency of waves is necessary to reveal thermal effects, and general mechanical engineering is not serviced at such frequencies. The other reason maybe that solving coupled thermoelasticity is too complex.

This work employs the Rayleigh series of Lamb's formal integral solutions as a basis set and thus expands the scattering fields in a thermoelastic half-plane. Solutions of Lamb's problem, regarding as the formal integral solutions, are described by various loading conditions at the surface of a half-plane. Additionally, the derivatives of the Lamb's integral formal solutions with respect to x-component are regarded as the Rayleigh series. Lamb's integral bases are the immediate solutions to expand such scattering field; however, their infinite integral intervals and their oscillatory terms restrict the usefulness of the basis sets. To overcome the difficulties, this study not only deforms the original infinite integral paths to the steepest descent paths, but it also adds integral paths around the branch cuts to reduce double-values conditions on the complex plane. Since the integrations are counted on the optimal paths with single-value results, numerical calculation of these bases has high accuracy and efficiency [11].

In the case, Rayleigh series obtained from Lamb's integral solutions is used to be the basis sets to calculate them numerically. Furthermore, Betti's third identity and orthogonal conditions are applied to form a transition matrix and to solve the boundary value problems of the alluvial valley [12]. Displacements and temperature gradient results at the surface of a semicircular alluvial valley are demonstrated. The results indicate that softer alluvial valley is associated with a substantially greater amplification at the interface of the alluvial valley.

2 Rayleigh Series of Lamb's Formal Integral Solutions and Matrix Method for Scattering

In linear isotropic thermoelasticity, the governing equations [5] are

$$\begin{aligned}&\mu\nabla^2\mathbf{u}+(\lambda+\mu)\nabla\nabla\cdot\mathbf{u}=\gamma\nabla T^*+\rho\ddot{\mathbf{u}}\\&\nabla^2 T^*-\dot{T}^*/\kappa-\eta\nabla\cdot\dot{\mathbf{u}}=0\end{aligned}\qquad(1)$$

where λ and μ are Lame's constants; the quantity $\gamma=\alpha(\lambda+2\mu/3)$; ρ is the mass density; T^* is a relative temperature; $\kappa=\lambda_o/C_\varepsilon$ is the thermal diffusivity; λ_o is a thermal conductivity; and C_ε is heat per unit strain. Assume that heat sources and body forces are absent. Set two potential parts, ϕ^* and ψ_y^* corresponding to longitudinal and transversal waves, satisfy the Helmholtz equations. So we have $\mathbf{u} = \nabla\phi^* + \nabla\times\psi_y^*$ and $\nabla\cdot\psi_y^* = 0$. Substituting into Eq. (1) yields

$$\begin{aligned}
&[\nabla^2\nabla^2 + (k_p^2 - q - \eta\,\kappa\beta q)\nabla^2 - i\omega k_p^2/\kappa]\ \phi = 0 \\
&(\nabla^2 + k_s^2)\psi_y = 0 \\
&T = [(\lambda+2\mu)/\gamma](\nabla^2 + k_p^2)\phi
\end{aligned} \tag{2}$$

in which $k_p = \omega/C_p$, $k_s = \omega/C_s$, $q = i\omega/\kappa$, $\beta = \gamma/(\lambda+2\mu)$, and the coupling efficient $\varepsilon = \eta\,\kappa\beta$. Furthermore, $k^4 + [\,k_p^2 - q(1+\eta\,\kappa\beta)]k^2 - qk_p^2 = 0$ is obtained, and the roots of this equation are defined by k_{p1} and k_{p2}. Applying Fourier transformation pairs $\tilde{\Phi}(k) = \int_{-\infty}^{\infty}\Phi(x)e^{ikx}dx$ to Eqs. (2), we have the results

$$\begin{aligned}
&\phi(x,z) = (1/2\pi)\int_{-\infty}^{\infty}(A^\alpha e^{-\sqrt{k^2-k_{p1}^2}z} + B^\alpha e^{-\sqrt{k^2-k_{p2}^2}z})e^{-ikx}dk \\
&\psi_y(x,z) = (1/2\pi)\int_{-\infty}^{\infty}C^\alpha e^{-\sqrt{k^2-k_s^2}z}e^{-ikx}dk \\
&\beta T(x,z) = \frac{1}{2\pi}\int_{-\infty}^{\infty}[A^\alpha(k_p^2 - k_{p1}^2)e^{-\sqrt{k^2-k_{p1}^2}z} + B^\alpha(k_p^2 - k_{p2}^2)e^{-\sqrt{k^2-k_{p2}^2}z}]e^{-ikx}dk
\end{aligned} \tag{3}$$

Then displacements $\mathbf{u}$ can be obtained. To develop the basis set as Rayleigh series in half-plane, let us consider the derivatives (m) with respect to x-component of the displacements and temperature of Fourier integral solutions.

$$\begin{aligned}
u_x^{\alpha(m)} &= \int_{-\infty}^{\infty}\frac{(-ik)^m}{2\pi k_s^m}[-ik(A^\alpha e^{-\sqrt{k^2-k_{p1}^2}z} + B^\alpha e^{-\sqrt{k^2-k_{p2}^2}z}) + v'C^\alpha e^{-\sqrt{k^2-k_s^2}z}]e^{-ikx}dk \\
u_z^{\alpha(m)} &= \int_{-\infty}^{\infty}\frac{(-ik)^m}{2\pi k_s^m}[-v_1A^\alpha e^{-\sqrt{k^2-k_{p1}^2}z} - v_2B^\alpha e^{-\sqrt{k^2-k_{p2}^2}z} - ikC^\alpha e^{-\sqrt{k^2-k_s^2}z}]e^{-ikx}dk \\
T^{\alpha(m)} &= \int_{-\infty}^{\infty}\frac{(-ik)^m}{2\pi k_s^{m+1}}[A^\alpha(k_p^2 - k_{p1}^2)e^{-\sqrt{k^2-k_{p1}^2}z} + B^\alpha(k_p^2 - k_{p2}^2)e^{-\sqrt{k^2-k_{p2}^2}z}]e^{-ikx}dk
\end{aligned} \tag{4}$$

Lamb's formal integral solutions are described by various loading conditions at the surface of a half-plane; each loading condition represents an independent set of the thermoelastic field. Herein isothermal surface boundary conditions are considered, and three loading cases are applied to yield the constants, A^α, B^α, and C^α. For the vertical loading case $\alpha = v$, the boundary condition are defined by $\sigma_{zz}^{(0)}(x,0) = -\mu\delta(x)$, $\sigma_{xz}^{(0)}(x,0) = 0$, and $\sigma_{xz}^{(0)}(x,0) = 0$, and we have the results

$$
\begin{aligned}
A^{v} &= -(k_p^2 - k_{p_2}^2)(k^2 + k^2 - k_s^2)/F(k) \\
B^{v} &= (k_p^2 - k_{p_1}^2)(k^2 + k^2 - k_s^2)/F(k) \\
C^{v} &= 2ik[\sqrt{k^2 - k_{p2}^2}(k_p^2 - k_{p_1}^2) - \sqrt{k^2 - k_{p1}^2}(k_p^2 - k_{p_2}^2)]/F(k)
\end{aligned} \tag{5}
$$

and the equation $F(k)$ called the Rayleigh function also can be obtained. Similarly, the boundary conditions $\sigma_{zz}^{(0)}(x,0)=0$, $\sigma_{xz}^{(0)}(x,0)=-\mu\delta(x)$, and $T^{h(0)}(x,0)=0$ for the horizontal loading case ($\alpha=h$) and $\sigma_{zz}^{(0)}(x,0)=0$, $\sigma_{xz}^{(0)}(x,0)=0$, and $T^{(0)}(x,0)=\delta(x)/\beta$ for the heat source loading $\alpha=h$ can lead the results of A^{α}, B^{α}, and C^{α}. Herein Rayleigh series turn out to be the series of Lamb's formal integral solutions, and this series are regarded as the basis set used in this work.

An incident plane wave propagating in z-direction with an incident angle θ_{p1} is considered to treat the scatterer as shown Fig. 1. Potentials of the predominantly longitudinal wave is $\phi^{i} = A_1 e^{-ik_{p1}(x\sin\theta_{p1} - z\tan\theta_{p1})}$, and $\psi^{i}=0$ is for the transversal wave. Apply isothermal boundary conditions to stress and temperature fields at the surface, the uncertain coefficients A_1 can be obtained numerically.

Let an alluvial valley inside S be filled with a material that differs from that of the surrounding medium, as shown in Fig. 1; all material constants of the alluvial valley are designated by the subscript "(o)". The wave field in the surrounding half-plane is denoted by $\mathbf{u}$, and the refracted filed inside the alluvial valley is $\mathbf{u}_{(o)}$, and we have

$$
\begin{aligned}
\mathbf{u} &= \sum_{n=0}^{\infty} \sum_{\beta=v,h,T} a_n^{(\beta)} \hat{\mathbf{u}}^{\beta(n)}(\mathbf{u},T) + \sum_{n=0}^{\infty} \sum_{\beta=v,h,T} \mathbf{c}_n^{(\beta)} \mathbf{u}^{\beta(n)}(\mathbf{u},T) \\
\mathbf{u}_{(o)} &= \sum_{n=0}^{\infty} \sum_{\beta=v,h,T} \mathbf{f}_n^{(\beta)} \hat{\mathbf{u}}_{(o)}^{\beta(n)}(\mathbf{u},T)
\end{aligned} \tag{6}
$$

Moreover, we have the transition matrix

$$
\mathbf{a}^{\alpha(m)} = \sum_{n=0}^{\infty} \sum_{\beta=v,h,T} \mathbf{Q}_{mn}^{(\alpha,\beta)} \mathbf{f}_n^{(\beta)} \qquad \sum_{n=0}^{\infty} \sum_{\beta=1}^{3} \mathbf{F}_{mn}^{(\alpha,\beta)} \mathbf{c}_n^{(\beta)} = \sum_{n=0}^{\infty} \sum_{\beta=1}^{3} \hat{\mathbf{Q}}_{mn}^{(\alpha,\beta)} \mathbf{f}_n \tag{7}
$$

where

$$
\begin{aligned}
\hat{\mathbf{Q}}_{mn}^{(\alpha,\beta)} &= \int_S [\, \hat{\mathbf{t}}_{(o)}^{\beta(n)}(\mathbf{u},T) \cdot \hat{\mathbf{u}}^{\alpha(m)}(\mathbf{u},T) - \hat{\mathbf{u}}_{(o)}^{\beta(n)}(\mathbf{u},T) \cdot \hat{\mathbf{t}}^{\alpha(m)}(\mathbf{u},T)]\, dS \\
\mathbf{Q}_{mn}^{(\alpha,\beta)} &= \int_S [\, \hat{\mathbf{t}}_{(o)}^{\beta(n)}(\mathbf{u},T) \cdot \mathbf{u}^{\alpha(m)}(\mathbf{u},T) - \hat{\mathbf{u}}_{(o)}^{\beta(n)}(\mathbf{u},T) \cdot \mathbf{t}^{\alpha(m)}(\mathbf{u},T)]\, dS \\
\mathbf{F}_{mn}^{(\alpha,\beta)} &= \int_S [\mathbf{t}^{\beta(n)}(\mathbf{u},T) \cdot \hat{\mathbf{u}}^{\alpha(m)}(\mathbf{u},T) - \mathbf{u}^{\beta(n)}(\mathbf{u},T) \cdot \hat{\mathbf{t}}^{\alpha(m)}(\mathbf{u},T)]\, dS
\end{aligned} \tag{8}
$$

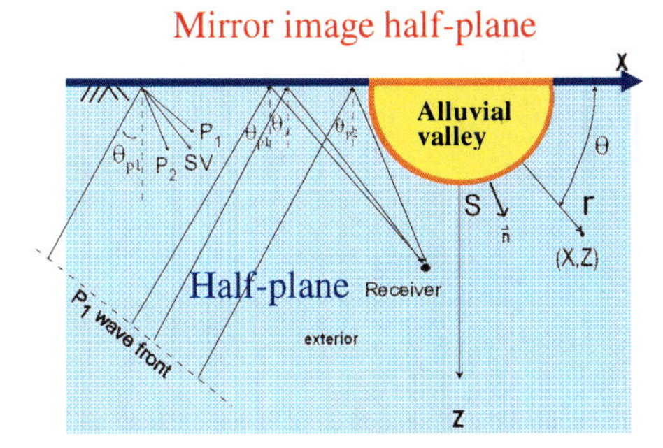

Fig. 1 Three continuum materials, including half-plane, mirror image half-plane, and alluvial valley, and three interfaces within the theromelastic scattering model

3 Modified Steepest Descent Path Method and Numerical Results

To evaluate the integrals in Eqs. (3) and (4), the modified steepest descent path method is adopted. Within the form of Lamb's formal integration, the general forms are written by $I = \int_{-\infty}^{\infty} [E(k)/F(k)]\, e^{-\sqrt{k^2-k_*^2}\,z-ikx} dk$, and the integrands can be separated into three parts within the presented branch cuts, $\sqrt{k^2-k_*^2}$ (k_* can be k_{p1}, k_{p2}, and k_s). Let $x = r\cos\theta$ and $z = r\sin\theta$. The exponent term is $f(k) = \sin\theta\sqrt{k^2-k_*^2} + ik\cos\theta$. Let $[df(k)/dk] = 0$ to obtain the saddle point located at $k^* = k_*\cos\theta$. Then let $f(k) - f(k^*) = \tau^2$ ($\tau \in R$), we have

$$\begin{aligned} I &= e^{-ik_* r} \int_{-\infty}^{\infty} [E(\tau)/F(\tau)]\, e^{-r\tau^2} [dk(\tau)/d\tau] d\tau \\ k(\tau) &= -i\cos\theta(\tau^2 + i\,k_*) + \tau\sin\theta\sqrt{\tau^2 + 2\,i\,k_*} \end{aligned} \tag{9}$$

For various positions of the receivers, typical position of a receiver is shown in Fig. 2a. However, the integral paths may cross the branch cuts, and the singular pole may locate inside the enclosing integral contours. When the integral path crosses the branch cuts, the integral paths or the branch cuts should be deformed to prevent any discontinuity to prevent double-values (Figs. 2b and 2c). When the including the singular poles, the residual values are considered in figures.

$$\begin{aligned} I &= e^{-ik_* r} \int_{-\infty}^{\infty} \frac{E}{F} e^{-r\tau^2} \frac{dk}{d\tau} d\tau + \int_{\Gamma P1\to P2} \frac{E}{F} e^{-\sqrt{k^2-k_*^2}\,z-ikx} dk \\ &+ \int_{\Gamma P1\to S} \frac{E}{F} e^{-\sqrt{k^2-k_*^2}\,z-ikx} dk - \operatorname{sgn}(\cos\theta)\, 2\pi i \cdot \left[\frac{E}{F'} e^{-\sqrt{k^2-k_*^2}\,z-ikx}\right]_{k_R \operatorname{sgn}(\cos\theta)} \end{aligned} \tag{10}$$

where $\Gamma(P1 \to P2)$ and $\Gamma(P1 \to S)$ represent the integrals along the branch cuts.

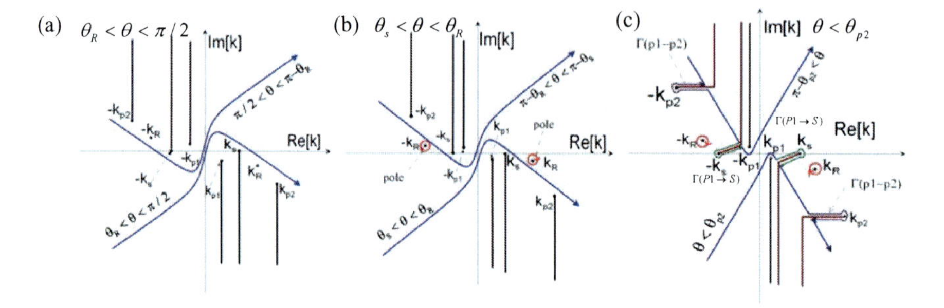

Fig. 2 The steepest decent paths with various receiver positions, (a) typical case, (b) Rayleigh pole included, and (c) two branch cuts and Rayleigh pole included.

The regular solutions are obtained by considering the combination of outgoing waves and incoming waves. The integration path Γ_{out} along $-\infty$ to ∞ at the real axis is deformed to Γ^{+} (Fig. 3a); and the incoming wave Γ_{in} integrating from $-\infty$ to ∞ is deformed to Γ^{-} (Fig. 3b). The integral becomes $\hat{\Gamma}=\Gamma^{+}-\Gamma^{-}$, and both infinite parts are canceled out as shown in Fig. 3c.

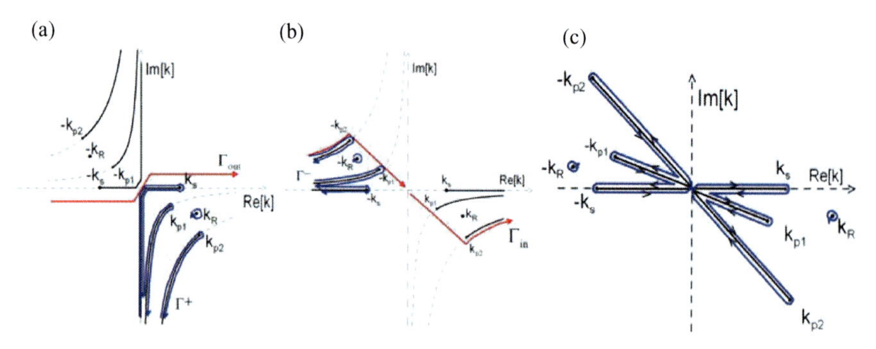

Fig. 3 Integration paths for standing wave (a) outgoing wave Γ_{out}, (b) incoming wave Γ_{in}, and (c) the deformed integral path $\hat{\Gamma}=\Gamma^{+}-\Gamma^{-}$.

In the study, scattering from a semicircular alluvial valley is considered. The outside material is assumed to be copper and the alluvial valley to be softer one. The radius of the valley is $a=0.5\pi/k_s$. The reference temperature is at 20°C. The coefficients are $\varepsilon \cong 0.02$, $\lambda/\lambda_O=1.1$, $\mu/\mu_O=1.6$, $v/v_O=1.04$, and $\alpha/\alpha_O=0.483$. The frequency is $\omega=2\pi\times10^9\ rad/\sec$. Figure 4 shows the Rayleigh series of Lamb's integral solutions of the singular type plotting along the interface of a semicircular. The results obtained using the basis functions for m=0~7 are perfectly stable within the double precision operation, and the accuracy of the results obtained is within 10^{-13}. A case with various angles of P1 wave incidences is considered. Figure 5 shows the normalized displacements and the normalized gradient of temperature plotted along the x axis. The results indicate that the displacements are amplified close to the interface inside the alluvial valley. This phenomenon is associated with the bounded refracted waves inside the alluvial valley, which amplify the wave.

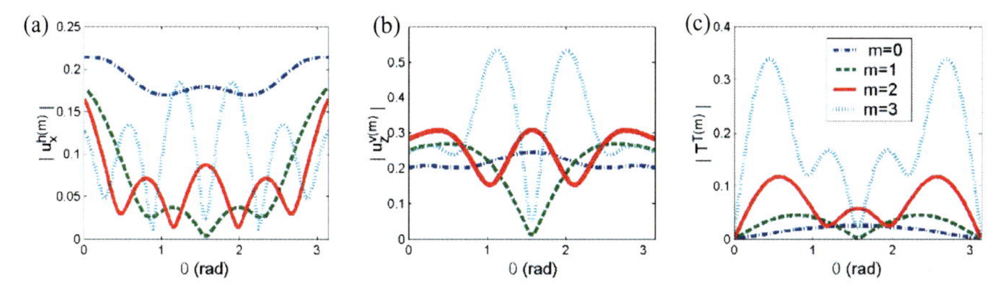

Fig. 4 Rayleigh series of Lamb formal integral solutions demonstrated by the cases, (a) $|\mathbf{u}_x^{h(m)}|$, (b) $|\mathbf{u}_z^{v(m)}|$, and (c) $|T^{T(m)}|$ plotted along the interface of the alluvial valley.

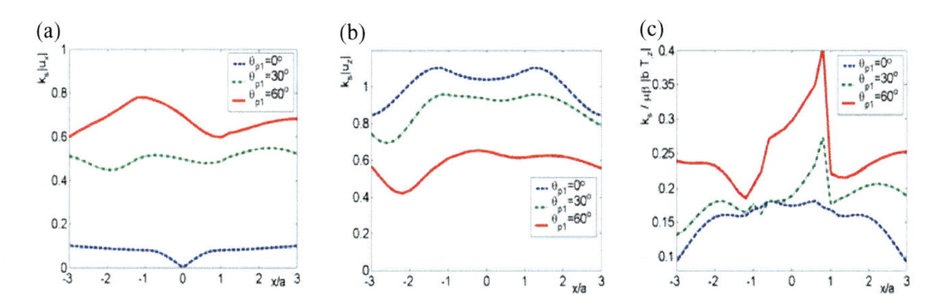

Fig. 5 Normalized displacements $k_s |\mathbf{u}_x|$, $k_s |\mathbf{u}_z|$, and temperature gradient $\dfrac{k_s |bT_{,z}|}{\mu\beta}$ plotted at the surface of the alluvial valley along x-axis.

4 Conclusions

A set of basis functions presented is obtained from Rayleigh series of Lamb's formal integral solutions, and three loading types are adopted to derive the basis sets. Each order satisfies the isothermal boundary condition. Though the basis functions are formed from integrals with highly oscillating terms, the modified steepest descent path method is utilized to improve the accuracy of the results obtained by numerical integration along the optimal integral paths. Different angles of receivers of the modified steepest paths in the specific cases are demonstrated. The solutions of the wave basis obtained by the steepest decent path method are also demonstrated. The example of scattering by an alluvial valley impinged by the P1-incidence is considered. The results show that the normalized displacements are amplified at the same side of the alluvial valley as that of incidence, while the normalized temperature gradient is amplified at the side of the alluvial valley opposite that of incidence.

References

1. Green, E., Lindsay, K.A.: Thermoelasticity. Journal of Elasticity 2(1), 1–7 (1972)
2. Lord, H.W., Shulman, Y.A.: A generalized dynamical theory of thermoelasticity. Journal of the Mechanics and Physics of Solids 15, 299–309 (1967)
3. Biot, M.A.: Thermoelasticity and irreversible thermodynamics. Journal of Applied Physics 27, 240–253 (1956)
4. Chandrasekharaiah, D.S.: Thermoelasticity with second sound: A review. Applied Mechanics Reviews 39, 355–376 (1986)
5. Nowacki, W.: Dynamic Problems of Thermoelasticity. Noordhoff International Publishing Leyden, The Netherlands (1975)
6. Lockett, F.J.: Effect of thermal properties of a solid on the velocity of Rayleigh waves. Journal of the Mechanics and Physics of Solids 7(1), 71–75 (1958)
7. Chadwick, P., Windle, D.W.: Propagation of Rayleigh waves along isothermal insulated boundaries. Proceedings of the Royal Society of London, Series A 280(1380), 47–71 (1964)
8. Sherief, H.H., Helmy, K.A.: A two-dimensional generalized thermoelasticity problem for a half-space. Journal of Thermal Stress 22, 897–910 (1999)
9. Sharma, J.N., Chauhan, R.S., Kumar, R.: Mechanical and thermal sources in a generalized thermoelastic half-space. Journal of Thermal Stress 24, 651–675 (2001)
10. Rajneesh, K., Tarun, K.: Propagation of Lamb waves in transversely isotropic thermoelastic diffusive plate. Int. J. Solids Struct. 45, 5890–5913 (2008)
11. Yeh, C.S., Teng, T.J., Liao, W.I.: On the evaluation of Lamb's integrals for wave in a two-dimensional elastic half-space. The Chinese Journal of Mechanics 16(2), 109–124 (2000)
12. Chai, J.F., Teng, T.J., Yeh, C.S., Shyu, W.S.: Resonance analysis of a 2D alluvial valley subjected to seismic waves. Journal of Acoustical Society of America 112(2), 430–440 (2002)

Author Index